THACH WEAVE

THACH WEAVE
The Life of JIMMIE THACH

STEVE EWING

NAVAL INSTITUTE PRESS
ANNAPOLIS, MARYLAND

Naval Institute Press
291 Wood Road
Annapolis, MD 21402

© 2004 by Steve Ewing
All rights reserved. No part of this book may be reproduced or utilized in any form or by any means, electronic or mechanical, including photocopying and recording, or by any information storage and retrieval system, without permission in writing from the publisher.

Library of Congress Cataloging-in-Publication Data
Ewing, Steve.
    Thach weave : the life of Jimmie Thach / Steve Ewing.
    p.   cm.
    Includes bibliographical references and index.
    ISBN 1-59114-248-2 (alk. paper)
    1. Thach, John S., 1905– 2. Admirals—United States—Biography. 3. Fighter pilots—United States—Biography. 4. United States. Navy—Aviation—Biography. 5. Fighter plane combat. I. Title.
    V63.T43E95 2004
    359'.0092—dc22

                                                          2004004448

Printed in the United States of America on acid-free paper ∞

11 10 09 08 07 06 05 04 9 8 7 6 5 4 3 2
First printing

Contents

	Preface and Acknowledgments	vii
1	A Razorback Goes to Sea	1
2	Test Pilot, Patrol Planes, and Scout Planes	15
3	Fighting Squadron 3 Readies for War	28
4	World War II and VF-3's First Battles	41
5	Battle of Midway	64
6	The Weave Validated and Named	83
7	The Thach Weave Gets Help from Hollywood	96
8	Battle of the Philippine Sea	108
9	Operations Officer: Destination Leyte	120
10	Operations Officer: Destination Tokyo Bay	140
11	Perspectives on Task Force Personalities	168
12	Hollywood Again, Pensacola, and Unification	180
13	Korean War Combat on USS Sicily	192
14	Inchon, Chosin Reservoir, and Final Korean Operations	209
15	Commanding Officer to Flag Rank	223
16	Task Group Alpha	239
17	Commander, Antisubmarine Warfare Force, Pacific	258
18	Deputy Chief of Naval Operations (Air)	271
19	Four Stars, Last Call, and Honors	283
	Appendix: Combat and Command Decorations	299
	Notes	303
	Bibliography	321
	Index	329

Preface and Acknowledgments

Thach Weave: The Life of Jimmie Thach joins *Fateful Rendezvous: The Life of Butch O'Hare* and *Reaper Leader: The Life of Jimmy Flatley* to complete the trilogy on carrier fighter tacticians in the Pacific War. In the dark days following the 7 December 1941 Japanese attack on Pearl Harbor, U.S. naval aviators discovered their aerial foe was far more formidable than expected. No American aircraft carrier-based fighter could match the Japanese Zero fighter in climb and maneuverability. With inferior fighters, and too few of them, tactics had to be developed to counter the proven capability of Japan's carrier airmen and their planes. Only U.S. Navy carrier pilots could keep the enemy from obtaining strategic objectives in the Pacific that potentially could push the U.S. Army and Navy back to the West Coast and lead to a negotiated settlement favorable to Japan.

Before bombs descended on Pearl Harbor, one American naval aviator had already developed an appreciation of the performance variance between Japanese and American carrier fighters. Lt. Cdr. John S. "Jimmie" Thach lost sleep for days on end before solving the problem theoretically and then testing it in the air with Lt. Edward H. "Butch" O'Hare. But pilots in the other squadrons had to know what Thach and O'Hare had learned. In time, Lt. Cdr. Jimmy Flatley, who would coin the phrase "Thach Weave," was the conduit for that.

In late 1941 and for much of 1942, Thach, O'Hare, and Flatley struggled with aerial combat tactics, training, and combat, never stopping to give any meaningful thought to the idea that history might remember them as the architects of sweeping changes in carrier fighter tactics. The three were occupied with thoughts of how their tactics and training might save their lives, save the lives of squadron mates, and keep a significant U.S. military presence in the Pacific. They approached their similar trials differently, with duty and urgency their only common attributes. Although he was pleasant and known to be a man who enjoyed life, the youthful O'Hare was shy and reserved on first meeting with others and all business in training others for combat. The religious Flatley was basically quiet, introspective, patient, and interested in all aspects of his pilots' lives, and he tried to train them to do what the ordinary pilot could do.

Thach was a friend of both O'Hare and Flatley, but he did not resemble either in personality. He expected all his pilots to perform as he did, a perhaps-unrealistic expectation but a standard never lowered. His gregarious demeanor, perpetually smiling face, and demonstrated excellence in the air formed the basis of a charisma enjoyed by few of his contemporaries. But Thach did not depend on charisma alone. He projected an aura of objectiveness that was on occasion demonstrated when he spoke in no uncertain terms in regard to inadequate policies, equipment, and personnel. Not given to being mean-spirited, most often he did not offend, and when he did, his smile warmed most philosophical opponents.

By the end of his career, Thach had won many friends in addition to his many combat decorations and had risen from test pilot to naval aviation's top billet. He even made a few stops in Hollywood, working for the film industry before and during the war and again as a consultant after the war. His World War II aerial tactical innovations overshadowed his status as an ace, his creative thinking and planning as task force operations officer during the last year of the war, his exemplary performance during the Korean War, and his outstanding work in antisubmarine warfare. Throughout his career he proved to be an innovator, constantly finding answers to challenges and following through to ensure Navy-wide familiarization and, often, implementation. From ideas to improve retrieval of target sleeves before World War II to developing a method to counteract kamikazes and launch planes faster during the war, Thach was widely recognized as an officer of superior talent. Decorated for combat leadership off Korea, he was tabbed by the Navy's top leaders to testify before Congress during the unification debate before being charged with designing an experimental antisubmarine task group. Successful in these assignments, he then became part of the Navy's top leadership before retirement.

In recent years, the Naval Historical Center has gathered for the secretary of the navy the names of officers and men whose roles and status qualify them to have ships named in their honor. One researcher for the center wrote that Thach was "truly a most interesting man who kept getting more interesting each day I did the research for our recommendations."[1] As a person and naval officer, Thach was interesting because he was deeply interested in both his profession and other people. Many who knew him recall that he had a rare ability to make strangers feel like an old friend. As readers of the first two volumes know, O'Hare's life ended tragically and Flatley's demise was premature. It is appropriate, then, that the final chapter in the trilogy of the three men who developed and championed the Thach Weave concludes with the one who lived a full lifetime, rose to the top rank of his profession, and made the most of every challenge and opportunity along the way.

Just as it was for the two earlier books in this trilogy, many individuals contributed to *Thach Weave*. James Harmon Thach III, a nephew of Adm. John S. Thach whose father attained the rank of vice admiral before retirement, was particularly helpful in providing family information. Officers who served with Thach during his career were equally helpful, including Lt. Gen. Robert P. Keller, USMC (Ret.), Rear Adm. William N. Leonard, USN (Ret.), and Capt. E. Timothy Wooldridge, USN (Ret.).

A number of friends who also write on navy subjects were again very helpful. John B. Lundstrom (my coauthor for *Fateful Rendezvous: The Life of Butch O'Hare*) read the full manuscript and offered relevant papers. Two other coauthors of earlier books, Clark G. Reynolds and Robert J. Cressman, contributed—Reynolds by reading a significant section of the manuscript and Cressman by answering several calls for details in Admiral Thach's life. Keith Gilbert was kind enough to allow me access to papers from his research on Adm. John S. McCain. And, Mike Brasseur, Rich Leonard, Owen Miller, Capt. Stephen T. Millikin, USN (Ret.), and Barrett Tillman provided information, interest in the project, and always welcome encouragement.

It is my understanding that some years ago a trunk filled with Admiral Thach's papers arrived unannounced at the National Museum of Naval Aviation, NAS, Pensacola. Whoever had the presence of mind to preserve his legacy deserves a salute from all who treasure the history of the navy and naval aviation. Museum director Capt. Robert L. Rasmussen, USN (Ret.), deputy director Buddy Macon, and historian M. Hill Goodspeed fully recognized the great value of the papers and have ensured their preservation within their Emil Buehler Library. Other members of the museum staff who assisted me during protracted hours of research were Daniel Clifton, Jim Curry, and Helen Watson.

Also offering assistance and encouragement was Capt. E. Earle Rogers II, USN (Ret.), vice president for communications with the Naval Aviation Museum Foundation.

While much of the research was completed in Pensacola, all of the writing was done in my office aboard the museum aircraft carrier *Yorktown* at Patriots Point Naval and Maritime Museum, Charleston Harbor, and Mt. Pleasant, South Carolina. While only my name is displayed on the cover, a number of colleagues within our museum made the writing a team effort. Executive director David Burnette ensured the necessary resources for the project, his assistant Laura Langston provided administrative support, and Eleanor Wimett was untiring in offering technical advice. Retired naval aviator Waring "Butch" Hills was frequently called upon for assistance on aviation matters as was former CEO, Rear Adm. James H. Flatley III, USN (Ret.). Julia Hammer assisted with research while Lori Livingston and Andrew Dombrowski contributed to the organization of the index. Retired navy submariners Tom Sprowl and Don Bracken read the two chapters on antisubmarine warfare and offered helpful suggestions. And all the maps for the book were crafted by Mr. Bracken.

As in the past, support from the staff of the Naval Institute Press was superb beginning with the continuing encouragement from director Mark Gatlin and senior acquisitions editor Thomas J. Cutler. Managing editor Rebecca Hinds, senior production editor Kristin Wye-Rodney, and copy editor Karin Kaufman were cheerfully helpful at every turn in the process, and Paul Stillwell and Ann Hassinger guided me to several valuable sources in their oral history section. Publicity manager Susan Artigiani has proved to be an indefatigable cheerleader in promoting the stories of the naval aviators around whom this trilogy was created.

In conclusion, final statements in the acknowledgments of the two earlier books bear repeating. The veterans who served with Jimmie Thach, Jimmy Flatley, and Butch O'Hare were especially generous in providing conversations, interviews, and correspondence during the research process. Their names are listed in the bibliography and, regrettably, many are now also listed on tombstones. They offered their lives and service to their country, and all seemed to believe that there would be others in subsequent generations who would do the same. At least unto this day, they have been correct.

THACH WEAVE

1

A Razorback Goes to Sea

World War II survivor accounts from both the United States and Japan reveal that many combatants gave little or no thought to the person in their gun sights, and even less thought to the name of their adversary. Had there been a requirement that they know their opponents' names before fighting, an aerial battle might have ended before John Smith Thach could explain why he was known by several names. He did not use John, except when he was required to officially write his name as recorded on his birth certificate. From his earliest days, he was known as Jack, the name used by his family throughout his life. But at age eighteen, he had the name Jimmie thrust upon him, and that was the name that stuck.[1]

During World War II, a number of Japanese pilots and ship gunners had an opportunity to end the life of John S. (Jack or Jimmie) Thach. But Jack almost ended his life at age two when he mistook some rat poison for cookies while on a visit to his grandparent's farm in Tennessee. Discovering the absent child in the attic happily devouring the deadly "cookies," the family quickly poured mustard water down his throat to induce vomiting. In an effort to stop the distasteful treatment, the struggling child demonstrated the quick thinking evident in his future roles as fighter pilot and diplomatic flag officer: he suggested that the family not waste all the mustard on him but save some for everyone else.

Born in Pine Bluff, Arkansas, on 19 April 1905, Jack was the third of four children born to James Harmon and Jo Bocage Smith Thach. James Harmon Jr.

(born 13 December 1900) was the Thach family's eldest child, followed by Josephine, Jack, and Frances. Jack's father was a school principal when he met the future Mrs. Thach, also a teacher. Mr. Thach's protracted presence in her classroom was the first indication he was interested in more than the world of education. In 1911, the household moved from Pine Bluff to Fordyce, the presence of Jack's maternal grandmother being a bonus to his education. With a standing invitation for afternoon tea, Jack's grandmother, Etta Bocage Smith, entertained him and his playmate guests with stories of life in Europe. Seldom mentioned was her late husband, Capt. John Smith, who died during the Civil War. In later years, Jack came to appreciate his grandmother even more as he understood how difficult it was for her to raise her three daughters and a son on a schoolteacher's salary and then put them through school.

Jack not only observed adversity in the life struggles of his grandmother but also experienced it directly. As a teenager, he was often frustrated and discouraged as he encountered situations he did not or could not master to his degree of satisfaction. On one occasion, while feeling sorry for himself at age thirteen, his mother stopped what she was doing and invited Jack to sit down with her on the back steps of their home. There ensued a conversation that stayed with him for the remainder of his life. His mother told him several stories of people she had known who had faced difficulties and challenges and then summed up her discourse with the thought that Jack could do anything he wanted. The keys, she said, were that he had to want his goal bad enough to do something about it and that he had to be willing to invest his time. Her formula could be reduced to two words: effort and patience.

Jack experienced his share of adversity in that transitional stage of life from boyhood to manhood, but he also experienced bountiful joys. Two favorites stood out. First were the vacations with his family to the Saline River some twenty miles from Fordyce, where camping, hunting, and fishing were the order of the day. In this setting, Jack learned to fish (often using bread rather than worms for bait) and hunt. Progressing proficiency with fly rod, rifle, and shotgun often translated into the evening meal.

While not necessarily intended as an educational endeavor, trips to the river and woods nonetheless were just that. With both parents being teachers before Jack's father went into the insurance business, it was only natural that instruction in camping, fishing, and hunting was presented in an educational context. Early on Jack learned that everyone needed to participate in setting up the camp so that preparations were complete before dark. Survival training in the early twentieth century was a normal expectation rather than the unique experience it is now. And along with many other boys of that generation, learning to shoot where birds and game would be rather than where they were when first sighted would serve as a considerable aid years later. In the future, the target would be humans, flying in machines that could maneuver as quick as deer, rabbits, or birds.

The second great joy of Jack's teen years was athletics. A man of all sports in all seasons, Jack played football and basketball and ran track. In addition to being fun and good exercise, athletics—especially football—was (and is) perceived as a stepping stone to manhood in the conscious or subconscious minds of youths. That football might develop teamwork skills was seldom a reason for reporting to the opening day of practice. The prospect of respect among peers and admiration from coeds often outranked mere interest in the game. At five feet ten inches and just over 120 pounds, Jack approached the football field his second year in high school determined to play. Given that his small school did not have the twenty-two players necessary to hold a full scrimmage during practice, making the team was not difficult. Playing end at only 120 pounds his first year was difficult, especially on defense.

The single-wing formation was a common formation on offense in the 1920s and remained so into the 1940s. Its basic strategy was to place as many players as possible in front of the ball carrier. Consequently, Jack usually saw at least two blockers bearing down on him nearly every play. On offense he was fast and could catch the ball, but too often he found himself still clearing his head from collisions from playing defense. Somewhat discouraged by his self-assessment as a defensive end, he was happy to be shifted to the backfield his senior year. Calling plays, kicking, running, and passing from his quarterback (tailback in the single wing) position, he was infinitely happier than he had been the year before, despite still suffering from inadequate numbers for full practice scrimmages. Indeed, during his senior year while playing safety on defense, he was flattened on one play and his coach trotted quickly onto the field and bent over Jack's prone body. Anticipating an encouraging or a sympathetic word, Jack was simply told to get up. There were no substitutes at his position. Although the team's pragmatic coach could not afford the luxury of sympathy for injured players, the Fordyce team was successful in 1922 and nearly won the state championship. The year following Jack's graduation, the team, coached by his brother-in-law Bill Walton, who married Josephine, did bring a state title to the small town of three thousand.

4

THE NAVAL ACADEMY

Athletics had been gratifying to Jack in high school and graduation did not end his enthusiasm. It was a factor in motivating him to attend college, and it was his association with sports that paved his way to Annapolis. Although they were not poor, the expenses of college for four children were beyond the reach of the Thach family. With World War I over in November 1918, there was little trepidation when older brother James Harmon Jr. entered the U.S. Naval Academy in the summer of 1919. His first choice had been West Point, but all congressional appointments from Arkansas were filled. Although unfamiliar with the Naval Academy, he did know that it offered the same free education as West Point. Not surprisingly, James Harmon Jr.'s positive experience there influenced Jack to bypass thoughts of West Point in favor of Annapolis.

Entrance to either of the academies was competitive, but Jack got in without taking an examination. Several prominent citizens leaned heavily upon U.S. Senator Joe T. Robinson to support the application of their local all-sports star while Jack negotiated an elevation of his math grades with the high school principal who, coincidentally, was also the football coach. Fortuitously, his history and science teacher was also the basketball and track coach. Regrettably for Jack, Senator Robinson and the Fordyce teacher/coaches were not present to help him when he arrived at Annapolis in the summer of 1923. Quickly it became evident that he was not as academically prepared as many of the 1,006 new plebes around him. Even physically he was not prepared. Before final acceptance, he had to submit to X-rays of the lungs because he appeared to have had tuberculosis. At five feet eleven inches and 130 pounds, he was nearly as tall as he would grow (six feet), and not until much later in life would he exceed 160 pounds—and then not by much. The examining doctor, assuming the somewhat emaciated lad before him had no previous experience with organized sports, unintentionally insulted Jack by recommending he build himself by participating in athletics. Calmly—on the surface—Jack related his recent past sports accomplishments and the consultation ended with the doctor telling Jack he was accepted into the academy, and to eat more!

For many people, things that are important take one to the top, and to the bottom, of the emotional scale. Jack loved the academy, the beauty of its grounds, the athletic facilities and the overall experience, but there were a number of challenges and moments that left him at or near the bottom of his emotional ladder. After surviving the entrance physical he made the mistake of laughing during the second day of plebe summer drill when a classmate marched in the opposite direction of the ordered command. Immediately Lt. Gerald F. Bogan (later vice admiral and an officer Jack would serve with several times) demanded to know who laughed and Jack confessed. While running up and down the field as punishment, Jack wondered why Bogan did not appreciate the humor of the moment. Soon, of course, he learned that a major reason for drill was the discipline it instilled.

Academics proved to be the most significant challenge, and Jack was immediately and constantly "on the tree," the academy expression for low or failing grades. Not fully knowing how to study, combined with time spent on the football field, put him in trouble. Football was tiring physically, and considerable academic study time was lost while he learned the plays of the varsity's next opponent. As signal caller on the B squad, Jack had to run Princeton, Notre Dame, Pennsylvania, or Army's plays in practices against his more senior teammates, who would actually take the field on the fall weekends. Occasionally Jack's practice squad would score against the varsity, thanks in large measure to his running and passing. In his 1923 hopes and dreams, he aspired to greatness on the gridiron. But before the season was over he had twice suffered a dislocated shoulder. The second time was the last time: the doctor told him he was through with football.

For the remainder of his life, Jack's thoughts returned to Annapolis and what might have been. Making the memory even more bittersweet was the fact that his graduation class won the 1926 national championship in football. And the classmate (Thomas J. "Tom" Hamilton) who did the running, passing, and kicking from the tailback-quarterback position for the undefeated 1926 Navy team won All-American honors. Even though Jack later understood his injury was a blessing in disguise, he likely would have traded his eventual four stars for a place on the 1926 All-America Team. Even though he did not play his last three years, the short bio in the 1927 yearbook, *The Lucky Bag*, noted that he "was well known on the football field."[2]

Another occurrence during his early days at the academy also stayed with him throughout the remainder of his life. Older brother James Harmon Jr., called "Harmon" in the family, graduated only days before Jack arrived. While at the academy, James Harmon Jr. was known as Jim or Jimmie, especially to members of the football team. With a place on the football team, Jack was privileged to eat at the training table with upper classmen who knew his brother, and they insisted that he was "a little Jimmie Thach." Later it was just Jimmie. Jack's adamant protestations availed nothing, even his argument that having two men named Jimmie Thach in the Navy might create problems. Years later it did, when surface officer James Harmon Jr. received orders to command an aircraft carrier and aviator John S. received orders to command a battleship.³ Protests to the contrary, Jack became Jimmie because his classmates addressed him thus. Even his future wife did not know his preference for the name Jack until after they were married, and even she chose to call him Jimmie.

Despite the absence of football for his last three years at the academy, life was still good. Wrestling and crew helped fill the athletic void, and socially he was "never known to miss a hop or to overlook any fair lady there. And seldom a Saturday passed that he didn't dash out to meet some sweet girl's train."⁴ Summers brought interesting midshipmen cruises, albeit on old coal-burning ships. Scrubbing the decks and shoveling coal was memorable if not enjoyable. Academics remained a "never-ending worry, constantly growing heavier."⁵ But on 2 June 1927, he graduated, standing 494th in a class of 579. Over 400 who started with him in 1923 had either resigned along the way or were forced out due to unsatisfactory performance. Indeed, there was a conscious effort to weed out some of the class to preclude having so many junior officers. In 1927, Jimmie was not overly concerned with his class standing, but in the years leading up to World War II, he came to understand that class standing was a matter of importance. It was not only a factor in promotion but also affected his standing on application for quarters, and there were occasions when he lost quarters to others several numbers above him.

Just before leaving the academy for his first assignment, Jimmie and the other new ensigns were provided with a short course in aviation. Three or four new officers were invited aboard Curtiss H-16 twin-engine flying boats. Though it was little more than an orientation course, Jimmie enjoyed the ninety-mile-per-hour rides, especially the bow gunner's cockpit, where he could lean forward to obtain an unobstructed view of the flight's direction. When the course was over, the man who would eventually rise to the top position in naval aviation did not foresee a career in aviation.

SURFACE SAILOR TO RELUCTANT AVIATOR

In 1927, the heart of the fleet was the battleship, and Jimmie's orders to the USS *Mississippi* (BB-41) was a choice assignment. Commissioned on 18 December 1917 during World War I, the thirty-two-thousand-ton second member of the three-ship *New Mexico* class was still in its original configuration (modernized in 1931) when Jimmie reported aboard on 14 July 1927 for an eleven-month tour of duty. Assigned to communications duty as one of several assistant radio officers, Jimmie encountered one of many practices that caused him to think critically and formulate ideas for improvement. The Navy could communicate with more than one addressee in 1927, but the Army could not. Over time that changed, but Jimmie began to take notice of problems and it became a habit for him to develop resolutions to them.

While attached to the *Mississippi*, Jimmie decided to follow up on the aviation orientation received at the academy by reporting to San Diego for an elimination course in flying. There he was given permission to fly solo after less than six hours of dual instruction, a surprise to the young ensign, as he was not overly confident of his ability on takeoffs and landings. Feeling very much alone on his first solo in the back seat of his NY trainer, Jimmie took off as ordered, flew around the field, and then made a rough landing that turned out better than he expected after the first hard bounce. The solo marked successful completion of the elimination course, at which point Jimmie was directed to fill out a form indicating whether or not he wished further flight training. Although responding affirmatively, he did not make a formal application and returned to his shipboard duties.

On 7 June 1928, Jimmie was transferred to another of the Navy's prime capital ships, the battleship *California* (BB-44), because of its need for additional junior communications officers. Aboard the flagship of the Battle Fleet, Jimmie continued his communications duties and became interested in cryptanalysis after completing a short course in code breaking. Assignment to the *California* brought another benefit: it was his brother's ship. Lt. (jg) James Harmon served aboard as assistant operations officer. The advantages were many for the two close brothers. In addition to enjoying each other's company, they could wear each other's clothes, and on one occasion Jimmie successfully substituted for his brother on a date after James scheduled two dates for the same evening. But somewhat unexpectedly in these happy days near the end of the Roaring Twenties, orders arrived in February 1929 for him to report to Pensacola, Florida, for flight training. Jimmie immediately sought to delay his reporting date. The aviation officer on Rear Adm. William V. Pratt's staff, future flag officer Lt. Cdr. George D. Murray, advised Jimmie that if he did not report as directed he probably would not be accepted later. Consequently, Jimmie, with some reluctance and resignation, packed for the trip to Florida.

The trip to Pensacola in March 1929 was more eventful than expected. His circuitous route to Florida included a visit to New York, where he had an acute attack of appendicitis. Declining an operation at an expensive civilian hospital he strapped on an ice pack, attended a few more parties, and then went on to Pensacola where he suffered another attack the day of his flight physical. Not acknowledging his pain, he passed the physical and then reported to sick bay.

Pensacola in the spring was a great place to be. The azaleas were in full bloom, and the days were warm and the nights cool. Air conditioning was all but unheard of, and by May the heat and humidity made ground school studies difficult. Flying, however, was a relief regardless of the altitude due to the movement of air. And by the time he arrived at Pensacola, Jimmie was much more favorably disposed to the prospect of flying—this despite the fact that one of the people he encountered was Lt. Gerry Bogan, his old nemesis from the parade ground at the academy. In 1929, Bogan was in command of one of the training squadrons and apparently he did not remember the young ensign who was smart enough not to stir his memory. Indeed, it was Bogan who bolstered Jimmie's heretofore-shaky confidence after a final squadron check. Also serving at Pensacola much of the time Jimmie was there were Lts. John G. Crommelin, Frank Akers, and William O. Davis, aviators who were already recognized as outstanding pilots and would later make notable marks on naval aviation. Although his direct relationship was limited with these instructors, he profited by observing them, and he profited by watching a flight demonstration team as they performed precision flying over Pensacola in their Boeing F2B fighters.

Like most student pilots—including academy classmate Ens. Paul H. Ramsey—Jimmie had his good days and not-so-good days throughout ground school and practice flights in primary seaplanes, primary land planes, service type scout-bombers, multiengine patrol planes, and combat planes. Nearly two-thirds of those who started with him in flight training did not complete the course, and at one point he was within one down (unsatisfactory grade) of being eliminated. But before reaching the halfway point of the training that officially began 4 April 1929, Jimmie was highly motivated to become a naval aviator and his performance equaled his desire. By the time he was designated a naval aviator on 4 January 1930, he had risen to the top of his class both in ground school and in flying, attaining a final average of 3.178 and 3.2 in combat flying.[6]

FIGHTER PILOT

In March 1930, Jimmie traveled to San Diego. He was happy to be in the Mediterranean climate of southern California rather than the heat and humidity of

Florida. But he was not happy with the amount of space on North Island being used by the Army Air Corps (Rockwell Field), and he was sorry to see that there was a lack of concrete on North Island just as there was too little at Pensacola. He was also disconcerted by the impact the Great Depression was having back in Arkansas. Soon both he and his brother began assisting the family there as their father's business began to feel the effects of the economic downturn. Promotion to lieutenant junior grade on 2 June 1930 was welcome more for the money than for any other reason.

From March into June 1930, Lieutenant (jg) Thach had orders for sea duty with the aircraft tender *Aroostook* (though still classified as a minelayer, CM-3). As engineering officer for VJ-1B, his main interest was the Loening OL-8 observation amphibian that featured a large single float under the fuselage instead of a flying-boat hull. For much of the spring, *Aroostook* was not in San Diego, and near the time it returned in June, he joined Fighting Squadron 1 (VF-1B), the noted "High Hat" squadron based on North Island, where he remained until June 1932. Squadron commanding officers (CO) during his tenure with VF-1B were Lt. Cdr. Arthur W. Radford, future chairman of the Joint Chiefs of Staff, and the seemingly ubiquitous Gerry Bogan.[7]

Prior to Jimmie's experience and for more than ten years following, new fighter pilots learned their trade and art in the squadron. Fresh from flight training at Pensacola, the new pilots were not as well trained as they might have thought, and in peacetime there was the luxury of utilizing time to develop skills as he trained with the squadron. Not long after Jimmie joined VF-1B, the squadron began trading in their single-seat Boeing F2B-1 fighters for the two-seat, in tandem Curtiss F8C-4 Helldivers. The F2B-1 first came to the squadron in January 1928, was powered by a 425-horsepower Pratt and Whitney R-1340B engine, had a maximum speed of 158 miles per hour, could operate up to twenty-one thousand feet, and carried one fixed .30-caliber and one .50-caliber gun.

The Boeing was strictly a fighter, but the Helldiver was intended to be primarily a scout and dive-bomber (maximum 500-pound bomb load). Still, with two fixed .30-caliber forward-firing guns in the top wing and a flexible .30-caliber gun for the rear seat, the Curtiss was survivable in a dogfight. The range of the Helldiver was double that of the Boeing it replaced, but its top speed of 146 miles per hour, lower ceiling (sixteen thousand feet), and markedly slower climb did not impress fighter pilots. By the time Jimmie left the squadron, the Helldiver's days as a first-line aircraft were near an end. And that was not too soon. While Arthur Radford, for whom Jimmie often flew as wingman, was in command of the squadron, he made it a point to call attention to how fast the fabric covered wings and tail deteriorated. When the commanding officer of the *Saratoga* (CV-3) came to inspect on one occasion, he initially did not believe the problem to be serious. But when he touched the fabric on a Helldiver and his hand went through, he immediately realized that appearances were misleading.

Although Jimmie held great respect and admiration for Radford and Bogan, most of his flying time with VF-1B placed him in formation with Lt. Herbert S. "Ducky" Duckworth (Thach's usual section leader) and Lt. (jg) Edward Page "Bud" Southwick. With Duckworth, Southwick, Radford, Bogan, and other pilots, Jimmie began his apprenticeship in the air. There were many lessons to learn, some easy and some requiring considerable practice. In time he became squadron navigator and had the responsibility of calibrating all the compasses. While serving in this capacity, he probably saved Bogan's life when he noticed the squadron commander was flying in the wrong direction after an exercise. Investigation on the ground revealed a misplaced magnet that rendered Bogan's compass useless.

Fear, according to Jimmie, was the unknown, and he believed in 1930 that he was sufficiently knowledgeable about aviation to not be preoccupied with thoughts for his personal safety. Although he had no desire to stand on the edge of a tall building, he did not feel that he was in a high place while flying. An emergency was always a possibility, but that did not worry him because he fully believed that he had sufficient training to handle a forced landing. Not anxious to experiment with a parachute he nonetheless knew he would hit the silk before trying to ride down in a burning plane. Death, however, was an occurrence around him often enough to keep him attuned to the need for remaining focused when flying. Fellow students at Pensacola died while he was in flight training, and even one of his elimination-training instructors was killed when he dove too steeply on a practice target and the wings of his plane separated from the fuselage.

The pilots of VF-1B during Jimmie's tenure were renowned for flying a nine-plane formation tied together by a manila line when they had an appropriate audience such as national air shows. The pilots took off together, gained altitude, performed maneuvers including loops and landed with all lines intact. As with any flight the most difficult periods were takeoff and landing. A mistake then could pull one plane into another or, preferably, break the line. Even when not tied together, formation flying required that the planes not stray too far out from the others. On some flights Thach and Southwick would fly so close to Duckworth that they would intentionally bump his plane just to be mischievous.

Although audiences were not present when aerial gunnery was practiced, pilots took a special interest in these exercises. As the Depression rolled on, ammunition became scarce, therefore valuable and therefore an item for appropriation or midnight acquisition by other squadrons also based at San Diego. All pilots desired recognition of efficiency in aerial gunnery, but to keep them honest, the aviation squadron version of the Navy Department training manual *Orders for Gunnery Exercises* (OGE) had to be updated and revised. Early loopholes allowed a pilot to fly up beside a target sleeve until his gun barrel was within inches of the sleeve, obviously a tactic that was not realistic in combat. The OGE was revised to require pilots to make firing runs by approaching the target sleeve from various angles without hesitating or dropping back. Failure to fly according to OGE prescribed patterns disqualified the pilot for that particular exercise.

Arguments that might develop over aerial gunnery practice did not find expression in individual battle practice (IBP) when pilots engaged in mock combat because a camera gun was synchronized with the firing trigger. The camera also had a small stopwatch that showed in the corner of every frame of the film. Watches in the camera guns were synchronized before the pilots left the ground, and upon the completion of the mock combat a viewing of the film left no doubt as to which pilot was first to fire. In addition to solving arguments, the film revealed exactly where the problems were and pilots benefited from the documented evaluation of their performance. Even from the beginning, Jimmie recorded outstanding gunnery scores. Without question in his mind, his ability to shoot in an airplane was a direct extrapolation from his experiences as a boy hunting quail with his father. Handed a shotgun at an early age, he learned to fire where a bird would be when it was hit rather than where it was at the moment he pulled the trigger. He understood deflection shooting long before he learned the term or heard it discussed as he trained to be a pilot.

That there would be arguments was expected, given that the essence of most any fighter pilot was a spirit of competition. The pilot's competitive spirit was evident in the air and on the ground. Athletic contests became miniature wars when pilots were involved, and some enlisted veterans of aircraft carriers recall that what was supposed to be a basketball game on carrier hangar decks were more football than basketball and that a sailor was safer in the boxing ring.[8] Looking back on his career, Thach believed that competition in combination with experience was perhaps the most common characteristic of the better fighter pilots.

When Thach graduated from the Naval Academy in June 1927, the United States Navy had only one aircraft carrier, the *Langley* (CV-1). Converted from a collier (coal ship), *Langley* was small and slow but nonetheless the test platform that nurtured naval aviation at sea from 1922. On 16 November 1927, the carrier *Saratoga* was commissioned, and a month later, on 14 December, the *Lexington* (CV-2) was commissioned. Both large (thirty-three thousand tons) and fast (thirty-four knots) carriers were converted from battle-cruiser hulls as a result of agreements at the 1922 Washington Naval Conference. While Thach was assigned to VF-1B, the squadron was attached to *Saratoga* and most of his early carrier landings were aboard it. Experiments aboard CV-2 and CV-3 picked up where they left off with *Langley*, with both air groups and ships company always looking for better techniques in handling and servicing aircraft. Not all good theoretical ideas proved functional, however. One example was Lieutenant Duckworth's idea to have all planes landing aboard the carrier catch the number eight or nine wire instead of the number three. But during his flight demonstration, Duckworth missed the number nine wire and went into the barrier. In theory the idea was good because planes catching the eight or nine wires could be moved out of the way faster to allow following planes to land. Of course all time saved was lost once a plane went into the barrier and the deck was closed. Still, it was beneficial for aviators to experiment with new ideas to improve carrier performance, especially given the fact that the two new carriers only had one large elevator to serve the hangar and flight decks and one small one. And the elevators on *Saratoga* and *Lexington* were very slow compared to later carriers (thirty-three seconds verses approximately thirteen seconds on an *Essex*-class carrier).

HIGH HAT IN HOLLYWOOD

In the summer of 1931, *Saratoga*, VF-1B, and Jimmie became movie stars. As the Navy cooperated with the Hollywood production of the movie *Hell Divers*, none of the VF-1B pilots received any extra pay. Nearly all of the scenes filmed received Navy's blessing, except the portion that showed *Saratoga*'s arresting gear and the tail hooks on the planes. That was then considered confidential

A Razorback Goes to Sea

and was therefore left out of the final version.⁹ The screenplay was written by Lt. Cdr. Frank W. "Spig" Wead, who had earlier commanded a fighter squadron before becoming paralyzed as a result of a fall down a staircase. Starring in the MGM production were Clark Gable, Wallace Beery, Cliff Edwards, and Conrad Nagel. The actual flying was performed by VF-1B with Ducky, Jimmie, and Bud flying most of the more exciting scenes. Jimmie recorded seventy-five flight hours during the production, some enabling cameramen to get footage that was later shown in the background as the actors recited and acted out their parts on a sound stage. Other flying, however, proved more challenging. One scene required Jimmie to depict a ground loop after landing on a sandy beach. He did it in one take, leaving the film's director greatly impressed and too afraid to ask that it be done again. On another occasion, an eighty-pound camera was placed near the wing tip of Jimmie's plane and he was asked to do a slow roll in front of a hangar and fly close enough to it to just barely miss. However, when he went into the maneuver he discovered that the weight of the camera would not allow him to roll as planed. Flying upside down, he rolled the only way he could, providing an even more exciting take than expected. Afterward he commented that he would not attempt that maneuver again for any amount of money.

Jimmie discovered that the actors were fun people to be around and was not surprised to see that they took their occupation as seriously as he and the pilots of VF-1B took theirs. He also discovered that they liked to laugh, poke fun, and be as mischievous as he and Bud when they "bumped" Ducky's plane from time to time. On 10 August 1931, Cliff Edwards wrote lyrics for a nonexistent song, crediting "Weary Berry" for the "Dreary music." One section read:

> We went out to the stable, met with Gibble Gable
> He really should have been down in the hatch
> I tho't that he was screwie, 'cause he was very spewey
> He had just been in a spin with Jimmie Thach[10]

The experience with the movie people was enjoyable, and ten years later it would pay dividends for both him and the Navy that neither could have foreseen in 1931. At age twenty-five, handsome and dashing, there were also some social dividends. *Hell Divers* was not primarily a romantic film, but one could not have been associated with Hollywood without meeting at least a few of the town's more alluring stars. One acquaintance became lifelong. Years later Jimmie shared his thoughts of Sally Rand, a 1920s film star later better known for her famous fan dancing. Somewhat surprisingly, he found her to be more interested in a quiet married life with family rather than the notoriety of Hollywood.[11] Others who remember seeing Rand perform recall that she had a face that could stop a steam locomotive and a body that could derail it.[12]

Sally Rand notwithstanding, Jimmie had romantic inclinations south of Hollywood. In San Diego he had met the daughter of Dr. Leland D. Jones, Madalyn, who was a senior at San Diego State College majoring in music and education. A wedding date was set for early 1932, but when Thach learned he was headed to sea for a protracted exercise, the two married in December 1931. Intelligent, articulate, and cultured, Madalyn brought all the desirable attributes to the marriage necessary to complement and further Jimmie's Navy career.[13]

In July 1932, Thach was detached from VF-1B and was on his way to Norfolk, Virginia. His time in San Diego had been rewarding both socially and professionally. Upon retirement he and Madalyn would return to San Diego, and in the intervening period they would think of it as home. Professionally he felt within that he had not only passed the apprentice period of his aviation development but also grown as an officer. Looking back, he credited Arthur Radford as the most influential commanding officer under whom he served. Radford always knew how to relate what the squadron was doing in relation to how it served the Navy and the nation. Anything that was not good for the country was not good for the Navy, the squadron, or the person. It was an axiom for Radford and was adopted by Thach to pass on to those who served with and under him.

2

Test Pilot, Patrol Planes, and Scout Planes

By July 1932, when Jimmie and Madalyn Thach packed for the journey from San Diego to Norfolk, Virginia, the Depression was having a significant impact on naval aviation and all other branches of the military. Flying time diminished as funds for fuel were expended, and squadron training suffered as a result. Naval aviation at Norfolk also felt the pinch, but flight testing was less affected than squadron training, and Thach enjoyed many hours in the air during his assignment as a test pilot.

TEST PILOT (1932–1934)

There is no better evidence of Thach's comfort while flying and his potential as a test pilot than his discomfort while flying a particular exercise on Lieutenant Commander Radford's wing with VF-1B. Although he knew he needed to visit the head (restroom) before taking off, he took off without attending to the matter. Once aloft, the urge to relieve himself exerted itself forcefully, and he told his rear-seat passenger to keep the Helldiver in level flight while he climbed onto the wing. The passenger appeared equally worried about the fact that he was not a pilot and that he could be on the receiving end of Thach's problem. Radford could not help but notice that one of his pilots had abandoned his cockpit in favor of a wing, that the plane was separating from the rest of the formation, and that it was flying erratically. He signaled the errant plane to rejoin the formation, and it did. But its pilot never again took off without first attending to the needs of Mother Nature. The rear seat passenger probably reconsidered his career choice, and when Radford learned Thach was tabbed to become a test pilot, he undoubtedly thought the Navy had made a wise decision.

The basic requirements of the new assignment were to conduct rough-water tests for experimental seaplanes and new planes proposed by numerous manufacturers. In 1932, manufacturers did not have to test fly their aircraft, but they had to get them to Norfolk in some manner. Many planes proved to be flying bricks and were never accepted for production, but the competition inherent to the testing process assured progress. Not surprisingly, Thach and other pilots assigned to this duty were especially interested in new planes intended for carriers. They conducted night-flying and gunnery trials and scrutinized new equipment produced for naval aircraft.

Thach's first experience as a test pilot with the experimental division at Naval Air Base, Norfolk/Hampton Roads was memorable, if not entirely enjoyable.[1] Even before he had time to get out of his blue service uniform, several fellow test pilots coaxed him into making a quick experimental flight so his opinion could be added to the team that had been working with it. The plane was a torpedo-bomber that apparently had a difficult horizontal stabilizer control making the plane nose-heavy. Paddle-type balances had been installed to correct the problems, and Thach caught a glimpse of the plane as another test pilot circled the field. Upon landing, Thach climbed into the cockpit and listened to instructions to take the plane up and then land it on the field where a replica of a carrier flight deck had been built. He was told to catch the one wire on the replica deck first with a three-point landing, catch the wire normally with the hook, and make a two-point landing keeping his tail up and not letting the wheels touch the deck until the hook brought the plane down. Soon into the flight, Thach realized the plane was unable to maintain level flight without constant effort on the controls. Still, his first three-point landing caught the wire. Up and away again, he came in, rolled as directed, missed the wire, went around again, and caught the wire. Then came the final test to catch the wire in free flight without letting the wheels touch. Though still gyrating up and down, Thach caught the wire—whereupon the plane slammed down to the deck and all but disintegrated.

Both wheels had exploded upward on either side of the plane, and their remains were still rolling around in the grass as one exceedingly unhappy Lieutenant (jg) Thach emerged from the aircraft ruins. Also rolling around on the grass, in laughter, were his newfound colleagues, some with the same rank he held but, he surmised, all senior in service time. Unable to threaten disciplinary action, he nonetheless let them know his state of mind when he bellowed, "What did you do to me? Here, the first airplane I get in I crash it. I've never crashed an airplane before in my life." Rather than apologize, they told him that the crash was fully expected and someone had to do it. The crash did indeed prove that the plane had serious longitudinal control problems, and despite the rough introduction, Thach later counted the pilots as friends.[2]

Being a test pilot was dangerous, and Thach had to spend most of his flight pay for costly insurance premiums. But there were rewards. In addition to helping determine which experimental planes would be rejected, the experience of having been a test pilot would pay dividends in future flying assignments for both him as an individual and those he would command. It was also helpful to all naval aviators to have an opportunity to test major flight concerns, such as the need for a carrier-based plane to have a controlled slow-speed tight-turn capability. And there was a considerable sense of satisfaction to be the first or among the first to fly any new airplane being considered for Navy adoption, as the first experimental type from manufacturers had to undergo Navy tests before there could be a production contract.

Some tests during Thach's tenure as a test pilot were especially memorable, the test marking the transition from one primary aircraft builder to another being near the top. The Boeing Airplane Company (later Boeing Aircraft) had maintained a relationship with the U.S. Navy since 1917 and produced front line carrier fighters from 1923 into 1938. The Army Air Corps equivalent was the Boeing P-12, and the company had such a corner on the fighter market that few manufacturers offered a challenge. The carrier-based Boeing F4B did have a major drawback, however, as its fixed landing gear adversely affected top speed. The Navy made an inquiry to the fledgling Grumman Aircraft Engineering Company in 1931, asking if it could resolve the F4B's problem by installing retractable wheels. Grumman, which had just developed a lightweight all-metal float with retractable landing gear for the Navy, responded that there was insufficient space for the stowed gear in the Boeing, but Grumman was in the process of designing a two-seat fighter that would have recessed wheels.[3]

The successful introduction of the new Grumman fighter to the Navy brought on a test between it and Boeing's experimental response, the XF6B. Thach was involved with both planes. The test required the two planes to line up on the field—there were no runways then—and at the given signal, both were to take off and see which was first into the air. Thach took off in the Grumman and not only was first into the air but was so far ahead of the Boeing that he had time to make a diving pass at the XF6B. Boeing representatives were incensed and blamed Thach for "showing off." So the pilots were reversed and Thach in the XF6B was still on the ground when his fellow test pilot in the Grumman roared over him in a diving pass. Equally negative for the Boeing was its tendency for a wing to drop into the ground when simulating a carrier landing. Rear Adm. Ernest King, then chief of the Bureau of Aeronautics (BuAer), flew down to witness the test, and upon seeing the XF6B's wing hit the ground before the wheels, he commented that he had seen what he came to see and retreated to his plane and Washington, D.C. Happily, Thach was on the ground with King rather than at the controls for this particular test. Boeing went on to build such war-winning bombers as the B-17 and B-29, but it was Grumman that would build the war-winning carrier-based fighters.

Although fighters were and would remain Thach's passion, he also had considerable interest in the Navy's large patrol boats. When the United States entered World War II, the Navy already had the Consolidated PBY Catalina, one of the greatest patrol planes ever built. Like many planes, however, its evolution was protracted, and in 1933, Thach had an opportunity to work with a forerunner of the PBY, the XP2Y. At Hampton Roads, crash boats and barges assembled near the big patrol bomber for rough-water tests. Finally the weather cooperated with strong winds that lifted waves to the minimum six-foot height required for the test. The shock of the landings popped rivets, caused leaks, and even broke bones in the feet of Thach's senior pilot, who had decided to observe the landings from down inside the hull. Ten takeoffs and landings were planned by the two test pilots and mechanic, but the big flying boat was not equal to all ten of the challenges. In other tests in other planes there were occasions when the crash boats were needed to fish Thach out of Chesapeake Bay.

It was not unusual for Thach to fly several different planes in one day, and at night. Morning might find him testing a dive-bomber for carrier landings with a full bomb load, in the afternoon he might fly a seaplane in rough-water tests, and at night he might run fighter gunnery tests, a particularly interesting experience when the instrument panel lights went out. Familiarization with a new plane, an absolute must for more contemporary pilots, was not the norm in the early 1930s when Thach was serving as a test pilot. Before rough-water tests he and other test pilots might enjoy one or at most two flights with a new plane. But there were many occasions when the failures of a new plane were not known until the test pilot discovered them during its initial flight. Instrument familiarization usually was not a significant problem; there just were not that many to view.

PATROL PLANES (1934–1936)

In June 1934, Thach was detached from his test pilot duties in Virginia and ordered to sea duty with Patrol Squadron 9 tended by the *Wright* (AV-1). Unlike the smaller *Aroostook* and "bird-class" *Gannet* (AVP-8) and *Sandpiper* (AVP-9) that served planes piloted by Thach, the 448-foot *Wright* was a true seaplane tender. With a beam of fifty-eight feet, the *Wright* could lift large patrol planes aboard without having parts of the aircraft extend over the sides. For nearly all of Thach's time with VP-9, *Wright* was based at San Diego and operated in the Pacific except for exercises in the Caribbean.

During his two-year tour with VP-9, Thach was tabbed for one special mission in part because of his experience as a test pilot. The assignment placed him on temporary duty back at Norfolk for a test flight in the XP2H-1, one of the largest airplanes built in the United States into the early 1930s and the first with four engines. Weighing close to twenty-four thousand pounds empty, the big experimental patrol craft could carry double its weight, and for the 15–16 January 1935 test flight from Norfolk to Panama, it would lift off the ground at fifty-two thousand pounds, most of that load being fuel.

There were multiple objectives with the XP2H-1 flight, even though the chances of Navy adoption were slim due, in part, to the reverse operation of the major controls in the cockpit. The first objective was to measure consumption to see just how far the plane could fly. A second objective was to test the best fuel-throttle settings given that the plane became lighter as a result of fuel consumption. But it was the third objective (discussed later) that caused King to once again travel to Norfolk to observe part of a test. Pulling Thach aside, the BuAer chief, who was renowned as something less than a sympathetic figure for much of his career, put his arm around Thach and told the young lieutenant in a father-to-son tone to put the plane down anywhere if he ran into trouble. King knew this was a pioneering experiment and Thach had to feel some pressure, especially as newspapers were covering the test.

With a crew of eight—Thach (in charge of the test) and two other pilots, two engineers, a chief mechanic, helper, and one radioman—the flight was successful despite some strong head winds. A major lesson learned, in addition to which power settings were most economical and the fact that the plane seemed to be able to fly forever, was that twenty-five straight hours in the air was too long for a crew. When the radioman passed out, no one could wake him and there was no one else aboard who could send messages. And Thach was in miserable shape at the end of the flight. His one attempt to get some sleep resulted in a cut on the head when he was suddenly awakened as two engines quit (the mechanics did not transfer tanks before they ran dry).

Happy to be at Coco Solo on the Atlantic side of the Canal Zone, Thach turned his attention to the third objective of the test. After King had reassured Thach prior to his departure from Norfolk, he specified that he wanted the plane in absolute overload condition for a test at Panama to determine if the big plane could take off in its ground swells. The paramount issue King wanted resolved was whether or not the existing breakwater at Coco Solo needed to be extended. The sea swells passing through the opening left a very restricted area for takeoffs, a persistent problem for the three patrol squadrons at the naval operating base. Those attached to the submarine base at the same place were adamant that the breakwater should not be extended because they wanted as much sea as possible to use for a quick diving escape should the need arise. After resting sufficiently, Thach and his crew climbed back into the XP2H with a full fuel load and attempted takeoff. Several attempts resulted in no successful takeoff, one lost wing-tip float, several leaks in the delicate hull, and bruises to the crew. Still, the test revealed things that were beneficial as well as what needed to be improved and marked one more step toward the new monoplane design that became the PBY. And the breakwater at Coco Solo was extended.

A second memorable mission while Thach was with VP-9 occurred in the summer of 1935, when he flew one of a dozen patrol planes to the Aleutians. The purpose of the mission was to test the squadron in deployment to a remote place for advance base exercises. The initial pioneering period for patrol boats in which emphasis had been placed on developing suitable and dependable aircraft to operate in the patrol function was passing. Sufficient progress had been made in aircraft development that by 1935 emphasis could be placed on operational exercises to prepare for the expected demands of a future war. Patrol planes had been used in World War I for antisubmarine work and coastal defense, functions primarily defensive and somewhat limited to the Atlantic. In the Pacific, Japan had invaded Manchuria in 1931, and in subsequent years the U.S. Navy began to think more seriously about a possible conflict with the Land of the Rising Sun. Looking at a globe it was immediately apparent that the shortest route from Japan to the United States was via the Aleutian Islands down the coast of Alaska and Canada. In 1935, the obstacles of seeking enemy ships and submarines or establishing forward bases in the Aleutians region were only partially appreciated.

The aircraft utilized by VP-9 was the Martin PM-1 biplane with two 525-horsepower Wright Cyclone R-1750D engines that propelled the big patrol plane at a cruising speed of 80 miles per hour and a top speed of 115 miles per hour. The open cockpit was not a desirable feature, especially for flights into the frigid waters of the north Pacific anytime of the year. Also undesirable, but absolutely necessary, were the bearskin flight suits. Although they kept pilots and crew warm, they were heavy, gave off a malodorous scent when dry and emitted an almost unbearably odiferous essence when wet. And they offered no protection for the face.

Knowing of the impending mission, Thach decided to end his chronic problem with appendicitis and have his appendix removed. That problem solved, he was able to concentrate solely on the problems of the mission. The cruise, as it was termed, was a series of flights up the coastline, a coastline often not visible and at the time poorly charted. Weather forecasting was inadequate and navigational aids were all but nonexistent. Flying solely by instruments was not an option because necessary instruments, particularly those that would help avoid mountain peaks that were not on charts, were not then available. For that time, the best navigational tool in a heavy fog was to descend until the white tops of the water could be seen over the sea or crashing onto the beach. White water was usually visible when nothing else was.

Mountain peaks were only the beginning of the pilot's problems. Weather was the foremost challenge, the cold temperatures taking third place behind storms and fog. Storms in the northern Pacific had characteristics not otherwise experienced by Thach. Totally calm seas and clear skies could disappear in as little as fifteen minutes and be replaced by near hurricane force winds and blizzards with zero visibility. On one occasion during the mission, the squadron encountered one such storm and discovered that an hour of flying into the blizzard had netted them only about seven miles. Even while the planes were riding at anchor, some winds registered fifty knots on air speed meters. Remaining anchored sometimes presented a major challenge as high winds threatened to blow the planes—and their anchors—upon rocky beaches. And if the weather and rocks were not enough to contend with there were also extreme tides that befuddled both the aviators and their tenders, once leaving the anchored planes high and dry for several hours when the twenty-five-foot tide ebbed.

Although remembering the Aleutians mission as fascinating and interesting, Thach also recalled the twenty-four-hour demands in the climatically hostile environment. Only a few years later the experience of having been under stress all day for many days proved beneficial, when weather concerns took their place alongside dangers from humans off both Japan and Korea.

SCOUT PLANES (1936–1937)

In June 1936, Thach was ordered to Scouting Squadron 6-B, where he experienced his first aviation unit command. Senior aviator in the squadron in June, a month later his promotion to full lieutenant brought status equal to his responsibility. Although reporting directly to the gunnery officer, he also had direct access to the commanding officer of his ship, the *Omaha*-class USS *Cincinnati* (CL-6), and it was during this assignment that he qualified to stand officer of the deck watch while under way. For all twelve months of this assignment, the cruiser division and Thach remained on the West Coast.

Scouting Squadron 6-B was based aboard a light cruiser division of four ships. The aviation squadron was comprised of twelve pilots, three allocated to each ship that carried two scout planes. The scout planes were the Curtiss-built SOC-1 Seagulls, small biplanes that were easy to fly and considerably faster than the patrol boats flown in VP-9. The single 600-horsepower Pratt and Whitney R-1340-18 engine provided a top speed near 150 miles per hour and a cruising speed of 130 miles per hour. Although capable of carrying up to 650 pounds of ordnance and mounting one fixed forward-firing .30-caliber Browning in the upper wing and one flexible mount in the rear cockpit, the SOC's primary function was scouting and main battery gunfire observation. Despite being blasted from zero to seventy miles per hour over *Cincinnati*'s sixty-five-foot compressed-air-driven catapults, which caused momentary blackout, Thach enjoyed flying the Seagull.

When first built, the ten units of the *Omaha* class did not have planes or catapults aboard. The addition of both, however, greatly enhanced the basic scouting function of the cruisers. To take advantage of the wind, the catapults were turned parallel to the bow to the greatest degree possible. This practice placed the catapult very close to the bridge and in full view of the CO and others whose duties placed them there. Blackouts, as noted, were the norm, but on one occasion the *Cincinnati*'s commanding officer was surprised to see Thach explode by his bridge with a board across his face. Wondering aloud why such an experienced pilot would pull such a stunt (which nearly put him in the water), he was told by the navigator that Thach was practicing blind flying. In truth, the gunnery officer had borrowed Thach's navigation board without knowing that Thach always tied it to the framework with string to keep it from getting loose. The gunnery officer had placed it back where he found it, under the SOC's instrument panel, but by not informing Thach of the temporary requisition, he nearly extinguished one of naval aviation's brightest lights.

Of course Thach could have suffered a mishap anytime he returned to the cruiser from a flight. The SOC had a large, but thin-skinned, float on the centerline that was nearly as long as the plane. To facilitate landing, the cruiser would make a sharp turn to level the sea and create a slick. Timing was critical for all phases of the operation. The slick did not last long as the SOC approached the ship and a small towed sled upon which the float would rise far enough to engage a hook. Vision was obscured during the entire effort by the engine and heavy spray kicked up by the prop. Once on the sled, problems were not over because the back seat observer/gunner had to connect the ship's crane cable to a ring on the upper wing of the SOC. That done, pilot and observer trusted the crane to lift them safely back aboard.[4]

The man Thach relieved when he came to Scouting Squadron 6-B was Lt. Warren W. "Sid" Harvey, one of the most promising officers in naval aviation. Harvey's outstanding work with the squadron was well known and appreciated but created both advantages and disadvantages for the new senior aviator. That standards and expectations were high was an advantage, but Thach knew he would be hard pressed to be better than the best. Nonetheless opportunities for improvement arose. Ship's gunners were constantly firing too soon at towed sleeves—even before they were within range—and Thach was asked by the CO to share his knowledge with them. During one catapult shot, a Seagull left behind the target sleeves on the rail of the ship. The dragging towlines forced the plane to land and precluded the normal procedure for retrieval therefore requiring the ship to stop. Responding to the CO's concerns, Thach and others worked together to develop a refined method that did not require an SOC to drag the unwieldy lines and sleeve down the catapult. The result was a practice bomb cut in half in which the sleeve and towlines were placed. As fittings on the practice bomb were already in place for bomb racks the entire operation was streamlined. The CO was sufficiently impressed to recommend that the innovation be submitted to the Bureau of Aeronautics. It was and BuAer accepted it.

Another improvement attributed to Thach was retrieval of target sleeves. Upon arriving in the squadron the routine was to have a Seagull drop the target sleeve near the cruiser immediately after a gunnery exercise. A small boat would be placed in the water to pick up the sleeve, a process that did not move quickly due in part to the weight picked up by the sleeve while in the water and the time required to get the boat back aboard. Thach's suggestion was to have a Seagull fly up the port side of the cruiser. When alerted by seeing the steam rise from the ship's whistle, the pilot was to cut over the bow and release the sleeve so it would land on the heavy wire running from the foretop to the bow. Happily, the experiment worked on its first try, Thach orchestrating all on the bridge immediately beside the CO. Although not always successful, it did work most of the time, and every catch kept the ship from stopping to place a boat over the side and wait for its return. Soon the enthused CO got into the spirit and wanted to give the steam whistle signal himself. Fortunately he got very good at it.

PATROL PLANES AGAIN (1937–1939)

In the late spring of 1937, Thach was detached from Scouting Squadron 6-B and *Cincinnati* with orders to report in June to Patrol Squadron 5-F, Coco Solo, Canal Zone, Panama. Initially he gave serious thought to protesting the orders. Having served with VP-9 for two years from June 1934 to June 1936, he believed he already had sufficient service and experience with patrol planes. In addition, he was very interested in climbing back into fighter aircraft, and he had no desire to return to the poverty-stricken Central American country of Panama. Happily his concerns were soon forgotten because the assignment proved far more interesting and beneficial than he expected.

When Thach reported to VP-5F in 1937, the squadron was flying Consolidated Aircraft Company's P2Y3 aircraft, but it was not long before the same company's newer PBY patrol planes, "the big boats" as they were then referred to, were available for operational use. The PBY was a refinement in aircraft technology with impressive capabilities. The early PBY-2s had a top speed of 178 miles per hour, a cruising speed just over one hundred miles per hour, a range up to twenty-one hundred miles, and a service ceiling of nearly twenty-one thousand feet. The increased range and ceiling were notable improvements over the earlier P2Y3. Four machine guns were mounted for protection, and the ability to carry up to 2,000 pounds of ordnance classified the big amphibian as a patrol-bomber.

Initially serving as the squadron's communications officer, Thach again found a training routine of scouting, search, horizontal bombing—with the new Norden bombsight in PBYs—and gunnery. Remembrance of World War I seemed far in the past as antisubmarine work received little emphasis with night flying being the major operations exercise for a significant portion of Thach's second assignment in patrol aircraft. One of the squadron's major objectives was to develop tactics for illuminating surface and shore targets with flares that descended under small parachutes. Several problems had to be solved. First, how many flares were required to provide sufficient illumination in both clear and inclement weather? Second, where should the flares be dropped in relation to various wind speeds? Third, at what altitude should the flares be dropped? And finally, how much time was allowed between adequate illumination and the arrival of bombers? The search for answers required months of trial and error. The squadron was flying so much in the darkness of night that newer pilots found daylight landings somewhat disconcerting, especially when they could see the obstructions hidden from view during night landings in the narrow harbor.

When the new PBY became available the squadron flew the older P2Y3s to Norfolk to turn them in and then traveled to San Diego to pick up the new planes. While in San Diego the pilots checked out in the PBYs and helped install bomb sights and other new equipment. After several weeks, the squadron was ready to fly fourteen of the new planes back to Panama. The squadron CO, Lt. Warren Berner, had directed Thach to develop a procedure for foul weather deployment to keep the planes from possibly flying into others when fog, snow, or rain erased all visibility of lights. Thach's system called for the CO to hold his course and altitude with the other planes—starting with ones farthermost out in formation—to turn out, fly for a specified time, and then climb to an assigned altitude above his former position. Once the objective of achieving lateral as well as altitude separation was accomplished, the respective planes would return to the original course and hold position until visibility allowed reformation. Development of the plan proved timely as the squadron encountered heavy weather en route to Panama. Not long after the foul weather spread was ordered, Thach—at Berner's command—attempted to fly over the front. Unsure whether or not the other thirteen planes were where they were supposed to be Thach made the attempt but at the top of the new PBY's ceiling he was still in the soup. That was only the beginning of other problems. An attempt to communicate with Berner was foiled by the discovery of ice on the antenna, followed by turbulent air and lightning strikes that burned his radioman's fingers and baseball-sized blue balls began drifting around in the cockpit. The blue balls were eerie and frightening and seemed to be from another world. Thach began to think that he was not long in being in another world, a thought amplified when he discovered that other unfamiliar lights turned out to be his wingman's plane nearly on top of his own. Concern only got worse when he learned that his wingman had been with him all through the dangerous flight rather than deploying as planned in the foul weather spread. After five hours of battling the storm, the weather cleared just about dawn and the formation reformed only to encounter yet another storm short of Panama. Again the foul weather spread was ordered, and when the formation passed through it, everyone was where he was supposed to be. After twenty-four hours of nonstop flight, the formation landed unnerved and exhausted but safe and satisfied with lessons learned and experience gained.

As earlier stated Thach was not initially happy when his 1937 orders to Panama arrived in part because of an addition to the family, son John S. "Jack" junior. Over the years he had developed an interest in golf that had risen to a level just short of passion. While Coco Solo offered little opportunity for him to remove clubs from the bag, Guantànamo, Cuba did. Flights to Guantànamo were welcome primarily to escape the relatively more primitive conditions in Panama even though in some respects—particularly sanitation—Guantànamo was as primitive as Panama. On one excursion to Cuba, Thach and several others exited the BOQ (Bachelor Officer Quarters) in response to some commotion on the pathway to the building. An exceedingly excited Chinese cook running at high speed with his pants down announced that while seated in an outdoor privy and reading the newspaper comic pages, he realized that what he thought was a bug nibbling on his ear was instead a six-foot snake. Along with others, Thach approached the privy and discovered the large water snake still apparently reading the comics. Returning to the BOQ, Thach reached for his golf bag and selected the short nine iron. Approaching the privy with the same train of thought that any golfer would have when ninety yards from the hole and needing a perfect shot for birdie or to save par, Thach entered. While his swing would not have received high praise from a professional golfer, he nonetheless executed a perfect shot. Reptiles were also known to crawl into the wings of planes, their discovery while in flight creating concern beyond consternation. As far as is known, however, Thach never recommended the installation of antisnake nine irons in Navy aircraft.

3

Fighting Squadron 3 Readies for War

In June 1939, at age thirty-four, Lt. John S. Thach was just about where he expected to be career-wise when he reported to Fighting Squadron 3 (VF-3) at North Island, Naval Air Station San Diego. Promotions had been slow for all officers since his graduation from the academy in 1927, so he was not disappointed to be a lieutenant after twelve years of service. Even if he had been overly concerned about rank, his assignment to a fighter squadron would have overwhelmed that concern. Fighting Squadron 3's "Felix the Cat" insignia was just as renowned with fighter pilots as was the "High Hat" insignia representing the squadron in which he had previously served. And like the earlier "High Hats," "Felix" also called *Saratoga* and San Diego home.[1]

The debate on whether the Army Air Corps and U.S. Navy would continue to share crowded North Island had been settled. The Army gave up Rockwell Field and moved to a new training site at Sunnyvale, California, late in 1935, thus allowing the Navy to expand on North Island and optimize available space. And in 1939, there were acres of concrete instead of the grass fields predominant during Thach's earlier tour with VF-1B.

TRAINING WITH VF-3

Soon after reporting to VF-3, Thach found himself assigned as gunnery officer, an assignment for which he was well qualified and to which he was intrinsically drawn. The pleasure of his new assignment was still in the incipient stage when a collateral duty mishap quickly cloaked his joy. Serving in rotational duty as fleet air officer-of-the-day (OD), Thach was responsible for determining the direction from which planes should land and designating which runway should be used. On the morning of the fateful incident, Thach determined the weather was too bad for planes to either take off or land. In his office some distance from the control tower—where his assistant, who implemented the OD's signals, stood duty—Thach did not know that Bombing Squadron 2 under the command of Lt. Cdr. Don Felt had taken off for a scheduled exercise. Felt's planes had to cross the mountains to reach the desert area for the exercise, and they attempted to fly through a pass as the weather worsened. However, at least three of the aircraft did not have sufficient altitude and they slammed into the mountainsides.

Along with others on duty and in positions of responsibility on the day of the accident, Thach was called as a witness during the investigation. Although everyone knew the regulations at the time invited just such a tragedy, someone had to be blamed. The regulations gave the officer-of-the-day responsibility to open or close the field, but each squadron commander had the authority to determine whether or not his squadron would fly. The obvious to the contrary notwithstanding, both Thach and Felt received letters of admonition from Capt. John H. Hoover, commander, Aircraft, Battle Force. Throughout his career, Thach came to know many people in the Navy's government and civilian sectors, and he liked and admired 99 percent. Capt. (later Adm.) "Genial" John Hoover was on Thach's short list of the other 1 percent.

The letter of admonition was not the only issue between Thach and Hoover. Hoover was particularly concerned about security for Navy installations following the 1 September 1939 outbreak of war in Europe, when Hitler's legions attacked Poland. After adding more sentries and implementing other changes, he declared North Island secure. Itching to spread an itch to Hoover, Thach seized upon another duty of the officer-of-the-day to demonstrate that Hoover's declaration of security was overstated. As the OD was also responsible for ensuring that sentries challenged all whom they did not recognize, Thach blackened his face, put on coveralls, and employed commando tactics soon after the sun went down. Successfully crawling across the field, he dashed unseen between sentries and placed sabotage tags on aircraft before gleefully making out his report for the following day. Not unexpectedly, Hoover did not accept the OD's demonstration as evidence of inadequate security in spite of

Thach's recommendation that sentries would be more effective if posted inside the hangars closer to the airplanes.

Happily, Thach did not spend most of his time on rotational OD duties. Flying fighters was the top priority and the new gunnery officer was in his element while preparing guns on the ground and especially when firing them in the air. Upon joining the squadron in June 1939, Thach found VF-3 flying the F3F-1, one of Grumman's last successful biplanes but known to be obsolete. Although Thach and other VF-3 pilots generally liked the F3F, they could not help but notice that the German fighters in the new war were fast monoplanes that carried much heavier armament than the Grumman. The two fixed, forward-firing .30-caliber guns of the 231-mile-per-hour F3F-1 would surely forfeit advantages against a German fighter not only in armament but also in speed and altitude. These characteristics for comparison were known, whereas the F3F's disadvantage in climb was only surmised in the early days of the European war. Consequently, there was considerable relief among VF-3 pilots when the nine Brewster Buffalo F2A-1 monoplanes were made available to the squadron in December 1939.

The Brewster Aeronautical Corporation had received a contract for a new monoplane fighter in 1936, but the Navy wisely did not discourage Grumman Aircraft from continuing development of their monoplane entry, the F4F Wildcat. Although there was little difference in climb or service ceiling, Fighting 3 pilots initially liked the F2A-1 Buffalo as it was fun to fly and could reach a top speed more than ninety miles an hour faster than the F3F-1 biplane and mounted four .50-caliber machine guns. European leaders also quickly saw the advantages of the new fighter and placed orders. Given that Western European countries were already in combat, priority was assigned to them rather than U.S. Navy and Marine Corps squadrons. This factor, plus Brewster's production problems, meant VF-3 would receive only ten F2A-1s instead of the required eighteen, thereby leaving the squadron operating with two different fighters for three-quarters of a year. In September 1940, the newer version F2A-2s began making their way to VF-3 and the squadron was then able to turn in their nine remaining F3F-1s in October.

Gunnery practice with the F2A-1 and F2A-2 Buffalos remained similar in many ways to that with the earlier biplanes, except that the Buffalo was not quite as maneuverable and, as an advantage, carried much heavier armament. As before, Individual Battle Practice (IBP) was similar, with two planes either beginning the exercise by having one fighter with an altitude advantage or with the two fighters approaching each other head on. The goal in either exercise was for one of the fighters to lock onto the tail of the other. Even though bullets were flying in earnest in Europe, they were still scarce in U.S. Navy carrier squadrons so some exercises were conducted with only twelve rounds of ammunition in one gun. While continually complaining over the scarcity of ordnance due to the effects of the Depression and more recently the need to transport it to Europe, the Navy pilots were nonetheless becoming particularly adept at making each practice-round count.

Although each round had to count due to the economics confronting each squadron, training nonetheless had to continue in realistic form. Greatly aiding in this practice was a camera gun, bore-sighted parallel to the guns with a small clock in each frame of the film. The gun cameras were synchronized to operate only when the fighter's trigger was pressed and by setting all clocks on the ground immediately before IBP. This not only aided in accuracy but also presented a realistic mode of combat. Further, it eliminated wasted time on arguments as to who shot first and served as an educational tool to hone skills and eliminate mistakes. Given its significance as a training aid, Thach was surprised to learn that neither the British nor the U.S. Army pilots utilized it.

The gun camera also greatly assisted pilots in their training for the major dissimilarity of training between the older biplanes and new monoplane fighters. When all U.S. Navy fighters and bombers were biplanes, some had rear-seat gunners that made a rear approach exceedingly dangerous and a deflection shot (approaching and shooting at an angle in the direction where another plane was heading) was necessary. Deflection shooting was not born as a result of the higher speed monoplanes, but the faster, relatively less maneuverable monoplane targets required deflection shooting more often than with biplanes.

Although gunnery was an interest Thach maintained throughout his career, he soon assumed duties as operations officer for the squadron and was serving in that capacity when Lt. Cdr. "Sid" Harvey (USNA 1924) took command of VF-3 in early 1940. In October 1940, Thach relieved Lt. Charles H. Quinn (USNA 1926) as executive officer (XO), an appointment that placed him second in command. Harvey, one of Thach's all-time favorites, as he was with almost all who knew him, had commanded *Cincinnati*'s scouting squadron immediately before Thach took command in 1935 and recorded an enviable performance. Like Arthur Radford, Harvey was an officer of considerable vision and understanding of how individuals' daily work translated into service to the Navy and nation. And like Radford, when he began to speak, all around stopped what they were doing to listen. Highly likable as a human being, he was equally respected as an efficient leader and administrator.

Soon after Thach assumed duties as executive officer of VF-3 in October 1940, Harvey received orders to England to observe the Royal Air Force during the Battle of Britain (10 July–31 October 1940). Thach was to command the squadron until Harvey returned in early 1941, at which time Thach was to be temporarily detached for the same duty. On 12 December 1940, however, Harvey suffered a fatal heart attack in Washington, D.C., while en route to England. Although already in temporary command of VF-3, Thach did not expect to retain command, but to his surprise and satisfaction, he became one of the first 1927 Naval Academy graduates to command a fighter squadron. His promotion officially blessed, he was assigned a new executive officer in January 1941, Lt. Donald A. Lovelace (USNA 1928), a man with whom he was highly compatible both personally and professionally. Together the two would mold a fighter squadron destined to achieve a combat record second to none.

BUFFALOS OUT, WILDCATS IN

While the Grumman F3F biplane overstayed its welcome in VF-3 until late 1940 due to the slow arrival of the F2A-1s and later F2A-2s, the Brewster Buffalos became less popular as time passed. Initial versions of the fighter were a step forward for U.S. Navy carrier fighters, but both the F2A-1 and F2A-2 were plagued with mechanical problems. For much of the spring of 1941, VF-3 alternated between North Island and *Enterprise* (CV-6) as *Saratoga* was at the Bremerton Navy Yard in Puget Sound for modernization.

On 21 April, *Enterprise* left for Hawaii with VF-3 on board. Although delighted to be on Oahu, the pilots were frustrated with the performance of their planes. VF-3 never had to endure the problems of the F2A-3, the last of the three versions of the Buffalo and a real challenge to fly because of all the additional weight placed on the same airframe as the initial F2A-1, but they still had significant mechanical problems. The landing gear of all the Buffalos was subject to failure, and their engine problems seemingly never ended. The F2A-2's Wright R-1820 Cyclone radial engines often suffered damage to the master rod bearings, particularly at high altitude.[2] When the bearings went, the engine froze, and the problem was occurring with such frequency that VF-3's engineering officer, Lt. (jg) Edward H. "Butch" O'Hare (USNA 1937), decided to fit oil and temperature gauges to one of the fighters and fly to twenty thousand feet, about two-thirds of the fighter's service altitude. Not unexpectedly, the engine froze and O'Hare had to ride the plane down and make a dead-stick landing. The test revealed improper oil seals supplied to the Bureau of Aeronautics, but still it took time to disassemble the troubled engines and make necessary repairs.[3] And the oil leaked onto the windshield, making carrier landings, especially at night, all the more difficult.

The aircraft mechanical problems so affected VF-3 that by late May, Thach had less than half his fighters available for tactical exercises and a scheduled Fleet Problem. On 24 May, he expressed his concerns verbally and in writing to Vice Adm. William F. Halsey, admitting that he and his pilots, particularly new additions to the squadron, desperately needed gunnery practice and other basic training. After asking several questions to fully understand VF-3's problems, Halsey not only acquiesced to Thach's request but also forwarded a message to the commander at Ford Island, asking that he have his supply department ready day and night to handle gasoline and supplies for VF-3. Back on Ford Island, the squadron spent a month working on their planes, flying night and day and restoring combat readiness before returning to San Diego in early July. Later that month the squadron was to turn in their Buffalos for the Grumman F4F-3 Wildcat.

Early expectations of solving all their mechanical problems and worries by turning in the Buffalo were only partially met as the squadron pilots began to familiarize themselves with their new fighter. Several of VF-3's pilots, including O'Hare, were designated to ferry the Wildcats from New York to San Diego. By late October, however, the squadron had only six F4F-3s, and on 7 December only eleven were available. But the Wildcats that arrived in the fall of 1941 were not ready for combat and certainly not ready to face the Zero.

The shortage of Wildcats was compounded by a shortage of spare parts and other essential equipment for the planes. In one of the small three-by-six-inch black books Thach carried in his shirt pocket in the years before and during World War II, he kept a list of the problems attending the new Wildcats. When fully equipped, the Wildcats would be a definite improvement over the Buffalo, but the first arrivals did not have illuminated gun sights. Oxygen equipment and masks were missing, bomb racks were not attached, and solid tail wheels were absent. When spare parts for both engines and airframes would arrive was not known. Even blueprints were not available.[4] And if things were not bad enough, VF-3 had to give up their .50-caliber guns to a Marine squadron the *Enterprise* was preparing to transport to Wake Island in late November. In short, the mechanical condition of the squadron was, in Thach's words, "deplorable." Given all the Wildcat shortages, even the overweight final −3 version of the Buffalo began to look good.

With the prospect of war looming, some shortages that had been pervasive during the Depression years took on a different perspective. In the final months of peace, the shortage of ammunition that caused gunnery exercises to be held with only twelve rounds in one of four .50-caliber guns or with just one round in a gun was sometimes solved in two ways. For firing during a record practice, every round was fitted into the gun that was to be used before being belted to ensure the gun would not jam. It only took one oversized round to cause a jam. The second solution found VF-3 pilfering ordnance from good friend and 1927 classmate Lt. Cdr. Paul Ramsey's VF-2 squadron. Not comfortable in authorizing any of his enlisted men to steal the ordnance on their own, Thach organized the unspoken loan. Into 1971 he had not shared with his old friend and then-golf partner that he had appropriated the shells nearly forty years earlier. He need not have worried as reports have since surfaced that VF-2 officers and men were just as active in creative ordnance acquisition as were Thach and VF-3. Regardless of where the available ordnance was acquired, incendiary shells were not available. And Thach began working the math to see how long it would take rated ordnance men to belt and prepare 32,400 rounds of .50 caliber. Unhappy calculations indicated extra men would be required to meet wartime conditions.[5] In sum, whether Buffalo or Wildcat, VF-3 was not ready for war.

EVOLUTION OF THE THACH WEAVE

Intelligence reports frequently crossed Thach's desk as 1941 wore on and combat became more intense both in Europe and China. Some, however, struck a particularly sensitive chord. Catching his attention were reports of a new Japanese fighter, the Mitsubishi Zero, that purportedly could outfly any fighter in the United States Navy and Army Air Force. Although the Zero had been in combat from 13 September 1940, intelligence on it was slow to make its way out of China. Confirmation of the Zero's remarkable attributes from sporadic reports was documented in the 22 September 1941 Fleet Air Tactical Unit Intelligence Bulletin. The bulletin described the new Japanese carrier fighter as having a top speed of nearly 380 miles per hour and armament of two 20-mm cannons and two 7.7-mm machine guns. This and other reports addressed the Zero's high rate of climb (purported to be five thousand feet per minute but actually about three thousand) and superb maneuverability. Thach was inclined to give credibility to the bulletin as it appeared to have been translated from a Chinese fighter pilot report.[6] He initially thought the earlier reports and perhaps the bulletin might be exaggerated, but even if they were, he surmised the Zero had flight characteristics no U.S. Navy carrier fighter could match. Indeed, the estimated top speed of the Zero proved to be exaggerated, as the first Mitsubishis topped out at about 322 miles per hour. But the early reports on armament, climb, and especially maneuverability proved correct.

Aware that relations between Japan and the United States were rapidly worsening in the fall of 1941, Thach began thinking how he might counter such an aerial threat. In the previous months, VF-3 and other squadrons had experimented with tactics used in the European war, one of which was to have weavers cover the retreat of formations as they headed back to base. These covering planes usually did not fare well, and the tactic did not gain acceptance in the carrier squadrons exercising in the Pacific earlier in the year. Thach realized that the distance the two weavers were trying to cover was too large to be effective, resulting in just one plane against one or more of the enemy seeking to down crippled aircraft lagging behind a formation.

The theoretical thoughts of how to counter the potential adversary became a near obsession as Thach sat at his dinning room table night after night. Losing track of time while lost in his thoughts, his wife usually had to remind him of the lateness of the hour and that he needed to go to bed and rest to be ready for the following day's flights. To Thach's thinking, the three most important considerations for a fighter pilot were speed, rate of climb, and a tight turning radius—the tighter the better. As he continued to consider the problem, he placed speed as the least significant factor as a pilot could gain speed simply by diving. However, the pilot was totally dependent upon the capability of his aircraft to rapidly gain altitude or maneuver in tight turns. For days on end he carried his thoughts from the dining table to the squadron where there was considerable give and take.

Several VF-3 pilots thought the intelligence reports on the Zero had to be exaggerated. Thach responded that they might be, but he noted that even if the plane's characteristics were cut in half, they still exceeded the attributes of both the F2A Buffalos and the new F4F-3 Wildcats. Reports indicated the Zero could climb at the rate of five thousand feet per minute. Cut in half, it would be twenty-five hundred, a capability greater than VF-3's planes. The same held true for turning ability.

Back at his dining-room table in the small rented house in Coronado, Thach poured out a box of kitchen matches. Some matches did not survive the evening, as both Thaches were heavy smokers, but enough remained to place two to eighteen—eighteen was still the authorized size for a fighter squadron at the time—to represent his squadron's planes with a dozen or more to portray enemy fighters. Thach called to memory his days on the football field in Arkansas and at the academy and how he usually countered a larger, faster opponent by baiting him into position where the opponent's superiority was either negated or mollified. Building on this premise, Thach determined that the one advantage his planes might have over the Zero was four .50-caliber guns and the marksmanship of his pilots. In fact he came to believe that the only chance his planes had would be to find a way to bring their guns onto the target even if only for one or two seconds. This conclusion became the first axiom in his developing formula.

The second consideration of the Zero problem was the formation used in fighter squadrons at the time. The value of two-plane sections as opposed to the standard three-plane section had been discussed at length for years, and after the beginning of the war in Europe, in 1939, it was put to the test. Two squadrons, VF-2 and VF-5, were directed to test the two-plane section. Despite enthusiastic reports favoring the two-plane section from both squadrons, in the summer of 1940 adoption for all carrier squadrons was rejected.[7] In part, a 29 July 1940 confidential memo to Halsey written by Lt. Cdr. Miles Browning, the OinC, Fleet Aircraft Tactical Unit contributed to the rejection. In his memo, Browning championed the continuing use of the "6-plane VF unit" by implying it contributed in part to British success in aerial battles early in the Battle of Britain.[8] However, in the summer of 1941 Halsey reversed his earlier decision and directed that each eighteen-plane fighter squadron be comprised of three six-plane divisions with each division organized into three two-plane sections.

Thach had been unhappy with the three-plane section for some time and was especially disconcerted with the fact that in a three-plane section, wingmen had to spend an inordinate amount of time concentrating on maintaining formation and preventing collisions. Of course, this took time away from looking for enemy aircraft. After the experiments of VF-2 and VF-5, Thach was even more convinced the two-plane section was the only way to go, and he kept it for squadron training after he assumed command of VF-3, despite the fact that it did not become official policy until nearly seven months later.

Freed from the three-plane section restriction in July 1941, Thach was able to concentrate on how a two-plane section could best counter the Zero. While aboard *Enterprise* for sea duty in April 1941, he had already experimented with the British weaver tactics but with modifications and without the continuous weaving practiced by the British. In a 12 May letter to Halsey, he noted the potential of the tactic and indicated he would continue experimentation.[9] The experiments involving the two and three plane sections in combination with the weaver tactic brought the second factor of "lookout" into his evolving formula.

The third factor required to bring his formula into place was how best to combine the first two factors of firepower and effective lookout protection. To find the third piece of the puzzle he experimented in the air with his pilots in a variety of formations. A four-plane division did not work out. Two sections flew with one behind the other. Distance was opened, then closed, and planes flew in formation with one plane in a section higher than another; he even tried sections flying in formation at different altitudes. Each experiment presented problems and did not allow for optimum lookout or the highest probability of bringing enemy aircraft into friendly gun sights. Finally, the combination of formation flying trials and matchsticks brought the elusive third component into the formula. On the evening that Thach worked out the theoretical problem with matchsticks, he shared the results with his wife ("Madalyn, I've got it!") and promised it would no longer be necessary for her to remind him when it was midnight.

The solution for Thach's "beam defense position"—his initial terminology for the tactic—was to have two sections fly abreast each other at a distance of approximately the tactical turning radius of the Wildcat. This provided optimum lookout for both sections as each had good vision above and behind the other, and with each section looking at the other, individual pilots only needed to look in one direction. When attacked, there would be no need for time-consuming signals or dependence upon then-undependable radios because the section first spotting the enemy would turn toward the other section, thereby alerting it to the impending danger. As the enemy came into range to attack one of the two sections—it could not attack both simultaneously—the other section would have already turned to bring their guns onto the attacker.

Then, if required, the two sections could continue to turn toward each other in a scissors maneuver which would not only bring one of the sections into firing position but also serve to disrupt the attention and aim of the attacker. The theoretical formula—firepower, optimum lookout, and mutual beam scissoring—was in place. What remained was to transfer the theoretical formula to the aerial laboratory. The final exam for the beam defense position was an aerial experiment with Thach leading a division of two sections against an "enemy" division of two sections.

Lt. (jg) "Butch" O'Hare was to lead the "enemy." O'Hare had been with the squadron for over a year, having reported to VF-3 in July 1940. Almost immediately the young man raised in St. Louis proved to be one of the best pilots in the squadron. Long before he became squadron commander Thach led the squadron humiliation team that took newly assigned pilots into the air, gave them altitude advantage, and then quickly defeated the neophyte. Positions were then reversed, and the veterans again defeated the neophyte by proving the newcomer could not evade the attacking veteran or gain firing position. The ultimate objective was to demonstrate to the pilots with their freshly minted wings from Pensacola that they were not as good or as seasoned as they thought they were. O'Hare, to the contrary, won his mock battles with Thach when he joined the squadron. Not one to let an opportunity pass, Thach then placed bets on O'Hare as he proved to other veterans on the humiliation team that they could not evade him or get favorable firing position. After collecting his winnings, Thach placed O'Hare on the team.

As important as it was to humiliate a young pilot into a more realistic frame of mind, O'Hare's assignment for the test of the beam defense maneuver would prove more significant. To allow for the Zero's alleged superiority and to give O'Hare's division a guarantee of superior performance, Thach had little marks placed on the throttle quadrants of the four planes in his two sections. This was to limit him and his other three pilots by not being able to advance their throttles more than half way. Instructed to attempt a number of attack approaches, O'Hare led his planes into the air and gained altitude advantage. Attacking from several different angles, he discovered in reality what Thach theorized. Every approach found the two-plane section he attacked maneuvering away from his line of fire as all but an astern attack required a lead or deflection shot. And every O'Hare-led approach against the constantly weaving Thach division found the noses of two Wildcats in one section or the other pointed directly at him as he prepared to shoot. Back on the ground, O'Hare happily announced, "Skipper, it really works!" At best, the final exam proved an attack by superior aircraft was not a death sentence. At worst, it kept the squadron from being demoralized and the pilots had a plan. Any plan was beneficial as there was little chance it would be the same as that of an enemy, and anything that an enemy pilot found distracting was advantageous.

At San Diego in the fall of 1941, Thach and O'Hare were pleased with the successful evolution of the beam defense position. But it would be another year before another fighter pilot and mutual friend of both would proclaim the tactic the "Thach Weave"—after it saved his life during the Battle of Santa Cruz.

NAVAL AVIATION READIES FOR WAR (1939–1941)

From the time Lieutenant Thach joined VF-3 in June 1939 to the first week of December 1941, momentous events affected naval aviation. At the beginning of the war in Europe, the U.S. Navy had five fleet carriers in commission (*Lexington, Saratoga, Ranger* CV-4, *Yorktown* CV-5, and *Enterprise*). The June 1933 Vinson-Trammell Act authorized construction of the *Yorktown* and *Enterprise*, with *Wasp* (CV-7) following the next year (commissioned 25 April 1940). These three carriers came to the Navy in part to build up to treaty allocation, but they were primarily built to help the stagnant economy during the Depression. *Hornet* (CV-8) was authorized in the 17 May 1938 Naval Expansion Act and commissioned 20 October 1941 in response to worsening international relations rather than as a spur for economic recovery. Additional fleet carriers were authorized before and after Pearl Harbor, but none would be available for the better part of two years.[10]

Although aircraft factories were turning out planes of all types in ever-growing numbers, so many were being transferred to Europe that the number of combat planes available to the U.S. Navy was not greatly different in late 1941 from what it was in the summer of 1939. The number of combat aircraft in inventory on 1 July 1939 was 1,316, and in July 1941, it was 1,774. While congressional authorizations for more aircraft were encouraging, Thach and other fighter pilots complained throughout the 1939–41 period that not enough new fighters were available for training, much less combat.

The one area of rapid development, however, was pilot training, with eight hundred men a month being admitted to training by December 1941, up from approximately forty in 1939. In addition to greater numbers of pilots being trained, the time required for instruction had been cut in half from one year to six months. Despite a strong industrial base moving toward economic recovery and wartime production on the eve of the U.S. entry into World War II, there were only nine fighter squadrons to operate aboard seven fleet carriers. The Marine Corps had four. Few of these squadrons had all the planes, spare parts, and experienced pilots they desired, but the hour was at hand.

4

World War II and VF-3's First Battles

In June 1939, when Jimmie and Madalyn Thach settled into their quarters in Madalyn's hometown, the city that Jimmie too would eventually call home, neither knew that the interlude between the two world wars was within three months of its end. Both followed the news closely, and both knew that problems in Europe and the Far East were serious. Both understood that another war was possible, even probable, but the beauty and serenity of San Diego, and the hope that diplomacy would succeed in preventing another massive bloodletting in Europe and resolve the conflict between Japan and China, had a calming effect.

ADVENT OF WAR IN EUROPE

Benito Mussolini's Fascist Italy had invaded Ethiopia in 1935 and the following year joined Adolph Hitler's Nazi Germany in the Rome-Berlin Axis. The Spanish Civil War, a dress rehearsal for World War II, with Hitler and Mussolini supporting eventual Fascist victor Francisco Franco, raged from 1936 to 1939. In 1935, Hitler renounced the Treaty of Versailles, which had ended World War I, and occupied the demilitarized German Rhineland in 1936. In March 1938, he occupied German-speaking Austria, and in September 1938, he took control of the Sudetenland in Czechoslovakia. Despite a promise to seek no further territorial concessions in Europe, in March 1939, Germany took control of the

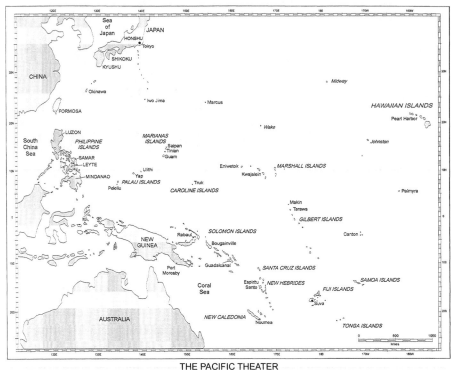

THE PACIFIC THEATER

remainder of Czechoslovakia and on 23 August of that year stunned the world with a non-aggression treaty with communist arch-enemy, the Soviet Union. With the immediate pressure of a two-front war held in abeyance, Hitler was free to set his sights on Poland and Western Europe. On 1 September 1939, his legions crossed the Polish frontier, and on 3 September, Great Britain and France declared war against Germany. In the United States, President Franklin D. Roosevelt immediately declared a limited national emergency. After protracted debate, the Neutrality Act of 1939, which allowed European democracies to buy war materials from the United States as long as they paid cash and provided the means for transportation, was passed by Congress. At the top of the list of needs in Europe was aircraft, and although Thach and others in naval aviation understood the rationale for the act, it nonetheless made it difficult for U.S. Navy carrier squadrons to obtain planes and spare parts to the degree desired for more than two years. In the spring of 1940, Germany again stunned the world by overwhelming Denmark, Norway, Holland, Belgium, and France.

With the British forced off the continent at Dunkirk, President Roosevelt quickly presented congress with proposals to strengthen an unprepared United States military. In July, the Two-Ocean Navy Act was passed authorizing additional carrier tonnage and an increase in the number of planes to fifteen thousand. On 2 September, fifty old World War I destroyers were transferred to Britain to help protect convoys in exchange for eight defensive bases stretching from the Caribbean to Newfoundland. And in the same month, the country's first peacetime conscription law was passed. In 1941, the last year of peace, additional legislation necessitated by the war in Europe was passed in Congress. The 11 March Lend-Lease Act allowed the cash-strapped British to accept war materials on a loan basis, not only aiding their military but also placing industry in the United States on a war footing and the general economy on a rapid pace to recovery. Germany and Italy knew the United States was not neutral but were not anxious to force Americans into the conflict, particularly after Germany invaded the Soviet Union on 22 June. But conflict with the United States was inevitable as German U-boats attacked convoys comprised of merchant ships flying the U.S. flag and protected in part by U.S. Navy warships. On 21 May, the American merchant ship *Robin Moor* was sunk, and on the twenty-seventh, Roosevelt proclaimed an unlimited national emergency. After the U.S. Navy destroyer *Greer* (DD-145) was attacked without damage in September, U.S. escorts were authorized to attack first if threatened. On 17 October, the *Kearny* (DD-432) was damaged by a torpedo and suffered 11 dead and 22 wounded. And on 31 October, the *Reuben James* (DD-245) was torpedoed and sunk with the loss of 115 dead, nearly three-quarters of its crew. Finally, in November the most serious restrictions of the 1939 Neutrality Act were abandoned and merchant ships were authorized to arm themselves. Although Hitler had ordered his U-boats not to fight unless in danger themselves, it was apparent that open warfare on the seas was only a matter of time, and surely a declaration of war was imminent on one side of the Atlantic or the other.

SLIDE TO WAR IN THE FAR EAST

Relations between Japan and the United States were generally favorable from soon after Commo. Matthew Perry's visit in 1854 until near the turn of the century. Victory in the 1898 Spanish-American War, which left the United States in possession of the Philippines, was viewed with some concern in Japan, but relations did not sour until 1905, when President Theodore Roosevelt served as mediator to end the Russo-Japanese War. Although the Japanese won decisive battles both on land and at sea, they were unhappy that they had to surrender some of their gains during the diplomatic endeavor to end the fighting. Soon after, the Japanese again felt discrimination when the San Francisco School Board decided it would not accept Asian students. Relations between the two countries declined to the point that Roosevelt's Great White Fleet arrived off Japan in 1909 not knowing whether they would have a fight or frolic. Happily, it was a frolic. Less than a decade later, both countries sided against Germany and the Central Powers, and after the war, Japan was awarded Germany's Pacific island possessions. Discrimination was formalized in 1922 when Japan agreed to the short end of a 5:5:3 ship ratio in the Washington Naval Treaty with the United States and Great Britain. And the Immigration Law of 1924 was received in Japan as a major affront. With visions of a "Greater East Asia Co-prosperity Sphere" and a stated desire to obtain more land for an already overpopulated country, Japan moved into Manchuria (northern China). Ignoring the argument that Java (Indonesia) had an even higher population density, the Japanese did not back away from their position. Their defense centered on the fact that Japan was devoid of many of the critical resources for industrial development (especially iron and oil). And they were not exaggerating when pointing to the intensive land use for agriculture, much of which was on terraced hills and mountain slopes. The bottom line was Japan's desire to create a feudal-type empire, and given the nineteenth-century and early-twentieth-century experience with imperialism, some Japanese leaders believed imperialism to be both their right and their destiny. Destiny in the 1930s appeared to support the Japanese expansionist desire given that the Depression-plagued Western democracies were too preoccupied with domestic economic problems to more directly confront challenges in the Far East. In September 1931, the Japanese military moved into Manchuria to compete for control of the region against both the Soviet Union and China. Letters of admonition from the League of Nations were initially obliquely answered but then ignored in 1936, when Japan withdrew from the league and disavowed the 1922 Washington Naval Treaty and 1930 London Naval Treaty. After signing the Anti-Comintern Pact with Germany in November 1936, the Japanese were in open warfare in China by 7 July 1939, although neither country officially declared war until more than four years later. The 12 December 1937 bombing and sinking of the USS *Panay* (PR-5) by Japanese planes was declared an accident, and U.S. attention largely passed back to the rapid and belligerent actions in Europe. Diplomatic overtures toward Japan were supplemented by the move of the Pacific Fleet from the West Coast to Pearl Harbor in May 1940. Two months later, in July, the Export Control Act prohibited oil, scrap metals, and other resources from being shipped to places other than the Americas and Great Britain. Japan's response was to further formalize its relationship to Germany and Italy on 27 September 1940 (Tripartite Pact). Less than a year later, the harshest measures short of war were taken when, in June 1941, the United States froze German and Italian assets and on 25 July took the same action against Japan, thereby eliminating the source of nearly 90 percent of that country's petroleum imports. There was oil in the Dutch East Indies (Indonesia) and other needed resources in Southeast Asia, but for Japan to seize the area would surely mean a greatly expanded war. While Thach realized the July diplomatic actions pushed Japan into a corner from which it either had to submit to international demands to resolve the China conflict or go to war, he and many others hoped for a peaceful resolution. On 25 October 1941, while Japanese diplomats traded proposals and counterproposals with their counterparts in Washington and Tokyo, all six of Japan's large aircraft carriers engaged in an at-sea exercise in preparation for a raid on Pearl Harbor. Ready to escort the carriers were two battleships, three cruisers, eleven destroyers, eighteen tankers, and supply ships. On 18 November, the armada headed into the open sea with orders to return home only if negotiations favorable to Japan were concluded by the twenty-ninth. The striking force was already halfway to Pearl Harbor on 2 December when the order came to "Climb Mt. Niitaka."

PEARL HARBOR AND WAKE ISLAND

On Sunday, 7 December, Jimmie and Madalyn Thach were riding in their Ford down Orange Avenue in Coronado on the way to have lunch with Don Lovelace, his wife, and several other members of the squadron when the program to which he was listening was interrupted with the news that Pearl Harbor was being attacked. Quickly he made a U-turn and raced back toward North Island.

Butch O'Hare also heard the news on his car radio while en route to have lunch with his wife of only two months. Everyone in the squadron had looked forward to the noon hour and afternoon on this Sunday having worked earlier on the day of rest to prepare for *Saratoga*'s arrival back from a five-week (left 30

October) yard period at Bremerton. The refitted carrier was scheduled to carry its own air group and Marine Fighting Squadron 221 (VMF-221) with fourteen F2A-3s to Hawaii. The news of war changed the order of loading to allow VF-3's Wildcats to go aboard last in case they had to be launched to defend the carrier, a fact that only exacerbated the chagrin of the squadron that had only half the number of Wildcats it should have had. If not for a strike by the company that produced a small electrical device needed for the electrically controlled variable pitch propellers for Wildcats, VF-3 would have had more planes available. But ready or not, the war was on.

After an afternoon, evening, and following early morning of purposeful loading, *Saratoga* pulled away from the dock and headed for the new war. Like Thach, many of his squadron mates were shocked by the surprise attack. It was particularly unsettling because numerous U.S. Navy exercises in Hawaii had demonstrated that surprise was possible and yet the Army Air Force—responsible for the defense of the harbor—was not prepared to ward off the attack. The nineteen pilots of VF-3 were anxious to get into the war and make right the things that had gone wrong on the Day of Infamy. Success in battle would be theirs, and soon. For some there would be career success and honors conferred. Three of the nineteen would eventually rise to flag rank, five would become aces, and one would win the Medal of Honor. But there would also be great sacrifice: twelve of the nineteen would lose their lives in less than four years, eight in less than a year.

For the week it took *Saratoga* to arrive in Hawaii, all aboard the carrier and escorting ships were a bit too anxious to fight. Reports of enemy carriers and submarines proved to be false alarms, but the several scrambles to find and attack the enemy at least provided some idea of how well prepared the men and their equipment were.

Time in Hawaii was very short, as *Saratoga*, *Lexington*, and *Enterprise* were ordered to prepare for operations in support of the relief of Wake Island, which had been under attack since 7 December (8 December Wake Island time). The plan was to land both aerial and ground forces and evacuate civilians from the American-held outpost some two thousand nautical miles from Pearl. Although at Pearl for only hours, what he saw there left an indelible impression on Thach. While not shocked to see that *Arizona* (BB-39) was still burning, he was surprised at the facial expressions of those he countered. Rather than the edgy bravado pervasive on the carrier, he found the people at Pearl still quite shaken from their harrowing experience the previous week. And he took a few minutes to analyze the attack concluding that the Japanese had done a commendable job in first striking the airfields before turning to other targets. That the new enemy would be a formidable opponent was confirmed when he talked with an Army pilot who had taken on a Zero and lost. The P-40 pilot related to Thach that he was on the Japanese fighter's tail and ready to shoot when it flipped over on its back, got on his tail and opened fire. For both Thach and his Army pilot acquaintance, there were no remaining doubts concerning the Zero's reputation for maneuverability.

There was one positive change in the postraid Pearl Harbor. Whereas previously even squadron commanders nearly had to beg—and sometimes did—for supplies, now they got what they wanted and got it without argument. Before 7 December, a squadron might need a part and there might have been three such parts, such as a carburetor, on the shelf. But regulations were interpreted then to mean that at least three parts had to remain on the shelf until others arrived. After the attack, if Thach needed three carburetors and there were only three on the shelf, he got all three. Before the war, even some carrier commanders did not fully accept the premise that a carrier's planes were the main battery. After 7 December, some of the same commanders vacillated on the concept, but at least the supply officers at Pearl Harbor had come to understand what was paramount to a carrier.

Even though the supply department at Pearl understood that those at the front had more need of supplies than warehouses, some of Thach's comments in his oral history indicate that he might not have been any more ready for war than others in the U.S. military. When orders came to abort the relief mission and return to Pearl, Thach was on *Saratoga*'s bridge with Rear Adm. Aubrey W. Fitch, the task group commander directly subordinate to Task Force 14 commander Rear Adm. Frank Jack Fletcher. Thach, Fitch, Fletcher, and everyone else on the relief mission was justifiably upset, Thach even suggesting that Fitch turn a blind eye to the orders. Although equally upset, Fitch responded that evidently someone in Washington knew something he and the others on the relief and support mission did not. And that was true.

Thach commented in his 1971 oral history that "we could have polished off that one little carrier they had and we could have knocked off all the transports and saved Wake Island." His comments echo those of many officers and enlisted men that were at the scene of events and battles but later did not have or gain access to the rest of the story. Indeed, that one little carrier was in fact both the *Hiryu* and *Soryu*, fast fleet carriers with a hundred planes accompanied by two heavy cruisers and two destroyers. Fresh from just pummeling Pearl, they would have been more than a match for the approaching and disjointed American forces.

Although Thach complimented several aspects of the Japanese attack on Pearl, he apparently never studied the attack in depth. Likewise, few of the Navy fighter pilots based at Pearl or attached to the three Pacific carriers wrote extensively on the alternate possibilities of the attack other than to patriotically state that they would like to have encountered the Japanese carrier planes in combat on that day. Even though the capability of the Zero was not fully known at the time, U.S. Navy carrier pilots should have known that three carriers against six off Pearl was not favorable odds. Too, many probably never fully appreciated that there were only forty-nine fighters aboard the three carriers, and seventeen of those were the overweight Brewster Buffalo F2A-3s. At least half of the forty-nine would have been retained as Combat Air Patrol to protect the three carriers, so there would have been far too few to fly escort for their attacking bomber and torpedo planes, all of which were slower than their Japanese counterparts. And it is significant to note that American carrier fighters would have had only about twenty-four rounds of ammunition to fire at each of the Japanese planes in the air on 7 December. More shells than that were usually expended just to charge the guns on Wildcat and Buffalo fighters before actually entering combat.

Aerial tactics then in use aboard *Lexington* called for attacking out of the sun, staying with section leaders, *and then reforming out ahead of the enemy planes before attacking again.*[1] Of course American carrier pilots discovered in the first hours and months of the war that pulling out in front of enemy planes was hardly possible when their planes could not even catch up to them. And just as it could have been at Pearl, so it probably would have been at or near Wake Island, and the U.S. Navy might well have lost one, two, or three carriers as the three Navy carriers were not steaming together.

Even before the recall of the Wake expedition force on 23 December (Wake time), Thach had reason to be upset. On the morning of the twenty-second, he lost his wingman, Lt. (jg) Victor M. Gadrow, in an operational accident. When Gadrow alerted Thach that something was wrong with his engine, Thach quickly returned to the carrier in order to give an emergency deferred forced landing signal. Sadly, Gadrow's plane got back to within sight of *Saratoga* but crashed into the sea.[2] That loss alone, which dropped VF-3's number of available Wildcats to twelve, would have been enough to trouble Thach. But there was also the recall on the following day and the knowledge that the .50-caliber guns that VF-3 had so meticulously prepared, primed, and polished before being ordered to transfer them to VMF-211 were lost. At least he could take some solace that some of the guns he reluctantly gave up did fire at Japanese targets.

VF-3 BOARDS *LEXINGTON*

Returning to Pearl on 29 December, *Saratoga* was ordered to sea on the thirty-first with Rear Adm. Herbert F. Leary on the flag bridge as both Fletcher and Fitch were ordered to other commands. January third was a good day for Thach, as word arrived that he had been promoted to lieutenant commander. But on 4 January he was in sick bay aboard the carrier with a nagging cold that held on long enough to put him in the hospital at Pearl when the ship returned to port on the thirteenth.

Fears were rampant that the Japanese might return to Hawaii, and so it was thought prudent to have the carriers outside the harbor as much as possible. Constant steaming, however, called for economy of fuel with consequent slower speeds than preferred in a potential combat environment. Concern for potential danger due to slow speed was realized on 11 January, when Japanese submarine I-6 put a torpedo into the port side amidships of the *Saratoga* while it was five hundred miles southwest of Pearl. Directed to the Bremerton Navy Yard for repairs, the big carrier also would have its 8-inch guns removed and a new antiaircraft battery installed. It would not return to Pearl until immediately after the June 1942 Battle of Midway.

Thach emerged from the hospital to find himself without *Saratoga* but he still had a full squadron of pilots and some additional new Wildcat fighters. Squadron tactical and gunnery training was interspersed with dawn and dusk patrols in conjunction with Army fighters, a routine necessary while the Army replenished aircraft lost the previous month. On 30 January, orders came for VF-3 to fly onto *Lexington* the following day as that carrier's fighter squadron (VF-2) was ordered ashore to trade their Brewster F2A-3s for Grumman Wildcats. Fighting 2's commanding officer, Lt. Cdr. Paul Ramsey, made sure his good friend Thach had the best Wildcats available. It was evident to both that VF-2 would need time to orient themselves to their new fighters and that Thach and VF-3 would almost certainly see combat first.

On 31 January, VF-3 flew aboard the *Lexington*, flagship of Task Force 11. Vice Adm. Wilson Brown was on the flag bridge while Capt. Frederick C. "Ted" Sherman stood on the captain's bridge. The mission initially called for the force to escort an oiler to meet Vice Admiral Halsey's Task Force 8 near Johnston Island southwest of Pearl and then continue south to cover another convoy. It was Thach's understanding that the force would be at sea only two weeks as the supplies on board were adequate for that period. And once aboard the carrier, Thach and his pilots were especially pleased with both the ship and the air department. Up to the beginning of the war, Thach had noted that not all carriers considered the air group as the main battery. This was due in part, he thought, because the top commanders often were not aviators and that led to certain restrictions that prevented adequate training at sea. On *Lexington*, however, he found that the carrier "was administered in every way to enhance the value of the air group. The training of the air group at sea in machine gunnery and bombing was given high priority," a practice any air group or squadron commander appreciated.[3]

LEXINGTON DISCOVERED OFF RABAUL

On 2 February, new orders arrived, and by the sixth the force was ordered to the South Pacific to protect islands maintaining the essential link to Australia. Although a few new ships joined Task Force 11 for the extended cruise, they did not bring additional food, a matter that was becoming a special concern on *Lexington*. By mid-February, the good food was all but gone, and soon only two meals a day—consisting mainly of canned beans and spinach—were being served. Especially missed was salt, particularly as the force moved toward the heat of the lower latitudes.

Before the middle of February, VF-3 knew there would be offensive action, probably in the area of the northern Solomon Islands. Soon it was known that the target would be Rabaul, the new and building Japanese base at the northern tip of New Britain. From the series of bases and fortifications in place and being built at Rabaul, the Japanese could direct forces throughout the Bismarck Archipelago, into the Solomons to the east and toward Australia to the south. Barring unforeseen circumstances, the aerial raid was planned for 21 February, Thach leading the attack with at least one division of VF-3, the intent being to destroy as many fighters as possible and keep the rest on the ground. The TBD

World War II and VF-3's First Battles

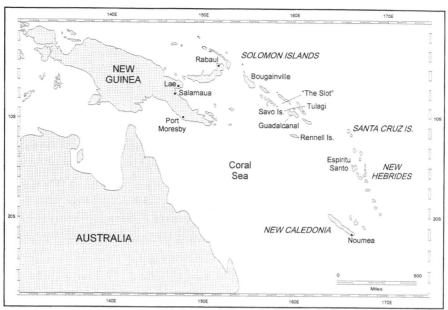

CORAL SEA and the SOLOMON ISLANDS

Devastators and SBD Dauntless dive-bombers would concentrate on ships in the harbor.

On 20 February, however, unforeseen circumstances—a constant variable in warfare—intervened. While *Lexington* was positioned more than four hundred miles from Rabaul in preparation for the next day's planned attack, it was discovered on the morning of the twentieth. As radio silence had become an absolute practice since the beginning of the war, Thach was nearly startled out of his seat when a loud voice broke in on his radio. The fighter director officer (FDO), Lt. Frank F. "Red" Gill, announced a vector to course 240 and stated that radar had picked up a snooper about thirty-five miles out. Thach and his wingman, Ens. Edward R. "Doc" Sellstrom Jr., turned as directed, leaving the other four Wildcats of Thach's six-plane 1st Division above the carrier in their two-plane sections. O'Hare, with his wingman in tow, attempted to follow Thach, but the squadron CO motioned him away. Thach knew that the contact was probably a patrol plane and would not be accompanied by fighters.

The discovery of Task Force 11 was made by the crew of a large, four-engine Kawanishi flying boat, and their message back to Rabaul ended all surprise for the planned raid. Now *Lexington* was to be the target instead of Rabaul. Thach and Sellstrom encountered a heavy cloudbank and rain as they neared the area where the FDO had placed them. With the FDO's further confirmation that the unidentified contact was near, Thach and his wingman plowed into the mist and, surprisingly, nearly on top of their quarry. Even the FDO was surprised, radioing a message to Thach that both planes had merged. Thach answered that the two planes were within thirty-five feet of each other, but the clouds were so thick he could see only one wing of the flying boat. Evidently gunners on the Japanese plane saw Thach or Sellstrom as the flying boat dropped enough to be noticed on radar. Breaking into the clear, the big patrol plane was not in sight, so Thach too descended to get under the cloud where visual contact was reestablished. Visibility was still poor, so Thach flew back into the cloud, made a couple of turns, and then resumed the base course the Japanese pilot was flying when last seen. Knowing the thunderstorm was isolated, he deduced the enemy would soon be in the clear. And soon he was.

Climbing for a little more altitude, Thach closed quickly, sliding to the right of the prey while Sellstrom, not immediately understanding his CO's directive, belatedly took up position on the left side. The idea was to bracket the big plane so it could not turn to place its defensive firepower at just one attacker. Following patrol aircraft doctrine also practiced by U.S. Navy patrol plane pilots, the Kawanishi held a steady course to allow its gunners a better opportunity to damage or discourage attacking fighters.

From above, Thach began his high side approach, thinking, as he later recalled, "With all of this training I've had over all these years, I'd better do it right this time. I even looked at my ball leveler while I was in the approach to be sure that I wasn't slipping or skidding, everything right, nice and easy, and took just what I figured was the proper lead on him, and waited until I got close enough, coming in from the side, and opened up . . ."[4] On his first pass Thach saw his shells hit the starboard wing and two engines at which he had aimed. However there was no apparent result. With Sellstrom now in position on the left and beginning his firing run, Thach began his approach from the right when suddenly the enemy's starboard wing exploded with flame. There was no need for further shooting for the two Wildcats as the flying boat burned, dropped its bombs, flew on a few minutes, dropped its nose and started down in a fatal descent. Smoke from the crash was visible back aboard the carrier.

Soon after Thach's section returned back over the carrier, the big CXAM-1 radar located another unidentified contact. This time the FDO sent Lt. (jg) O. B. "Burt" Stanley's section to investigate while Thach and O'Hare's sections remained over the force. Not unexpectedly, the contact proved to be another four-engine flying boat, and its fate before Stanley's and Ens. Leon W. "Lee" Haynes's guns was the same as Thach and Sellstrom's.

With surprise lost, Vice Admiral Brown decided to cancel the raid on Rabaul, but he changed course toward the objective in the hope that it would disconcert if not disrupt ongoing successful enemy operations in the region. As his ships were low on fuel, the risk of a running sea battle was unacceptable, and to close on Rabaul might bring his force under attack. Still, he had come too far not to affect the enemy in some manner.

By 1400, all six Wildcats of Thach's division were back aboard the carrier. Four of the pilots happily accepted congratulations although they were more anxious to find food, water, and cigarettes. O'Hare and his wingman not only had nothing to celebrate but also were agitated in not having been given an opportunity to engage either snooper. While Don Lovelace with his 2d Division circled the task force on CAP, Thach and his division smoked, quenched their thirst, and unhappily stuffed bland-tasting beans down their throats. Also aboard while Lovelace's division was on patrol were the six pilots of Lt. Noel A. M. Gayler's 3d Division, and they along with O'Hare and his wingman, Lt. (jg) Marion W. "Duff" Dufiliho, pressed Thach and the other three new combat veterans on what happened when they made their attacks. What kind of attack did they make? Were they the same as when shooting against a sleeve? Thach, Sellstrom, Stanley, and Haynes answered in the affirmative that their training translated directly to actual combat and that those who had not yet pressed the firing button were ready and able. Sooner than expected, they would join the ranks of combat veterans.

VF-3 DEFENDS TASK FORCE 11

After leaving the wardroom, Thach went to the squadron ready room to research charts and an intelligence manual to see what he could find on the two enemy bombers destroyed in the late morning hours. Years later he stated that he felt sorry for the crew of the enemy patrol plane because they no doubt believed the large number of guns aboard would allow them to successfully defend against fighters. Thach remembered similar propaganda-like statements while he was flying patrol planes, and he knew that Army bomber crews had heard the same assessments. Having served in fighters both before and after his assignments in patrol planes, he never bought into the idea that one swiveling gun could overcome the sighting problem and heavier armament of a fighter with an adequately trained pilot.

About 1630, Thach was still researching in the ready room when the call sounded for fighter pilots to man their planes. Racing to the flight deck he climbed into his Wildcat and took stock of the situation. Lovelace's 2d Division, low on fuel, was still in the air but preparing to land. Gayler's 3d Division had just been launched but was still climbing to altitude. Two F4F-3s were not operational; his wingman was siting in his cockpit several planes behind on the flight deck and deep in the distance enemy bombers were heading in.

Approaching Task Force 11 in two separate formations (nine and eight) were seventeen Mitsubishi G4M1 land-based Betty bombers. The two-engine, cigar-shaped bombers capable of more than 260 miles per hour could carry torpedoes, the most destructive weapon against ships, but these were armed with bombs for a horizontal attack. Still on deck, Thach could see the three sections of Gayler's division begin attacks on the tight division of nine bombers and very soon he could see planes falling in dark red flame and brown smoke. Unknown to Thach at the time, Gayler was talking to *Lexington* as furiously as he was fighting the bombers. The time-fused 5-inch antiaircraft shells were exploding above the bombers and peppering Gayler's planes as they gained altitude over the Bettys for attack.[5]

After what seemed an eternity, despite an exemplary effort by the flight deck crew and an actual time of less than fifteen minutes since the alarm, Thach, Sellstrom, O'Hare, and Dufilho were launched. Thus all sixteen of VF-3's Wildcats that could fight were in the air attacking, maneuvering for firing position, or climbing to meet the closing enemy formation. Helping the fighters immeasurably were Sherman's course changes, although the first maneuvers were to head into the wind to launch Thach's fighters and other fueled aircraft on the flight deck. Gayler's 3d Division caught the brunt of the attack, but the several flaming planes Thach could see before and after his launch were the result of the 3d Division's efficiency. And before he and the three pilots of his division could engage, Lovelace's 2d Division was in position, and they caught the remaining half of the enemy formation as it closed from directly astern of the carrier. Four Bettys survived Gayler and Lovelace's twelve Wildcats long enough to drop their bombs, but none came close to *Lexington* or its escorts. One bomber attempted to crash the carrier and failed, and the other survivors attempted to speed away.

World War II and VF-3's First Battles

Thach did not have altitude over the surviving Bettys immediately after their bomb run, but his Wildcat had plenty of fuel and ammunition. Knowing he had little chance to engage before any surviving bombers could drop their ordnance, Thach flew in the direction the enemy planes were headed. That his wingman and the other section had not joined on him was a matter of no concern since the targets were bombers rather than fighters. Chasing enemy planes that no longer had ordnance was not a priority for the FDO, and he ordered Butch O'Hare and his wingman to remain over the carrier, again much to the aggravation of O'Hare, who was beginning to think he might never be allowed to fight. Thach, however, had a green light to pursue the fleeing Bettys.

Within shooting range of the surviving bombers just after they missed their target, Thach was just below as the Bettys descended to pick up speed. Their mission now was to escape, and Thach's mission was to see that they did not. Coming up from below and behind, he lined up a side approach, instantly ran through the math and science of sighting a target, fired, and watched his shells disappear into the engine and wing. There was little time to watch the flaming, doomed bomber head down as his view shifted astern to watch in anger and horror as one of the 3d Division's F4Fs took a cannon shell in the windshield from another bomber just below him.

Thach's anger was simultaneously directed toward his squadron mate for disobeying a training axiom never to attack a bomber from directly behind and toward the Betty, whose tail-mounted cannon had taken the pilot's life. Quickly the anger directed at his dead squadron mate dissipated, but not so for the green and brown bomber and its crew. His rage channeled into determination, Thach decided to move up ahead of the Betty so he could get in close. Moving in, he noticed the gunners firing at him even though he was still out of their range. Diving from above he began his firing run and saw shells hit the wing and fuselage. Somewhat surprised that the bullets apparently had no effect, he pulled up, rolled over, and made another run. Looking back, there was no smoke or any other apparent damage, but suddenly the bomber that had just killed Ens. Jack Wilson spewed flame from a wing, blew up, and disintegrated.

Later Thach stated that his firing run on the second bomber had been too concentrated on the fuselage, a tactic acceptable against a fighter when there was the possibility of killing the pilot. Against a bomber, however, the aim had to be "target sleeve" precise and directed at a "vital spot" such as an engine.[6] Further, he deduced that a hole in an engine would ignite oil that was already under pressure and very hot, especially when being emitted as a spray.

At day's end, VF-3 could take credit for downing seven of the nine Bettys of the first attack and sharing an eighth with the guns of the carrier. The ninth was damaged by VF-3 Wildcats but finished off by a Scouting 2 SBD that abandoned scouting duties when the luckless, wounded Betty failed to escape notice some distance from Task Force 11. Thach clearly destroyed one Betty by himself and shared credit for the second Betty and the flying boat. Always concerned with squadron morale and the critical need for teamwork, he was perhaps too magnanimous, as the information now available indicates that both shared-credit targets would have been downed without the two assists he acknowledged. Years later, when recounting the battle, Ens. W. E. Eder, the pilot to whom he gave half a credit for the "Wilson" bomber, is not mentioned. And the only mention of his wingman (Sellstrom) was that they did not have time to join up after take off and "[I] couldn't see him."[7]

BUTCH O'HARE INTERCEPTS THE SECOND ATTACK

After downing the second twin-engine bomber, Thach continued on with others to chase the last of the nine bombers that had attacked in the first wave.[8] Soon after the battle, he would direct his pilots not to chase empty bombers too far from their carrier. On 20 February, however, he and several other pilots were moving away from the task force and Lovelace's 2d Division was attempting to land on *Lexington* when a second wave of enemy bombers appeared near 1700 hours.

As one Japanese bomber had engine trouble at takeoff, only eight were in the second wave. Whether eight or more, Red Gill had no option but to send the only two fighters available to meet the new threat and hope that other VF-3 fighters could return or climb in time to help. Realistically he knew help could arrive only after these bombers had released their bombs.

Aboard the carrier, Gill would have been even more worried had he known that Marion Dufilho's guns would not charge. Section leader Butch O'Hare saw his wingman's signals that his guns would not fire and ordered him to return to the relative safety above the carrier. To his everlasting credit, Dufilho refused to leave Butch, as he knew that his presence would draw fire away from the one fighter that did have live guns.

If Thach had to choose one pilot from his several outstanding pilots to press an attack alone, that pilot most likely would have been O'Hare. He was the neophyte fresh from Pensacola who had held his own against the humiliation team and was then invited to join it, the same pilot who had proven in the air that Thach's weave worked in reality as well as in theory. And he was the only pilot with functioning guns between eight enemy bombers and the single friendly carrier deck within two thousand miles. Butch had been left out of the fighting earlier in the day as a matter of chance and circumstance, and he was available to meet the new threat for the same reasons. The right man at the right time at the right place, he nonetheless had to meet the challenge. Unseen until his tracers and shells flamed two bombers on his first overhead pass, Butch appeared and disappeared too fast for enemy gunners to effectively retaliate or spoil his aim.

On his first overhead pass, two bombers flamed and fell out of formation. Recovering on the other side, he dropped two more out of formation on his second pass. Recovering again back to the side from which he began his first run, he attacked again and two more flamed, one suffering a violent explosion that tore the engine off its left wing. One last pass was abridged as his eighteen hundred shells were expended. Although O'Hare had shot six of the eight attackers out of formation, two were able to recover sufficiently to drop their bombs along with three others. In retrospect it is known that O'Hare was directly responsible for downing three and damaging a fourth enough to cause it to ditch. Still, his greatest feat was to disrupt the formation of the attackers. That contribution was made more evident by the near misses of the three bombers that survived long enough to drop ordnance against the carrier and one so far out of position that it dropped—and missed—over a cruiser.

Thach and Sellstrom, still not flying together as a section, chased and fired on the Bettys fleeing the second attack. Sellstrom got sole credit for one, and Thach emptied his guns on another. It did not get home, but its pilot survived.

The battle had been so intense and close that many on *Lexington* had a difficult time trying to accomplish their respective tasks. Even Captain Sherman was surprised to learn that Lovelace's fuel-starved Wildcats had slipped back aboard during the few moments he had ordered the carrier into the wind while maneuvering to frustrate the bombers' aim. After everyone was back aboard the carrier following the late afternoon battle, Thach immediately debriefed his pilots and then asked them to join him in drawing sketches of the bombers they had just destroyed. Intelligence officers claimed that the twin-engine Betty was a type already known to the Navy, but Thach and his pilots were convinced it was one they had not seen before either in the air or on an identification chart. Each pilot drew a part of the plane that they were sure they remembered and then a composite final drawing was made to take back to Pearl.

Butch O'Hare added his sketches to what eventually became the final composite and then warded off the compliments and congratulations coming from all in the ready room and wardroom. Soon everyone in the United States would learn of his feat.

ATTACK ON LAE AND SALAMAUA

Alerted to the U.S. carrier's presence, ships in Rabaul Harbor had taken to sea and Japanese fighters would no doubt be in the air before dawn on the twenty-first. Consequently, the plan to strike Rabaul on 21 February was off, but the idea was not dead. Task Force 11 steamed southeast during the last week of February to refuel, but there was nothing available to supplement the spinach and beans that had already worn out any lingering appeal. Weight loss among the pilots was apparent, complaints were uttered in ever more eloquent adjectives, and Thach even had nightmares about (more likely as a result of) the beans.

The Japanese presence at Rabaul, formerly held by the Australians until late January 1942, was still in an incipient stage in late February 1942, but there were sufficient facilities, Zero fighters, and supplies to hold the new acquisition. Landings at Rabaul, on Bougainville in the northern Solomon Islands, at Balikpapan in Borneo, and at Kavieng on New Ireland were strategic positions from which the Japanese could move to isolate Australia and defend the resource-rich areas of Southeast Asia either already captured or being assaulted. Capture of Port Moresby on the southwestern side of New Guinea's southern peninsula would not only expel Australian forces there but also secure the desired perimeter. Before attacking Port Moresby, however, landings were planned for Lae and Salamaua across the peninsula.

Wisely, Brown did not wish to attack Rabaul with just one carrier because that bastion was alert to the possibility of a carrier raid. En route to join Brown was Task Force 17, with Fletcher commanding aboard *Yorktown*. Refueling complete, both carriers and their attending cruisers and destroyers headed north toward Rabaul, but sightings of the Japanese invasion force heading for Lae and Salamaua brought orders to attack there rather than Rabaul.

By the time *Lexington* and *Yorktown* arrived in the Gulf of Papau and in position to attack on 10 March, Japanese forces were ashore at Lae (northernmost) and Salamaua, the two only some eighteen miles apart in a horseshoe shaped harbor. Their transports, however, were still unloading supplies, and at anchor the ships were inviting targets. The plan adopted by Sherman included the best thinking Thach and other squadron commanders could offer. *Lexington*'s air group comprised of Thach's VF-3, Bombing 2 (VB-2), Scouting 2 (VS-2), and Torpedo 2 (VT-2) would launch first followed by *Yorktown*'s air group (VF-42, VB-5, VS-5, and VT-5). Only VT-2 would carry torpedoes and fighters from both carriers would provide combat air patrol.

Given the distance to the targets, Thach's eight fighters launched from *Lexington* on the morning of the tenth and waited for the strike aircraft to launch, after which they returned to CV-2 to top off their fuel. Then they would catch up to the slower SBD Dauntless dive-bombers and TBD Devastator torpedo planes, the *Yorktown* air group following shortly after. Rather than approach from the open sea to the east or south where Japanese scouts would surely discover them, the attack route was from the southwest. However, the Owen Stanley mountain range that might hide their approach from enemy scouts posed a navigation hazard, especially for VT-2's thirteen torpedo-laden TBDs. A mountain pass seventy-five hundred feet high was located before the attack, but when Thach caught up with Lt. Cdr. James H. Brett's TBDs, they were having trouble climbing even to that altitude. For a few minutes it appeared the TBDs would not be able to surmount the peak of the pass and Brett informed Thach by radio that his planes might have to abort. Fortunately, Brett had glider training and knew that a sunlit area in the broken clouds could warm the ground enough to give him a thermal updraft. Happily, a sunny area was spotted and the Devastators circled it while rising and then washed over the illuminated pass. Once on the other side, Brett and the TBDs fell in behind Thach and his two divisions as they headed for their targets.

Arriving over the two target areas and their totally surprised occupants, Thach quickly determined that there were no enemy fighters to contest the raid. Still, he left Butch O'Hare at altitude to intercept any intruders while he and his three planes plus Noel Gayler's division descended to strafe. Only two floatplanes rose to challenge throughout the morning, one making the fatal mistake of flying in front of Gayler.[9] The absence of enemy fighters gave all the attackers freedom to pick out targets at will. Post-attack claims were inflated, as usual, but three transports were sunk, several other ships were damaged, and gun crews—both on ships and ashore—were killed or wounded. But the physical damage to ships, buildings, and supplies was far outweighed by the impact on Japanese planners. The few ships lost and damaged were already tabbed for the assault on Port Moresby and that attack was delayed. And to date this was the most damage enemy forces had sustained in a single attack thereby sending notice to the enemy that future acquisitions in the region might not be obtained without considerable cost.

As with every early engagement, there were lessons to be learned. While most factors in preparing for a sea battle were known, to climb high mountains a considerable amount of additional geographical and meteorological data was necessary. And Fletcher and Sherman were concerned about how near the coast they had to bring their carriers to allow for the short range of their planes. Thach shared these concerns but was especially interested in the lack of value of carrying two 30-pound bombs and the impact of his four .50-caliber guns on ships, particularly smaller ships such as destroyers and minelayers. The small fragmentation bombs had only impeded range and had probably caused no meaningful damage. Plus they would have had to be jettisoned early had there been enemy fighters over the targets. Too he was especially distressed at the performance of the aerial torpedoes. It appeared to him that only one exploded against an enemy ship and that others hit and did not explode or passed under targets and hit the beach. Just as live ammunition was not available in sufficient amounts for fighters before the war, the expenditure of live bombs and torpedoes had been inhibited by budget restraints. Now in combat the price of economy was sadly demonstrated. Although appreciated, a letter of commendation from Adm. Chester W. Nimitz, commander in chief, Pacific for the raid did not erase Thach's concerns for how prewar austerity was affecting the Navy's combat effectiveness.

TO PEARL AND THE BREAKUP OF VF-3

Lexington arrived back at Pearl Harbor on 26 March. Waiting were fresh food and a week at the Royal Hawaiian Hotel. While Pearl, good food, and the Royal Hawaiian were welcome, Thach was not thrilled over the loss of his squadron in the following weeks. First to go was Butch O'Hare—elevated to VF-3 executive officer after Lovelace received orders to command another squadron. Butch was ordered back to Washington to receive a Medal of Honor from President Roosevelt plus a promotion to lieutenant commander in a White House ceremony (21 April 1942). Although sorry to learn O'Hare was leaving, Thach continued to shepherd his younger protégé. First, on 27 March he faced reporters with Butch at his side on a piano bench in a press conference to address the particulars of Butch's heroism over the *Lexington* on 20 February. Uncomfortable with the attention and press, the modest and shy O'Hare deferred to his squadron commander who was not the least uncomfortable in detailing Butch's exploits. On 30 March, Thach and O'Hare sat down for a radio broadcast from KGU in Hawaii to NBC. The lengthy broadcast gave Thach an opportunity to describe his own air battles before Butch was invited to speak. On 11 April, Thach and O'Hare were ordered to make an aerial photography flight and pose for photographs on the ground near or in their Wildcats. Over time these pictures became historically significant, but at the time they were valued for public relations, support of the war effort, and the sale of war bonds.

Thach expected to go to sea again with *Lexington* when it departed on 15 April, but Rear Admiral Fitch personally informed him that he would not be aboard. New pilots arriving at Pearl needed someone with combat experience to teach them how to fight, and for the short run at least, this would be Thach's job. Fitch also told Thach that most of his VF-3 pilots would need to be loaned to Paul Ramsey's Fighting 2 and that a few others would go back to the mainland to train new pilots. The expectation was that in time Thach would get back his loaners and the new pilots he would train would comprise an expanded VF-3 of twenty-seven planes rather than eighteen. The loaners soon acquitted themselves admirably in the forthcoming 7–8 May Battle of the Coral Sea, but their expected return did not eventuate due to events unforeseen in mid-April by both Fitch and Thach. That Thach would miss them is perhaps best summed up by his remark "I wouldn't trade a one of them for the biggest ace in Germany."[10]

Replacements for VF-3 were to come, but on 16 April, Thach had two dozen brand-new F4F-4 Wildcat fighters with only himself to fly them—which he did. Before the end of the month, however, new pilots began to report and others arrived in May. It was clearly apparent that Fitch was correct in his assessment that the new graduates of the Advanced Carrier Training Group were not ready for combat. Although Thach would have preferred to take his new charges into the air for tactical training, he first had to teach them gunnery as they were extremely deficient in that most critical of fighter pilot requirements.

From 11 April to 19 April, Thach had one pilot assigned to him for temporary duty who could both shoot well and understand fighter squadron tactics. Lt. Cdr. James H. Flatley Jr. was a 1929 Naval Academy graduate whose career path had thus far been similar to Thach's. When the war began, he was Paul

Ramsey's executive officer with VF-2 on *Lexington* and in April had orders to report aboard *Yorktown*—then at sea—as VF-42's commanding officer. Flatley would catch up with *Yorktown* just before the Battle of the Coral Sea but a mix-up in orders resulted in having him fly into that two-day battle as VF-42's executive officer. Immediately after the battle, in which he shot down one enemy fighter and claimed another, he was ordered to take command of Fighting Squadron 10 (VF-10), which would go into action in August over and off Guadalcanal from *Enterprise* as the first carrier replacement fighter squadron.

Flatley and Thach had known each other for some time, but they became especially close in the few days they shared together in mid-April. While only orders survive in Flatley's papers for the period, Thach recorded the value of their brief Hawaii association in his memoirs. One of two special remembrances was Flatley's assistance of Army pilots who wanted to practice firing on a towed sleeve rather than, as was usual, firing at static targets on the ground. Lacking gun cameras and equipment to tow sleeves, the Army pilots were most appreciative of the assistance Thach offered, and their first experiences firing at the sleeves confirmed their expectations. One problem the Army pilots encountered, however, was an overhead approach, as they believed their P-39s were too fast for the maneuver. Flatley volunteered to test the maneuver, and after a checkout ride, he demonstrated that the P-39 could make an overhead approach. Then he led Army pilots through the maneuver. They were so appreciative that some checked out in Wildcats and began to copy the Navy pilots' practice of making field carrier landings. And one Army pilot asked Thach to assist in having him assigned to a carrier to fly a Wildcat as an exchange pilot. In time the exchange came about, but not in time for the Navy's next battle, in which the need for fighter pilots was critical.

Another valuable result of the time Thach and Flatley spent together in Hawaii was their discussions of Thach's new beam defense tactic. In April, Thach was already totally convinced that the optimum division size was four fighters, while Flatley had not yet rejected the six-plane division. The pros and cons of both tactics were discussed well into several nights, and when Flatley left on the nineteenth to find *Yorktown* at sea, he told Thach he would continue to experiment with a three-plane section and Thach could continue with two. Time and combat, they both agreed, would reveal the optimum number of fighters in a section and the best tactical solution.

After Flatley left, the pace of new pilots reporting to VF-3 accelerated. From late April to the middle of May, Thach picked up nearly a dozen pilots, mostly new ensigns from ACTG. Emphasis was placed where it had to be placed: gunnery at the top of the list, followed by field carrier landings. Sufficient progress was made in aerial gunnery that Thach felt good about the results of the training and the justifiable confidence building in his young charges. While he could fly on the wing of a neophyte to help him learn how to shoot effectively at a towed sleeve, he knew he could not fly beside a new pilot while making a true carrier landing. With no carrier available for such training, Thach had to resort to psychological ploys to convince his pilots that they could place the narrowly spaced wheels of a Wildcat on a moving carrier deck. Given the frequent wobbly touchdowns on land, occasional ground loops and at least one severely bouncing ground loop that narrowly missed several parked planes, Thach was having some difficulty buying into his own psychology.[11]

Time was needed to mold a new team for VF-3. In late May, however, Thach was informed that he would be going aboard *Yorktown* as soon as it could be repaired for damage received two weeks earlier in the Battle of the Coral Sea. He was told that something big was brewing but not given any details. His superiors at Pearl knew he needed more time to prepare his young pilots, but they, and now he, knew they had to fight again before they were totally prepared.

5
Battle of Midway

From the second week of December 1941 until early May 1942, Lt. Cdr. Jimmie Thach did not have to read information bulletins to know what was happening in much of the Pacific theater, as he was physically present for many of the more salient combat actions in that period. Disappointed though he was when *Lexington* sailed toward the Coral Sea on 15 April without him, he thought he would not be away from the action for long as his air group's host ship, *Saratoga*, was due back at Pearl in June. On 11 May, however, he learned that Butch O'Hare would soon be returning for duty and would relieve Thach as commanding officer of VF-3. There was no one else Thach would have chosen over Butch. But in the few days between learning of his approaching relief and receiving word to prepare to board *Yorktown*, he thought his days as a fighter pilot were probably over.

Reports from the Coral Sea battle brought both good and bad news. The good news was that at least one enemy carrier had been sunk (light carrier *Shoho*) and at least one other (*Shokaku*) had been significantly damaged. The bad news was that *Lexington* had fought its last battle and the *Yorktown* was damaged. Loss of the *Lexington* was not announced to the public until later, but the absence of its name from planning and its absence when *Yorktown* limped in was a meaningful clue for sailors in and around the harbor.

Secrecy was the norm during the war, and in May only a handful of officers knew what ominous peril was again heading toward the Hawaiian Islands. Although a fighter squadron commander, and one who would soon have the responsibility of leading pilots into combat, Thach was not one of them. It did not trouble him that he was not presented detailed information. Indeed, he felt it was somewhat of a blessing. As events unfolded and he did find out what was before him and his squadron, the challenge proved greater than expected.

Before Thach and his pilots flew out to *Yorktown* on 30 May, scores of Japanese ships were steaming toward Midway Island. After the war, critics would fault the Japanese for having multiple objectives in their plan. But time was critical to enemy fortunes, as some of their leaders, including Adm. Isoroku Yamamoto—architect of the Pearl Harbor raid and Midway plan—realized they had to secure their defensive perimeter. If that objective could be achieved, the United States might agree to a negotiated peace rather than pay the high price in casualties required to reclaim the Pacific. Gamble it was, but in May 1942, the accomplished card player Yamamoto had confidence he was holding the high cards. However, there was a wild card in the deck he did not know about, and that card—the breaking of his secret code—evened the odds to a considerable degree. Admiral Nimitz and his intelligence section that had broken the code knew the critical elements of Yamamoto's plan. But the brilliant Japanese admiral still had a much larger force than Nimitz, and the expectations at Pearl were that the Japanese would have from one to three carriers more than the U.S. Navy could muster in time to meet the assault.

Known to Nimitz was that only four of Japan's six large, fast fleet carriers were steaming toward him. All six had participated in the Pearl Harbor raid, but the two newest were not available for this battle. *Shokaku* had indeed been too badly damaged at Coral Sea to fight, and *Zuikaku*, though undamaged in the battle, had lost too many planes and pilots to be ready for another sortie. The flagship *Akagi*, along with *Kaga*, *Soryu*, and *Hiryu*, were showing some faded paint as a result of constant operations since the opening of the Pacific War. Still, their pilots were experienced, their planes had proven equal or superior to their American counterparts, and only success had attended the previous combat sorties of all four carriers.

TO THE *YORKTOWN*

Thach knew he could not go aboard *Yorktown* with only the handful of pilots he was busily training. This concern, however, was being handled at a higher level, where a decision was made to add pilots of VF-42 already aboard CV-5 to those few assigned to Thach's VF-3. This would bring Yorktown's fighter squadron up to twenty-seven, the new complement of fighters authorized for carriers. Despite the fact that most of the twenty-seven were loaners from VF-42, the squadron would fight under the VF-3 designation. Yorktown's attack squadrons had suffered significant losses at Coral Sea so adjustments also had to be made for those squadrons. Torpedo 3 commanded by Lt. Cdr. Lance Massey would take TBDs aboard Yorktown in place of diminished Torpedo 5 and Bombing 5 would be supplemented by Bombing 3. Understandably Bombing 5 was not the least bit pleased with its temporary redesignation as Scouting 5, but deep down all understood the exigencies of war occasionally required improvisation. And in the end, all understood that duty was paramount.

Duty was challenged on several occasions in the few days after Thach learned he was going back to sea. First, he would board a damaged carrier that might or might not be sufficiently patched up to fight. Second, he would have to combine a half dozen fresh ensigns with less than a dozen experienced pilots in Hawaii and then combine them with experienced pilots of VF-42 he had neither trained nor flown with. Third, he had to determine which experienced flyers to place in position as division and section leaders, and some would have to come from VF-42 despite his unfamiliarity with them. Fourth, the sixteen pilots of VF-42—the majority of his amalgamated squadron—had no familiarity with the F4F-4 Wildcat. All these challenges had to be addressed in less than three days.

Working day and night, shipyard and ship's crew repaired the most serious damage to Yorktown, and the crew was still working on the ship when it departed on 30 May. Thach could do nothing about the wounded carrier, but he set about to address the other challenges in a positive manner. On the twenty-eighth, he got an unexpected but exceedingly welcome assist from former executive officer and close friend Don Lovelace, who volunteered to again serve as Thach's XO. Lovelace had orders to take command of VF-2 and no one suggested he step into the Yorktown squadron dilemma except for himself. Indeed, permission from superiors was required, and they, like Lovelace, realized his VF-2 pilots would not be able to assemble for at least a month. Temporary duty approved, he joined Thach and a squadron greatly in need of all the leadership it could find. Lovelace immediately made a difference not only in helping teach the rookies the essentials of combat but also by adding the warmth of his personality to Thach's as they welcomed VF-42 pilots into the VF-3 squadron. Too, Lovelace assisted in the decisions as to who would be the other two division leaders and who would make the best section leaders.

Introducing the six-gun, folding-wing F4F-4 Wildcat to the now-former VF-42 pilots did not go as well as the personal introductions. The folding wings allowed more fighters to be stowed in carrier hangar decks, but the –4s were heavier than the –3 and –3As, and the six guns in place of four meant that firing time was reduced from approximately forty seconds to approximately twenty. As the former VF-42 pilots had faced the Japanese Zero in combat three weeks earlier in the Coral Sea, they were justifiably concerned with further limiting factors to a plane that could not match the Zero in maneuverability or climb.

Unknown to Thach during the three days he pieced together his squadron was that the biggest challenge to duty would occur before he was informed where the impending battle was to be fought. Flying out to the *Yorktown* on the morning of the thirtieth, he set down and watched the deck crew unfamiliar with the F4F-4 damage the flaps as they folded the wings. Quickly out of his plane, he calmly taught the crew how to properly fold the wings. Satisfied that the crew now knew how to fold the wings prior to pushing the planes forward, he left the flight deck for the wardroom and lunch. A few minutes later, however, he was back on the flight deck to witness the aftermath of a tragedy. Don Lovelace had led his division to the carrier, had landed, and was taxiing forward when his wingman—one of the new pilots just in from ACTG, Ens. Robert C. Evans—landed too hard, jumped the barrier, and crashed on top of Lovelace's Wildcat. The propeller of Evans plane exploded into Lovelace's cockpit, and Thach's close friend, the man upon whom Thach was depending to bring cohesion to the squadron, was dead. Thach's concerns about new pilots making a carrier landing were disastrously confirmed now that one singularly important leader and two planes were gone. The only solace Thach could find was that he had allowed Lovelace to chose his own wingman, but that could not erase a lifetime of grief. Immediately after lunch, Thach called the squadron together in the ready room and told them that no one present had suffered a greater loss than had he. However, duty required that he, and they, remove the fatal event from their minds and concentrate on defending a task force relying on the few souls in that one room to do just that.

TO MIDWAY

Leader of the 1st Division as any squadron commander would be, Thach had chosen Ens. Robert "Ram" Dibb to be his wingman as he was the first to report in early May, thereby allowing the two more training time together. With Lovelace gone, Thach restructured the squadron again, elevating former VF-42 pilot Lt. (jg) William N. "Bill" Leonard to executive officer and naming him as leader of the 3d Division. Leaders of the 2d Division, Lt. (jg) Richard "Dick" Crommelin, and 4th Division, Lt. (jg) Arthur J. "Art" Brassfield, were also former VF-42 pilots. Leonard, Crommelin, and Brassfield had not only fought enemy planes either off Tulagi or at Coral Sea but also claimed half a dozen aerial victories. All but two of the section leaders were also former VF-42 pilots.

On the first night out *Yorktown*'s air officer Lt. Cdr. Murr Arnold briefed Thach on the upcoming mission, the place, and the estimated strength of the enemy. He did not need to tell Thach how significant the battle would be to the fortunes of the United States and the Navy. Task Force 16 comprised of *Enterprise* and *Hornet* with Rear Adm. Raymond Spruance on the bridge in place of an ailing Vice Admiral Halsey had most of the aircraft that would fly into battle even though Rear Admiral Fletcher on *Yorktown* was in overall command. Of course, *Yorktown* in Task Force 17 would fight, but it was in a more precarious situation due to the carrier's recent damage, its hastily assembled air group, and the fact that CV-5 was the only carrier in TF-17. On the voyage toward Midway Thach assumed the three carriers would operate close enough that combat air patrols from each task force could support the other if necessary, but that topic was not at the center of Arnold and Thach's discussions. Early on the two agreed in principle that Thach would take eight planes divided into two four-plane divisions to cover the dive-bombers and torpedo planes with the remaining seventeen Wildcats left behind to defend TF-17. At least for a couple of days that was the plan.

Although he would have preferred to have spent every available moment practicing gunnery and aerial tactics as the force steamed west, Thach had to prioritize meetings with the other two squadron commanders and his own squadron mates. Both VT-3's Massey and VB-3's Maxwell F. Leslie suggested Thach's few escorts fly with the other into battle. Thach broke the tie by deciding the slow torpedo planes needed him more than the faster, higher flying SBDs. Although slow and obsolete, the TBDs had placed several torpedoes into the one Japanese carrier definitely sunk at Coral Sea, and Thach reasoned that the enemy most likely would not repeat their mistake of concentrating protective fighters only at high altitude. After meeting with Massey and Leslie, Thach then directly addressed the pilots of both attack squadrons with advice on how they might best protect themselves from Zero fighters that were nearly twice as fast as SBDs and three times faster that TBDs.

Recalling one of these briefings, then-Ens. Robert M. "Bob" Elder of VB-3 found Thach to be a serious student of aviation and a leader who obviously wanted people to understand the totality of a given aerial situation. Although he found Thach quiet and unassuming, he also observed that the VF-3 squadron commander had a direct, instructional approach and demanded air discipline. Wasting no time or words, Thach's lectures ended with an unmistakable "let's go" attitude that was later reinforced as it became apparent that Thach not only generated ideas and direction but also followed through.[1] Meetings with the attack squadrons emphasized how SBD and TBD pilots could save themselves defensively. In brief meetings with his fighter pilots, Thach emphasized the offensive nature of the fighters as the best defense but did not have occasion to discuss the rudiments of his weave with the former VF-42 pilots. That was something he had expected Lovelace to do, but that, of course, did not happen. Still, the significance of the battle was such that Thach even stressed the possible necessity of using their Wildcat's propeller to cut off part or all of the tail of enemy torpedo planes. If done correctly, by approaching from below—no easy task when a torpedo plane is already low in preparation for dropping its ordnance—and then rising up to the target, the spinning prop would not only damage the enemy's tail but also throw the Wildcat away from it.

By the night of 3 June, Thach had made the best use of the four days he had to prepare for the battle. Telling all to get a good night's rest, he himself could not. The decision to defend the torpedo planes was correct, so he did not worry about that. But how well it could be done was a concern that kept sleep from overtaking his consciousness. After the battle he would find himself frightened about it all, but not before. There just was no time for fear to work its way into his mind.

TO BATTLE

Early on the morning of 4 June, Thach found his charges in a state of high morale. Neither they nor Thach had any idea that the Japanese code had been broken, but they did know that the element of surprise should be on their side. It also was a beautiful day for flying. PBYs had seen the several different and widely spaced Japanese naval forces coming toward Midway, and VF-3 believed none of the American carriers had yet been spotted. But in war there are almost always surprises for both sides, and only thirty minutes before launch Thach learned he could not take eight fighters split in two divisions of four. He would have only six, giving him one full division of four with two others basically just following along on their own. Thach made one last attempt to change the decision on the reduction of fighters. He had not seen Fletcher or Capt. Elliott C. Buckmaster, and he was not supposed to see them. The best he could do was ask Arnold to go back up the chain of command to get the order rescinded. It was not. Years later Thach was sympathetic to the decision and whoever made it because he understood that every fighter available was needed to protect *Yorktown* and its attending ships. But on the morning of 4 June, he was very unhappy about it. And now that the enemy had been seen, Thach and his pilots knew how big and strong their opponent was, and that too was nothing to be happy about.

Hiding his disconcerted state of mind, he rounded up the five other pilots who would fly with him—with the others left behind for CAP listening in—and told them to stay together and for there to be no lone wolves. He reminded them of enemy tricks, such as having one fighter place itself in a vulnerable position only to draw fighters toward it so other Zeros could descend on the lone wolf. Dibb would still fly as Thach's wingman, with Lt. (jg) Brainard T. Macomber and Ens. Edgar R. Bassett (both former VF-42 pilots) rounding out the division. Mach. Tom F. Cheek and Ens. Daniel C. Sheedy would fly together, behind and below the other four pilots to give the six fighters a layered formation in defense of the TBDs.

Nearly two hours after planes from *Enterprise* and *Hornet* launched in hope of catching the enemy carriers by surprise and with most of their planes on deck, the six VF-3 Wildcats lifted off *Yorktown*. Soon the small escort division caught up with Massey and the other eleven TBDs flying with him in a running rendezvous. Even though it worked earlier off New Guinea and worked on 4 June, this was a practice he did not like because it was too dependent on good weather and excellent visibility.[2] Excited and nervous, Thach and the others in his division occupied themselves with the necessities of watching their navigation charts, switching on and testing their guns, and settling into the slow S-turns to remain near the much slower cruising TBDs (forty-knot cruising speed variance). Massey's squadron was flying in a formation of three planes per section, stepped down for maximum visibility and mutual free-gun support, and would continue until each section split and fanned out to attack.[3] The four V formations held steady at about fifteen hundred feet, with Thach, Dibb, Macomber, and Bassett maintaining an altitude near five thousand feet with their two sections also stepped down. Cheek and Sheedy brought up the rear and stayed lower than the other fighters but still slightly above the TBDs.

Massey's TBDs flew through the small, puffy clouds that were layered at about fifteen hundred feet, but Thach's visibility of the Devastators was not impeded. Although not concentrating on Leslie's SBDs flying well above, he could not help but notice when several bombs exploded in the water below. Thach correctly deduced that while arming their bombs, some electrical arming devices in the planes had inadvertently released them. Having lost his bomb, Leslie quickly alerted other pilots not to make the same mistake.

Unbeknown to Thach, the *Enterprise* and *Hornet* dive-bombers were having difficulty locating the Japanese carriers. The *Hornet* SBDs did not find the enemy on the morning of the fourth, and only some very clear thinking by Lt. Cdr. C. Wade McClusky eventuated in the *Enterprise* SBDs finding the carriers after fuel gauges indicated they might not get back to the "Big E." But the torpedo squadrons from all three American carriers found the Japanese as expected. *Hornet*'s VT-8 was first to make contact, followed by VT-6 from *Enterprise*. All fifteen TBDs from *Hornet* were shot down by defending fighters before they could gain favorable launching position, and ten of VT-6's fourteen Devastators fell into the sea without injury to the enemy.[4] *Yorktown*'s VT-3 was the last to approach the enemy, and Thach thought Massey might have received some pertinent information when he observed VT-3's TBDs make a small course change. Very soon, however, he could see the Japanese outer screen some ten miles away. Shortly after, he wondered why colorful bursts of antiaircraft fire were being directed at him while he was still out of range. It quickly became apparent that the enemy guns were alerting Japanese fighters to the approach of the American planes.

It appeared to Thach that most of the Zeros the Japanese had brought to Midway were coming just for him. He estimated about twenty fighters were streaming in one by one from astern and overhead, and although he gave them only a quick glance, he did not have time to initiate evasion or counterattack before Edgar Bassett's Wildcat was set afire and dropped from sight. While some twenty enemy planes were concentrating on his small escort group, he saw other Zeros diving toward the torpedo planes.

Being outnumbered at least three to one was only the beginning of Thach's problems. The last scenario he or any other Wildcat pilot wanted was to try to fight a Zero at low altitude. With their greater climb and maneuverability, Zeros would dictate the terms of battle under these circumstances. Thach immediately knew he was in no position to help Massey's torpedo planes and he tried to radio Macomber to open the distance between them so they could initiate the beam defense (weave) maneuver. His radio dead from battle damage already received, Macomber did not hear the message, and the former VF-42 pilot had not been with Thach long enough to automatically know that the current dilemma necessitated the beam defense tactic. It was ironic that the creator of the life-saving tactic was now in a life and death situation alongside a teammate who did not know how to execute it. Indeed, Macomber was being faithful to Thach's last words before leaving *Yorktown* to stick together.

For the first attacks Thach and the two pilots trailing behind him could only jink and move back and forth to dodge enemy shells. Following in line, Dibb and Macomber emulated Thach's defensive moves that for the most part called for exact timing as to when to quickly pull away from the path of the constantly descending Zeros. The abrupt turns to the right resulted in the first several attackers not being able to adjust their aim, and they pulled up and climbed back to altitude to continue the circle pattern of attack. Although Thach had turned back to his left to fire off a few bursts as the enemy planes bottomed out in their dives, his first efforts brought no results. Finally, however, one Zero slowed during his pullout—a fatal mistake, as Thach had just enough time to hit the vulnerable target.[5]

Knowing Macomber could not help and not knowing where Cheek and Sheedy were, Thach radioed his wingman Dibb and ordered him to assume the position of a section leader so they could initiate the weave. Dibb's radio was working and he enthusiastically pulled away from behind Thach to a couple hundred yards abreast. Ram Dibb was a rookie in his first combat while Macomber was a combat veteran. But for this occasion, the training time Dibb had with VF-3 made him the only partner Thach could count on for this tactical maneuver. With so many enemy planes in the air, Thach was not sure anything would work, but the answer came when a Zero followed Dibb during one of his turns. While counseling his charges that emotion had no place in combat, Thach found himself angry that the young, inexperienced Dibb was the target of this Zero. Wisdom called for a short burst of shells to hopefully cause the Zero to break off the pass, but it was apparent this Zero was not going to break off. Anger rising, Thach continued straight ahead, the firing button depressed, rather than ducking under the Zero. At last the Zero broke off, and as he passed close by, Thach could see flames pouring from its underside. Continuing the weave now discouraged the Zeros from following the Wildcats in their turns, but one made the same mistake as Thach's first kill, and when he was too slow in his pullout, Thach shot him down and added a third mark on his kneepad. Soon after, Dibb erased another enemy fighter converging astern of Thach and Macomber. Macomber did not then realize exactly what his two teammates were doing, but Thach and Dibb knew the weave was working up to expectations.

No more than a quick look at the torpedo planes could be spared. But no more than a quick look was necessary to see that one Devastator had exploded and some others were hurt.[6] By this time, the remaining TBDs and Wildcats were over the outer screen and therefore within range of antiaircraft fire. Although some TBDs were already down, the others split to attack from two sides. While such an attack seemed to guarantee that the target would take torpedoes from one side while turning from the other, the split also deprived the TBDs of the larger mutual support from back seat gunners. Without fighter support and the fewer number of shells being fired from the smaller formation, the Zeros pumped 20-mm cannon rounds into the surviving Devastators while the TBDs' .30-caliber guns attempted to respond in kind. The long run to the target became protracted as the enemy carriers turned their sterns to the approaching torpedo planes. At this stage the odds against the Devastators were overwhelming, and all but two slammed into the ocean, most—like those of VT-8 and VT-6—before their torpedoes could be released.

There was no time for Thach to watch the planes he shot at fall, except for his third kill, which was close to the water, but he knew that when he saw profuse red flames issuing from a plane, it was finished. Marking his kneepad for each plane downed, Thach realized that he himself would most likely not survive the battle and stopped the tally. He still had not seen Cheek or Sheedy, and he knew Bassett was down. Suddenly it appeared that the attacks began to slacken, perhaps to concentrate more on the torpedo planes. Then, while looking up for more pursuers, Thach glimpsed a flash of light, another, and then another as a descending streak of SBDs poured down from on high toward the Japanese carriers.

For the moment Thach did not know which squadrons were attacking, but later he learned that Leslie's SBDs placed at least four bombs into *Soryu* and that McClusky's SBDs inflicted fatal wounds on *Akagi* and *Kaga*. He did know during the dive-bombing attack that the carrier nearest him would not survive the day and suspected the other two would suffer the same fate. Too he would later learn there was a fourth enemy carrier, but he saw only three during his flight.

Flying between the screen and the burning enemy carriers, Thach looked for Devastators. Finding one he escorted it clear and then returned for another. Off to the side he spotted a Zero, and looking up he saw others just waiting for an opportunity to jump the first Wildcat to attack the decoy. Thach left the bait to dangle and continued the search for other TBDs while assuming that the declining numbers of Zeros could be attributed to some chasing SBDs or climbing to altitude in anticipation of more SBDs. Regardless, there were not as many around as there had been, and for the first time in the very long twenty minutes he had been within sight of the enemy ships, Thach thought he just might survive the battle.

Later Thach understood that his good friend Lem Massey and the other torpedo plane pilots and gunners missing from VT-3, VT-6, VT-8 (detached and flying from Midway), and even two torpedo-armed Army B-26s did not die in vain. They drew the protecting Japanese fighters so far down that the *Yorktown* and *Enterprise* dive-bombers had an unobstructed flight to the enemy carriers and encountered no opposition in their dives. But on 4 June and soon thereafter, Thach was upset that he had not been able to do more.

Flying beside one of the two surviving TBDs, Thach attempted to communicate with its pilot (apparently Mach. Harry L. Corl) to fly higher than twenty feet off the water. Thach needed altitude to be in position to move toward or defend against an attacker. And he needed more distance between them to allow for the slow S-turns while he kept the TBD in sight. Soon, however, it was apparent that the radio in the TBD was dead and Thach had to pull up and away.

The other Devastator survivor, former enlisted pilot, later commander Wilhelm G. "Bill" Esders, had Thach with him even though the fighter squadron commander was not physically present for all of his flight home. Exiting the Japanese carriers and screen with his mortally wounded gunner, ARM2c Robert B. Brazier, Esders was under fire by Zero pilots determined not to let the TBD escape, even though its torpedo had already been released. Esders recalled a Thach visit to VT-3's ready room and the words the fighter squadron commander had shared. Thach told Devastator pilots that to survive against a Zero when there were no friendly fighters present they must fly as low as possible to keep the enemy from flying underneath. Forced to maintain some altitude, the

Zero would then have to make overhead passes from either astern or the side. To further reduce the Zero's effectiveness, the TBD pilot should fly as slow as possible to shorten the amount of time the Zero had to fire and perhaps force the Zero into a stall if it fell below its minimum speed. The final portion of the tactic, according to Thach, was to watch the Zero during its approach while holding a straight course. The TBD pilot would then see the enemy's shells splashing in the water ahead and to the side as the attacker walked his shells into the TBDs path. At the last moment the TBD should cut in toward the Zero, thereby not allowing the attacker to correct his deflection-firing angle. Timing, of course, was the critical factor. And it is interesting to note that the sharp turns toward a Zero that Thach counseled for TBD pilots was the opposite direction of his sharp turns to the right in a Wildcat while he was evading Zeros in the same battle.

Esders was greatly relieved to find how well the tactic worked when first tried. However, he was greatly concerned that the enemy might find a way to compensate after seeing the maneuver. But physics and Thach's words were on Esders side, and his evasive movements worked for every attack. After what seemed an eternity, the Zeros—probably out of ammunition—departed. The last one flew up beside Esders, waved, and seemed to say, "You've earned it" before pulling up and heading away.[7]

BACK TO THE *YORKTOWN*

Heading away from the burning enemy carriers and the beehive of Zeros that had been attacking, Thach was exhilarated to know that the odds were better that he would live rather than die. During the twenty-minute fight, he was too busy for emotion other than for the one Zero he shot down when his temper got the better of him. Now he found himself scared—a delayed emotion, as the major danger was behind or at least in abeyance. Mixed with fear was admiration for his opponents, so much so that from that day forward he described their coordinated attacks as beautiful. And the individual Zero pilots had proven to be superb professionals. The Zero itself had also earned admiration, enough so that Thach was left in envy of its performance and wishing his Navy had a fighter equal to it.

Still attempting to keep the one TBD in sight, Thach radioed his Wildcats to follow him home. He knew Bassett was gone and Cheek and Sheedy were still unaccounted for. Later he learned that both had taken a number of hits, held their own by accounting for a couple of Zeros, and made it back, Cheek to *Yorktown* and Sheedy to *Hornet*. With fear and necessary cockpit functions competing for time in his mind, fear took an edge when he felt liquid on his legs and feet. Afraid to look, he moved both legs and feet and, after confirming that everything worked as it should, he looked. To his great relief the liquid was oil.

After nearly three hours in the air, Thach landed back aboard *Yorktown*.[8] Leslie's SBDs were overhead, but Arnold knew he needed to get his fighters back aboard first. Thach's mechanic noticed a hole in the fuel tank and investigation revealed a bullet. Further investigation revealed that an oil line had been shot away, nearly all oil was gone and a few more minutes in the air would have put him in the water rather than aboard. He had suspected that might be the case as his oil gauge dropped all the way to empty. The Zero pilots had even better aim than he had credited them with.

While mechanics made necessary repairs to the Wildcats and began to refuel and rearm them, Thach went to the VF-3 ready room.[9] There he evaluated who was available to fly other than the pilots then on combat air patrol. Cheek was aboard, but his damaged Wildcat was further battered when it flipped on landing. Sheedy was not aboard, and for the moment his fate was not known. Someone would have to fill Bassett's position in his division, and other changes would have to be made because he would soon need to relieve the current CAP. Sooner than expected, however, the call came for the ship to prepare for attack just a few minutes before noon. Procedure was for the fighter pilots to remain in their ready room rather than be exposed to enemy ordnance or friendly shells during an air attack so Thach and his squadron mates saw the combat with their ears. Audio vision was revealed as the 5-inch guns began to fire at targets ten miles away. The quad 1.1s began when the targets approached inside four miles, and at two miles the smaller-caliber weapons chimed in.

Defending fighters threw themselves against eighteen approaching enemy dive-bombers. Despite downing eleven, seven got through. Within a five-minute period Thach felt some thuds that seemed some distance from his location just under the island structure. But the thuds were closer than he realized. One bomb exploded on the flight deck, leaving a large hole just aft of the number two elevator, killing five and wounding fourteen in the number three 1.1 mount and killing twelve and wounding four in the number four mount. Reserve crews got both mounts back in operation, but at a reduced rate as loss of power affected water pressure cooling and several water jackets were breached leaving the number four gun on each mount frozen and jammed.[10] A second bomb entered the ship in the island and descended to the second deck, where the explosion damaged uptakes disabling firerooms below. A third bomb struck the forward elevator and passed into the elevator well damaging it and compartments adjacent and below. While all three bombs killed and maimed crew and temporarily shut down flight operations, the ship was not mortally wounded. From other ships nearby, the large clouds of thick, black smoke rising from *Yorktown* seemed more ominous, a perception magnified when the carrier slowed and then went dead in the water about 1230 while repair parties worked to put out fires amidships before moving to light off the boilers again.

The damage to *Yorktown* prevented planes from coming aboard. Many of Leslie's SBD pilots and the only two TBD pilots to survive were within sight of the carrier when the enemy bombs hit. Both Esders and Corl landed their TBDs in the water and were picked up while Leslie and other SBD pilots landed on *Enterprise*. After an hour of hard work, all fires were not out but were under control aboard CV-5. Auxiliary steam lines were rigged, allowing the ship to come alive again, and Thach was informed that the carrier could probably work up to sixteen knots, a speed at which his fighters could launch.

Given damage near the main fuel station, Thach decided to have his pilots man only those Wildcats that had twenty or more gallons of gas. With a nearly empty flight deck, there was plenty of distance for the fighters to get a long deck run to compensate for the slower-than-usual speed of the carrier during launches. But even with the long deck run, Thach dipped slightly as he cleared the deck at approximately 1440. Only a few fighters were already in the air for combat air patrol as Thach rolled down the patched deck, and they were moving to intercept another attack that was on the way in. Following behind Thach, Dibb, and Leonard was Ens. John Adams, also short of fuel and not sure how many shells were in his guns. Like Thach, Adams had to ensure a safe takeoff, switch his guns on and test them, crank up his landing gear (over thirty turns with his Wildcat lurching from side to side during the exercise), dodge the carrier's antiaircraft gunfire, and keep an eye on the incoming attack. Altitude was not going to be a problem, as these attackers were torpedo planes, but intercepting them before they released their torpedoes would be difficult at best.[11] The Wildcats following Thach had no choice but to accept combat as it was developing at low altitude rather than in the manner training had prescribed. One, who did not have time to crank up his wheels, claimed a kill but was himself in the water about a minute after takeoff, his plane downed by the carrier's guns or one of the Zeros escorting the torpedo planes. Another was killed under similar circumstances.

Out ahead, Thach was closing fast on the enemy formation as it split to launch torpedoes on either side of CV-5. One enemy craft with unique tail markings was crossing in front low on the water and Thach had a near perfect side approach. His six .50-caliber guns found the port wing and apparently a fuel tank as the fuselage burned away to reveal the plane's internal ribs. Though doomed, the enemy pilot—later confirmed to be the strike leader, Lt. Joichi Tomonaga—kept his plane in the air long enough to release a torpedo that Thach thought might have slammed into the amidships port side of *Yorktown*.

A second aerial torpedo also found the port side of *Yorktown* less than fifty feet from the first (frame 92 for the first and 80 for the second), and the design flaw that did not alternate boiler and engine rooms was about to cost the U.S. Navy dearly. With power gone, the carrier could not supply water to fight fires or, more important in this instance, evacuate water. Although only in the first couple of minutes of his flight, Thach could see his carrier begin to list. His gaze at CV-5 was interrupted when he mixed it up with a Zero before it headed back toward *Hiryu*, the fourth enemy carrier and the only one to escape damage earlier in the day. Thach thought he might surprise the Zero but the nimble enemy fighter pulled up and over Thach and then quickly ended up immediately behind. Fortunately a cloud was directly in front. Entering the cloud, Thach did a split-S and then pulled out ready to fire, but the Zero was gone. That short episode over, Thach again flew over *Yorktown* and sadly noticed its list had increased to the point that it might capsize. Obviously neither he nor anyone else could land on the ship.

TO THE *ENTERPRISE*, *HORNET*, AND HAWAII

The launch off the *Yorktown* was too quick to organize in sections so Thach and the other seven fought alone during the last attack on her. Nearing *Enterprise* and *Hornet* some thirty miles away, he was angered to see more than enough planes over the two carriers to handle any emergency. Just a few of them over *Yorktown* might have prevented the last attack. Unknown to Thach at that moment was the fact that several *Enterprise* fighter pilots were also hot under the collar that they had not been released earlier to help defend the *Yorktown*.

Landing on *Enterprise*, Thach had to suppress his misgivings as he was ordered to report to Rear Admiral Spruance. Spruance wanted his opinion on how the battle was going, and Thach repeated what the admiral had already heard, that three enemy carriers were burning. By that time, the location of the fourth enemy carrier had been confirmed, and twenty-four SBDs launched from the Big E were winging toward her. Fourteen *Yorktown* SBDs left homeless when CV-5 was first damaged were merged with ten *Enterprise* survivors from the morning attack. Led by Lt. W. Earl Gallaher of *Enterprise*'s Scouting 6, the unescorted Dauntlesses found *Hiryu* without a significant combat air patrol and left the last enemy carrier burning from stem to stern. *Soryu* and *Kaga* sank before sundown on the fourth, *Akagi* went down before daybreak on the fifth, and *Hiryu* sank after daybreak allowed spotters an opportunity to photograph the burning hulk.

The fighter squadron commander of *Enterprise*'s VF-6, Lt. James Gray, and his escort group did not encounter opposition during the morning flight, and VF-6 Wildcats on combat air patrol during the day had emerged basically intact. However that was not the case for *Hornet*'s VF-8. Although not finding the enemy on the fourth, ten of their number had to ditch for lack of fuel off Midway; included with the missing was VF-8's squadron commander, Lt. Cdr. Samuel G. Mitchell (later rescued). Consequently, on the morning of the fifth Thach was ordered to take nearly all his VF-3 (and VF-42 veterans) survivors to *Hornet* and assume command of the fighter squadron aboard that carrier. Flying CAP on 5 June from *Enterprise*, Thach landed on *Hornet* and organized combat air patrols for the composite fighter squadron before flying another CAP for a total of 5.9 hours on the day. While SBDs from both *Enterprise* and *Hornet* chased the retreating Japanese and sank one heavy cruiser (*Mikuma*) on the sixth, there was no need for a fighter escort as the enemy had no aerial attack capability after the fourth and distance was a problem for the Wildcats.

On 10 June, a message from Admiral Nimitz congratulated the victors, expressed his regret for those lost and ordered *Enterprise* and *Hornet* "to proceed North and drive the enemy from the Aleutian Islands." He added, "I know you will overwhelm the enemy there as you did at Midway thus insuring his complete defeat."[12] Within twenty-four hours, however, Nimitz decided the enemy presence in the Aleutians was such that they could be contained without having to commit his carriers. Thought by some to be a diversion, the Japanese thrust to the Aleutians was a concurrent campaign with the Midway operation rather than a diversion. Admiral Yamamoto agreed to the Aleutian campaign in combination with Midway to placate the Japanese Naval General Staff that preferred to concentrate on the South Pacific to isolate Australia. Yamamoto, as is well known, wanted to draw the American fleet into a decisive battle. To him, any move of the U.S. carriers toward Alaska would have been aborted once Japanese presence was known off Midway and when they came was not as important as just having them come to meet the challenge. The Americans could tolerate Japanese presence on remote Attu and Kiska, just as Nimitz decided on the tenth, but not Midway.[13]

On 13 June, the American carriers and attending ships began their return to Pearl Harbor. Not present were *Yorktown* and *Hammann* (DD-412). On the morning of 6 June, *Yorktown* was riding steady with a 26-degree list, and it appeared it could be saved. Repair parties were put back on board, and the destroyer *Hammann* came alongside to furnish temporary power. Considerable progress was made, but about 1330 two torpedoes launched from I-168 exploded against the starboard side of *Yorktown* and another erupted against the thin side of *Hammann*. The destroyer sank quickly with heavy loss of life, and work aboard the carrier was terminated. The extensive watertight compartments within the carrier kept it afloat until dawn on the seventh, when it slipped beneath the surface. After the war, Thach and others learned another submarine, the USS *Scamp* (SS-277), sank I-168 on 20 July 1943.

Back off Hawaii, Thach led his merged fighter group to Ewa Field and from there he flew on to Ford Island. Never passing up an opportunity to train, he incorporated gunnery exercises in both flights and again on the eighteenth while flying round trip to Kaneohe. On the nineteenth he was detached from VF-3, Butch O'Hare arrived to take command on the twenty-fourth, and Thach then headed to San Diego for leave before going on to his next assignment.

Thach and the pilots from the three carrier air groups were happy to be back in Hawaii, but they were most unhappy to find newspaper accounts of the battle giving credit to Army B-17s flying from Midway for much of the damage to the Japanese carriers. Thach and the others knew that the B-17s' contribution was limited to disrupting enemy flight-deck operations and the premature death of scores of fish. All knew not a single bomb dropped from the high-flying bombers had found an enemy ship, and all knew that high-altitude bombing against shipping was an exercise in futility. In sum, the exhilaration of victory at Midway among naval aviators was mitigated by distorted accounts of the battle. Given wartime censorship, it was a long time before carrier aviation's role was revealed, and even then it was not fully disclosed. *Yorktown*'s loss would not be announced to the public until 26 September 1942, but the pain of its loss and the pain of too many lost fellow naval aviators were indelibly etched in the mind of survivors forever.

THACH ON MIDWAY

When Admiral Thach conducted his oral history with the U.S. Naval Institute in 1971, he acknowledged that nearly thirty years after the Battle of Midway he had a greater understanding of why some decisions were made, but he nonetheless was still dismayed about other decisions. He could accept the reasoning of reducing his fighter escort from eight to six so that the one-carrier Task Force 17 could protect itself. Perhaps he could have had eight had the tragic loss of Lovelace and the two Wildcats not occurred, but looking back, he knew that two more Wildcats against such overwhelming odds would not have altered Torpedo 3's fate.

In his 26 August 1942 interview at the Bureau of Aeronautics, Thach stated that "positioning of our forces was a brilliant piece of strategy," a tip of the hat to Admirals Nimitz, Fletcher, and Spruance.[14] Although he respected the three after retiring with four stars on his own collar, he could not help but note that none of the three flag officers responsible for Midway was aviation trained. Specifically, Thach was still upset that Task Force 16 and Task Force 17 were either not instructed to remain closer or that the admirals did not realize the importance of not separating to such a degree that there could not be mutual air support. In defense of the two task force flag officers at Midway, Thach perhaps did not fully take into account that Spruance's two carriers still had to maneuver for landings and launches after *Yorktown* went dead in the water. While Spruance was not an aviator, he did have all of Halsey's aviator staff with him in flag shelter aboard *Enterprise*. And Thach may or may not have taken into account that Fletcher transferred tactical command to Spruance after *Yorktown* was initially abandoned (1823 hours). As will be seen later, by 1944 Thach was in a position to see that the problem of carriers being too far apart for mutual air support did not again occur.

Another matter Thach reflected on was a disagreement with Marc A. Mitscher, already promoted to flag rank but still commanding officer of *Hornet* at Midway. When Dan Sheedy's shot-up Wildcat landed aboard *Hornet*, all six guns went off for two seconds with the shells spraying the island, killing six and wounding over a dozen more. After Thach came aboard as temporary commanding officer of the mixed-bag VF-8 squadron, an investigation of Sheedy's plane indicated that his master switch had been turned off but combat damage had welded across the wires leaving the circuit on. Mitscher recommended installation of an automatic switch in the tail hook so that when the hook was lowered it would ensure no accidental discharge. Thach disagreed, stating that it was an unnecessary addition that would require difficult maintenance due to its location, and the possibility of dirt accumulation would counteract the desired intent. Mitscher was not happy that Thach did not agree with him and let him know it.

In the years after Midway, Thach, Scott McCuskey, Bill Leonard, John Adams, and others who were over *Yorktown* when it took the two torpedoes on 4 June felt a natural sorrow and regret that they were unable to intercept and down all the Japanese torpedo planes before they released. Each has talked openly of those few minutes as being at or near the top of their lifelong list of regrets, and each seems to have accepted the carrier's loss as both a personal and professional responsibility. Reminders that the ship would have survived if not for the submarine torpedoes two days later have failed to soften emotions, as each could justifiably note that without the aerial torpedoes on 4 June, CV-5 could have steamed to safety. Like others who dwell on real or perceived failures rather than bask in the glory of victories, VF-3 pilots lived with the regret of not having just a few more minutes' notice on 4 June 1942.[15]

In his early days as an aviator, Thach flew OL-1 seaplanes similar to this one attached to seaplane tender USS *Gannet* in Alaskan waters in 1934. A fighter pilot and test pilot before transferring into patrol planes, he flew most of the aircraft in the pre–World War II U.S. Navy. *U.S. Navy*

Thach (*second from left*) poses with (*left to right*) Cliff Edwards, Clark Gable, Wallace Beery, and fellow officers Lt. (jg) Herbert S. Duckworth and Lt. (jg) E. P. Southwick during the making of the movie *Hell Divers* in 1931. NA photo 80-G-450865

Fighting Squadron 3 aboard *Lexington*, 5 March 1942. Squadron commander Thach sits in the middle of the front row. To his right is Don Lovelace, and to his left is Noel Gayler, who sits next to Butch O'Hare. *U.S. Navy via O'Hare Collection*

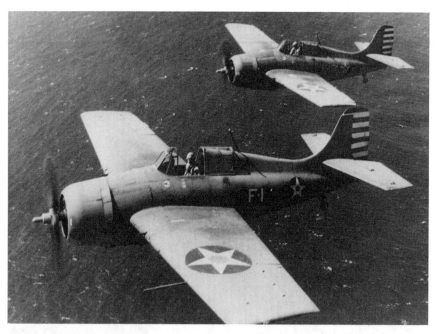

Thach in F4F-3 Wildcat F-1 looks toward the camera, with Butch O'Hare in F-13 on his wing off Pearl Harbor on 11 April 1942. *U.S. Navy via O'Hare Collection*

Thach (*right*) and Butch O'Hare pose in front of an F4F-3 Wildcat marked with O'Hare's kills off Bougainville, for which he earned the Medal of Honor. Thach received a Navy Cross for the same action. *U.S. Navy via O'Hare Collection*

Thach in the cockpit of an F4F-3 Wildcat on 5 May 1942 while training pilots who would fight in the Battle of Midway the following month. *NA photo 80-G-64822*

While serving as Vice Adm. John S. McCain's operations officer in 1944–45, Thach (*right*) marks Japanese airfields for the task force commander and recommends locations where his "Big Blue Blanket" could keep kamikazes on the ground. *NA photo 80-G-308561*

Lt. Cdr. Jimmy Flatley, who coined the term "Thach Weave," stands on the wing of an F6F-3 Hellcat just after making the first landing aboard the new carrier *Yorktown* on 6 May 1943. *U.S. Navy*

Damage-control personnel work to repair the flight deck and Thach's operations office, which was directly beneath the Avenger torpedo-bomber that exploded aboard Hancock on 21 January 1944 while off Formosa. Thach Papers

Task Force 38 commander Vice Adm. John S. McCain (left) and recently promoted Captain Thach (second from left) share pleasantries with Assistant Secretary of the Navy John Sullivan shortly before the end of the Pacific war in August 1945. Thach Papers

Thach transfers from a destroyer to a carrier via high line. Judging from his smile, he enjoyed this arrival just as much as flying aboard a carrier. *Thach Papers*

Quickly changing roles from antisubmarine carrier into attack carrier with the outbreak of the Korean War, USS *Sicily* was the second carrier to send planes into combat over Korea on 3 September 1950. It operated twenty-four F4U-4 Corsairs of Marine Fighter Squadron 214. Usually only eight Corsairs launched into combat at a time, as shown here. *Thach Papers*

Captain Thach on the bridge of *Sicily* early in the Korean War. *Thach Papers*

A VMF-214 Corsair prepares for launch from *Sicily* bound for a target in Korea. Note the 5-inch rockets and open cockpit—standard procedure for launch and landing to allow for a fast exit in case of emergency. *Thach Papers*

During combat operations off Korea, Thach could be found in the VMF-214 ready room nearly as much as on the bridge. Squadron commander Lt. Col. Walter E. Lischeid (*front right*) and pilots fully appreciated Thach's combat experience and welcomed his presence. *Thach Papers*

The marine pilots of VMF-214 made a difference with their close air support and interdiction sorties against the North Koreans. Shown here is an enemy tank that was destroyed by one of VMF-214's Corsairs near Seoul in September 1950. *Thach Papers*

Lieutenant Colonel Lischeid (left) talks to Thach on the bridge of *Sicily*. Lischeid became commanding officer of VMF-214 immediately before the squadron flew aboard and was killed in action on 25 September 1950. *Thach Papers*

Thach was never removed from public relations duties while off Korea. Here a correspondent, armed with camera, canteen, and well-stocked purse, arrives via helicopter to sit in on ready room briefings, interview pilots and crew, and talk with the carrier's CO. *Thach Papers*

Thach (middle) believed Assistant Secretary of the Navy (Air) John F. Floberg (right) was one of the best civilian servants the Navy ever had. NA photo 80-G-441265

Change of command ceremonies aboard the carrier USS Franklin D. Roosevelt on 22 May 1953. Thach relieved 1927 Naval Academy classmate Capt. George Anderson (right), who later became chief of naval operations. Thach Papers

Franklin D. Roosevelt at sea while Thach was commanding officer. Thach was the carrier's last CO before the ship was modernized with an enclosed bow, angle deck, and numerous other significant changes. Note F-2H Banshees on the flight deck beside and forward of the island. *Thach Papers*

Portrait photograph of Rear Admiral Thach soon after his promotion to flag rank and assignment as senior naval member of the Weapons System Evaluation Group with the Department of Defense. *Thach Papers*

Rear Admiral Thach on the bridge of his flagship, USS *Valley Forge*, during duty with experimental antisubmarine Task Group Alpha in 1958–59. An AH Seahorse helicopter rests on the flight deck near the island, with a Grumman S-2 Tracker visible on the port side of the flight deck. *Thach Papers*

A Soviet submarine on the surface with a supply ship. In the late 1950s, Soviet submarines had to surface to fire missiles, and it was during those few minutes or when being resupplied at sea that they were most vulnerable. *Thach Papers*

Thach speaks at the entrance to his headquarters at Pearl Harbor upon assuming the position of commander, Antisubmarine Warfare Forces, U.S. Pacific Fleet on 1 March 1960. *Thach Papers*

Thach (*fourth from right*) greets guests soon after assuming command of the Antisubmarine Warfare Force, U.S. Pacific Fleet. To his left is his wife, Madalyn, and to his right is Naval Academy classmate and former fighter pilot Rear Adm. Paul H. Ramsey. *Thach Papers*

There were many social occasions throughout the Thaches lives. They enjoyed the festivities, but by late 1966, both were more than happy to delegate such responsibilities—and joys—to younger officers. *Thach Papers*

There were few things Thach enjoyed more than golf. Here he views a hole-in-one trophy for his shot on the eleventh hole at the Pearl Harbor Navy and Marine course. *Thach Papers*

Thach was not overly impressed with Hawaii during World War II but came to appreciate it more during his 1960–63 tour. Madalyn, shown here with her husband as they prepare to cross the ocean to their next assignment, dutifully liked every place they were assigned but always considered San Diego home. *Thach Papers*

Having just left the Pentagon, where he was deputy chief of naval operations for air, Thach (*left*) visits with Adm. Charles D. Griffin on 25 March 1965, when Thach relived Griffin as commander in chief, U.S. Naval Forces, Europe. *Naval Historical Center photo 92748*

Battle of Midway participants George Gay (*left*), Thach (*second from right*), and Rear Adm. Max Leslie, USN (Ret.) discuss their respective Midway roles during a seminar at the National Air and Space Museum, 20 September 1979. *Thach Papers*

6

The Weave Validated and Named

Although combat in the Battle of Midway was over in June 1942, it was still being fought in Jimmie Thach's mind for many months thereafter. The *Yorktown* and too many good friends and squadron mates had been lost. Thach could not do anything to bring the carrier, officers, and men back to life, but he was determined to do whatever he could to ensure that mistakes at Midway were not repeated. At the top of his list were concerns pertaining to the F4F-4 Wildcat and training. He would have ample opportunity to address all aspects of training in his next assignment, but before reporting to Operational Training Command at NAS Jacksonville, he devoted his time and attention to fighting squadrons' most pressing problem.

THE NAVY FIGHTER PROBLEM

No one connected with the 4–7 June Battle of Midway believed the F4F-4 Wildcat had the wings to carry naval aviation to victory in the Pacific. In addition to the pilots who fought in the battle, operational planners, air officers, carrier COs, and flag officers were upset. Capt. Miles Browning, Halsey's chief of staff on loan to Rear Admiral Spruance aboard *Enterprise*, wrote, "Our F4F-4 . . . Performance greatly inferior to Jap 'zero' VF. . . . Range and endurance totally inadequate."[1] Other messages from both task force leaders were no less critical.

Improved fighters were coming, but not immediately. The Chance-Vought F4U Corsair first flew on 29 May 1940 and demonstrated potential to be a great fighter. But the fast gull-wing fighter experienced numerous development problems and as a result did not enter combat until 13 February 1943, with land-based marines. Corsair suitability tests revealed serious problems when landing on flattops, a situation that benefited the marines but was a disappointment for carrier squadrons.

Grumman had started design of the Wildcat's replacement in 1941. A Navy contract for the F6F Hellcat was received in June 1941, and the first experimental model flew on 26 June 1942, only three weeks after Midway (and well before the acquisition of a Zero found in Alaska on 10 July 1942). The new Hellcat did not enter combat until 31 August 1943, but it proved highly successful not only for carrier operations but also for performance again the Zero. Still, in mid-1942 all in a position to know or care realized it would be many months before either the Corsair or Hellcat could work its way from the factory to squadrons in training and finally to combat. Consequently, those with vested interests in the summer of 1942 knew the Wildcat had to continue to hold the line for months to come.

Nearly every major participant in the Battle of Midway complained about the Wildcat's inferiority to the Zero. Thach voiced *Yorktown* and VF-3's concerns in his 4 June action report:

> Six (6) F4F-4 airplanes cannot prevent 20 or 30 Japanese VF (fighters) from shooting our slow torpedo planes. It is indeed surprising that any of our pilots returned alive. Any success our fighter pilots may have against the Japanese Zero fighter is *not* due to the performance of the airplane we fly but is the result of the comparatively poor marksmanship of the Japanese, stupid mistakes made by a few of their pilots and superior marksmanship and teamwork of some of our pilots. The only way we can ever bring our guns to bear on the Zero fighter is to trick them into recovering in front of an F4F or shoot them when they are preoccupied in firing at one of our own planes. The F4F airplane is pitifully inferior in *climb, maneuverability* and *speed*. The writer has flown the F4F airplane without armor and leak proof tanks. Removal of their vital protection does not increase the performance of the F4F sufficiently to come anywhere near the performance of the Zero fighter. This serious deficiency not only prevents our fighters from properly carrying out an assigned mission, but has a definite and alarming effect on the morale of most of our carrier based VF pilots. If we expect to keep our carriers afloat we must provide a VF airplane superior to the Japanese zero in at least climb and speed, if not maneuverability.[2]

Two months later, in Washington at the Bureau of Aeronautics, he was still fuming. "I think you cannot over-emphasize the importance of having a fighter superior to the one the enemy is flying," he later stated. "Control of the air over any location is gained by fighters, and it's maintained by fighters. If you have a better fighter you have control of the air. Our aircraft carriers can be kept afloat only by fighters."[3]

In addition to climb, maneuverability, and speed, another of Thach's specific complaints on the Wildcat was the gun sight, which he deemed completely inadequate. However, he did compliment Grumman for the 6.5-degree angle on the nose and expressed concern about the long nose of the F4U Corsair.[4] To the question pertaining to the optimum number of guns for a fighter, Thach stated, "Increased fire power is not a substitute for marksmanship. The pilot who will miss with four won't be able to hit with eight."[5]

As noted, Thach was not the only unhappy commentator on the F4F-4's performance at Midway. The *Enterprise*'s VF-6 squadron commander, Jim Gray, had written a letter to BuAer nearly two months before the battle complaining about the poor performance of the F4F-4 compared to the F4F-3 but was still unhappily surprised at Midway. On 6 May 1988, at the second annual Naval Aviation Symposium, the theme was the Battle of Midway and its implications. A blue-ribbon panel of both Japanese and American luminaries sat on the stage at the Pensacola Civic Center. They and the audience were surprised to have another luminary, Gray, rise from well up in the seats and ask if he could speak to the F4F-4 Wildcat's performance during the battle. Over the years several written accounts of Gray's fighters not answering a call to come down and fight had taken a toll on the retired captain. In a tone of voice that revealed lingering pain, Gray stated that he had never had an opportunity to explain what happened, and the moderator, quickly realizing the significance of the man and his place in history, gave him the opportunity.

Gray explained that he knew he was over Torpedo Squadron 8 as he could count their fifteen planes and knew that VT-6 had launched fewer (fourteen). Approaching the Japanese carriers at twenty thousand feet in a running rendezvous, Gray's ten Wildcat's never heard a call to descend to protect Torpedo Squadron 6. Over the enemy carriers he lost sight of VT-8 under the clouds, but what he could see startled him. Looking at his fuel gauge, he found he had only a quarter of his gas remaining instead of the three-quarters he expected. Obviously the heavy F4F-4s had consumed more fuel than anticipated during the climb to twenty thousand feet. Commenting that "only an idiot runs out of gas in an airplane" and that *Hornet*'s ten escorting Wildcats did run out of fuel and ditch, Gray decided the only option open to him was to take his planes back to the carriers.⁶ With complaints rising from so many that had taken the Wildcat into battle. the Navy, especially the Bureau of Aeronautics, acknowledged the problem and looked for ways to address it. Butch O'Hare had visited the Grumman factory soon after receiving his Medal of Honor in April and stressed to the Wildcat and Hellcat builders the same thought he shared with President Roosevelt: naval aviation needed "something that will go upstairs faster."⁷ Grumman responded by placing a more powerful engine (the Pratt and Whitney R-2800) in their second Hellcat prototype in place of the Wright 2600 installed in the first. When production began, the R-2800 was the engine of choice. And the Navy had upped the number of fighters assigned to carriers from eighteen to twenty-seven while plans were being prepared to go to thirty-six when sufficient numbers of pilots and planes—and carrier decks—permitted.

THE FLATLEY LETTER

After leaving Thach on 19 April, Jimmy Flatley caught up with *Yorktown* in time to fight in the Battle of the Coral Sea. With orders to take command of VF-42 in his hand, he was surprised to find that CV-5's brass had already promoted from within the squadron, as they were unaware that Washington had cut orders naming him as the squadron's CO. Flatley helped resolve the issue by volunteering to serve as executive officer, knowing full well that engagement with the enemy was only a few days away. Needing all the experienced fighter pilots available, *Yorktown*'s CO was only too happy to accept the offer. On 7 May, Flatley led an escort mission when SBDs and TBDs sank the *Shoho*, and on 8 May, he led fighters in defense of *Lexington* and *Yorktown*. For the two-day battle he claimed two enemy fighters and tangled with numerous others. After the battle, Flatley left for San Diego to take command of Fighting Squadron 10, the first replacement fighter squadron to enter combat later in the year.

Flatley had fought the two-day Coral Sea battle in two different F4F-3 Wildcats but he was quite familiar with the experimental F4F-4. Logging more than twenty hours in the experimental XF4F-4 in February and March, he gave Lt. Cdr. Dale Harris from BuAer an earful after a dinner invitation to the Flatley home in San Diego in late March.⁸ After facing enemy planes in two days of battle a little over a month later, he knew that neither version of the Wildcat could equal the Zero in climb and maneuverability. Like so many other fighter pilots during the first six months of the war, he was surprised at how fast a Zero could climb away from trouble. But unlike Thach and Gray, both of whom were headed for noncombat assignments, Flatley was going back into combat with a squadron that would undoubtedly include many rookies. With the responsibility of preparing so many new young pilots, the last thing he needed among them was low morale emanating from a perception of inferiority. Consequently he looked for any and all positives that could be found.

The result of Flatley's experience and thinking was first committed to a series of notes that by early July had been converted into a small manual titled *Combat Doctrine: Fighting Squadron Ten*. Flatley's cover statement noted that the contents pertaining to combat data for fighters was the only available information at the time addressing actual war conditions. In the manual he discussed formation flying, fighter director orders, methods of attack—both advantages and disadvantages, when to shoot and when not to (not over 250 yards), and how long to shoot (not over three seconds). Everything from radio failure to fuel supply and altitude advantage to fighting in bright sunshine to clouds was discussed. Prominently absent, however, was any mention of the Wildcat, either the –3 or –4 version, and nothing was noted about the Zero. All mention of fighters was in generic terms.

Combat Doctrine was not completed until July, as Flatley in mid-June pushed that manuscript and drawings aside to write a letter in response to the negative reports circulating on the Wildcat. None of the fighter pilots who fought at Midway arrived in San Diego before his nine-page epistle, titled "The Navy Fighter," was sent to commander, Carriers, Pacific Fleet (ComCarPac). Not having dialogue with those pilots resulted in a few details not being correct (such as a statement that no fighter pilot had fought enemy fighters more than once), but that did not subtract from the essence of the letter. From ComCarPac, copies were forwarded to the Bureau of Aeronautics, Admiral King—the Navy's commander in chief—and Admiral Nimitz. Nimitz, commander in chief of the Pacific Fleet, was well aware of naval aviators opinions of the Wildcat not only from written reports but also from face-to-face discussions with principals returning to Pearl from Midway. Indeed, Nimitz was sufficiently concerned about the problem to have written a letter to King on 21 June suggesting modifications of the Army's P-40 to allow them to be used aboard carriers.[9]

Flatley had no knowledge of the depth of Nimitz's concerns and most likely did not expect his letter to find its way to the desks of Nimitz, King, or the secretary of the navy. With only six pilots assigned to him when he wrote his letter, Flatley's primary audience was the twenty to thirty fighter pilots who would join his squadron in the coming weeks.[10] Still, the essence and tone of his written thoughts struck a resonant chord with people in high places and the timing could not have been better.

In "The Navy Fighter," Flatley's goal was to convince fighter pilots that the Wildcat, when properly used, was not only equal to enemy fighters but superior. Fighting 10's skipper had not abandoned his reservations on the Wildcat because he knew neither the –3 nor –4 Wildcat could outperform the Zero one on one. Therefore Flatley wrote, "Our fighter squadrons must develop formation tactics which will eliminate the necessity for individual combat."[11] The answer was to fight as a team and not get into individual dogfights. The last line on page nine of his lengthy letter ended with "Let's take stock of ourselves and get down to work and quit griping about our planes."

On 27 June, two days after "The Navy Fighter" went into the mail, Thach and his wife joined Flatley and his wife, Dorothy, and a few other mutual friends for dinner at the Flatley home. As informal protocol dictated, the men and women separated before dinner and during dinner discussion centered on anything but the war. But both before and after dinner, when the men and women had again largely separated, sadness was shared as Thach related to Flatley the last minutes of Lem Massey and Edgar Bassett's lives. Massey was a 1929 Naval Academy classmate of Flatley; the two had sailed together to Pearl aboard *Enterprise* less than a year earlier. And in May, Bassett shot down two planes in the same actions in which Flatley had fought at Coral Sea. "The Navy Fighter," the Wildcat problem, and other mutual business matters were postponed that day, but both agreed to spend as much time as possible discussing tactics and strategy before Thach had to leave for Jacksonville.

Most of these meetings occurred in Flatley's office in the hangar serving as VF-10's home. Capt. Stanley "Swede" Vejtasa recalled that Flatley's quieter, low-key demeanor served him well in meetings with the more excitable Thach. Flatley did not discourage his pilots from jumping other squadron's fighters or any other planes in the training areas. Thach did not think this practice was a good idea, and when he complained "about those wild and dangerous pilots of VF-10," Flatley sat quietly, listened intently, and interrupted only to pass cigarettes to his vociferous friend. Having to interrupt himself while drawing on a cigarette, Thach lost some ardor and most of the rest when Flatley suggested they continue their visit while having lunch at the Officers Club. As opportunity arose, Flatley delicately moved the discussion from his "wild" pilots to tactics. According to Vejtasa, who had already distinguished himself in combat as a dive-bomber pilot and would again as a VF-10 fighter pilot, this tactic always worked because Thach never passed up an opportunity to discuss tactics.[12]

The two squadron commanders had known each other for nearly seventeen years. Their careers in naval aviation had been parallel in type of assignments and with the number of prewar fighter pilots being so small they could not help but frequently cross paths in the tight-knit naval aviation fraternity. Good friends on the basis of similar interests and outgoing personalities both well understood that their relationship was not based on professional agreement. Although both were easily approachable and friendly, there were marked differences in their respective personalities. Dorothy Flatley remembered Thach as a delight to be around despite an air of superiority. Two years ahead of Flatley at the academy contributed somewhat to the "air," but Dorothy recalled that no one was offended "because he was indeed superior."[13]

On most issues and matters, the two were on the same page, and neither had to remind the other that the heaviest burden of how to conduct the current war with carrier fighters had fallen on them and three or four others who had led fighters into combat. Over the next three years, Thach made entries into one of the little black books he always carried, some of them items to share with Flatley, who did the same, except for his reliance on looseleaf paper stored in files rather than pocket-sized notebooks.[14]

THACH RESPONDS

The strength of the Thach-Flatley relationship was not strained in the least in the days when the two were together in San Diego and Thach was asked to officially respond to "The Navy Fighter" by Capt. J. M. Shoemaker. Shoemaker, in the administration office of ComCarPac, appreciated Flatley's letter but was concerned with the several references to the six-plane formation (division). Consequently, on 15 July he passed it to Thach, "the most experienced fighter pilot in this area, for comment." In the same letter, Shoemaker proposed that Thach revise Flatley's letter and integrate pertinent sections in a revision of chapter 3, part 2, *United States Fleet 74 Revised* (USF-74), the "book" on carrier squadron operations. Once Thach completed his revision, the changes were "promulgated to the Service."[15]

Thach's four-page response was undated but undoubtedly was submitted back to ComCarPac before the end of July, when he left for his next assignment at Jacksonville.[16] He opened with an acknowledgment that "The Navy Fighter" was "an excellent approach to a subject, which in the past has been given too little attention by those responsible for the overall training of carrier-based pilots." Continuing, he agreed that "it is possible by increasing each squadron to thirty-six fighters and by exploiting to the fullest extent the personnel and material (F4F-4) now on hand to successfully attack enemy carriers and to effect complete annihilation of an enemy carrier attack group before their arrival at the point of dropping torpedoes or bombs." Completing his introductory comments he wrote "that an aircraft carrier under attack by enemy dive-bombers and torpedo planes can escape damage only so long as her fighter squadron is adequate in number and performance."

Leaving the opening comments in which he was in agreement with Flatley's letter, Thach moved to the areas of disagreement. He may have misunderstood what Flatley intended to say on page seven of his letter, as Thach apparently thought Flatley suggested that the weight of armor and self-sealing tanks might be traded off for better performance. Flatley may well have clarified that sentence for Thach, as he later stated that he differed with Flatley on only one issue, and that was not the weight versus performance issue.[17]

The issue Thach did disagree with Flatley on in "The Navy Fighter" was the same one the two men had been discussing since their time together at Pearl from 11 to 19 April. In late June, the two were not alone with their cigarettes and bourbon at Pearl talking well into the darkness of night but were now standing conspicuously in the light before all of naval aviation. The issue, the only issue, was tactical organization. Flatley had followed up on his April statement to Thach that he would continue working with a six-plane division while Thach continued with his four-plane division. In "Navy Fighter," Flatley had carefully listed all the advantages he saw with the six-plane formation, even what degree angles formations should be flown. Both men agreed that the two-plane section should be retained, so the disagreement was only on division composition.

At Coral Sea, Flatley had joined VF-42 too late to train with them and that lack of training resulted in Flatley's wingman losing track of him on his very first combat maneuver with things not improving much during combat the following day. On the other hand, Thach knew for a fact that it was all but impossible to keep track of a third section as he never found them once the fighting started at Midway. And he had every reason to believe that the mutual beam support maneuver had saved both he and Ram Dibb in that battle when they were so heavily outnumbered. Consequently, his rejoinder was written with conviction as much as logic.

After noting that a third section would get lost in combat and that six planes were insufficient for a combat escort (how well he knew), he presented the logic that a thirty-six plane squadron lent itself to four-plane divisions (nine). Somewhat surprisingly, he did not note that six could not divide twenty-seven. For escort missions he advocated a minimum of twelve planes in three divisions stacked to provide close support, close cover and intermediate or high cover. When awaiting enemy attack groups (and assuming twelve fighters would be flying with his own attack group) he championed having twelve planes on combat air patrol and the last twelve of thirty-six on deck ready for launch as soon as the enemy was known to be closing in.

Throughout his letter it was apparent that Thach was calling upon his recent experience at Midway. Although altitude was desirable, enemy torpedo bombers or dive-bombers had to be met from whatever angle was available, he stated. The desperation he felt immediately before Midway, when he told his pilots they might have to ram an enemy plane before it dropped ordnance, was reiterated in this letter after the battle. In it, he described in detail how a fighter "should move in to make contact, across to the right, up and away" (to use the fighter's propeller to cut off the tail of an enemy plane).

Letter finished and presented as directed, Thach, Madalyn, and son completed packing and prepared for their move to Florida. Flatley meanwhile continued to collect pilots and increase the intensity of training. He had already heard back from Dale Harris at BuAer that his letter "received most favorable comment," and Cdr. J. B. Pearson from the Engineering Department at BuAer echoed that thought.[18] In the first week of August, he received a 29 July 1942 letter from Secretary of the Navy Frank Knox congratulating and praising him for the essence and tone of "The Navy Fighter." Noting that "that is the kind of spirit that wins wars," Knox went on to say he too knew the Navy had to "do the best with what we have." And Flatley was preparing to do just that: take his VF-10 "Grim Reapers" and their F4F-4 Wildcats first to Hawaii, then aboard *Enterprise*, and then into combat.

O'HARE CONVERTS FLATLEY TO THE WEAVE

Everyone in the United States and Japan old enough to understand knew the battle that had opened 7 August on Guadalcanal in the Solomons could determine the war for both countries. On 8 August, VF-10 left San Diego bound for Hawaii. On the twentieth, the squadron arrived on Maui, and by the twenty-fourth, training was under way. The "Grim Reapers" did not have Maui to themselves. In addition to some Army planes, Butch O'Hare was assembling and training fighter pilots for his VF-3. O'Hare had not had the opportunity to serve with the "Grim Reapers" squadron commander as he had with Thach, but Butch and Jimmy Flatley were not strangers. At Pensacola in May 1940, Butch had attained his highest training scores in a Boeing F4B fighter under the tutelage of Flatley. In August and September 1942, the thirty-six-year-old Flatley and the twenty-eight-year-old O'Hare were equals in rank and command, but the modest O'Hare looked upon Flatley as an older brother who did not demand respect but deserved it.

By virtue of age and personality, Thach had been more direct in his discussions with Flatley regarding their tactical organization variance of opinion. Butch had been with Thach on the day before the war when the mutual beam defense tactic was first tested, so he was a believer without reservation. In his own non-assuming manner, Butch picked up the debate with Flatley on Maui as the two squadrons trained side by side. O'Hare first got Flatley's attention by relating how Thach had modified the controls on the planes in his division while allowing Butch's division to retain full power. As Flatley retained seventeen hundred revolutions per minute (cruise setting) in battle while others often went to twenty-two hundred, the potential to conserve fuel was a matter of special interest.[19]

After a few training flights, Flatley began to experience some of the problems Thach had emphasized—especially losing sight of the third section—and began to experiment with the four-plane division. "It took little practice to note the many advantages," and before the end of September, Thach's cogent oral and written arguments in combination with O'Hare's continuing support of the tactic converted Flatley.[20] Once convinced, Flatley was a whole-hearted convert. He not only adopted the four-plane division but also began teaching the weave to his squadron. Not concerned in the least that naval aviation peers might wonder why he had abandoned his previous detailed defense of the six-plane division, Flatley became an outspoken champion of Thach's tactic.

Flatley had no illusions as to where he and his "Grim Reapers" were going once they boarded their carrier. On 16 October, *Enterprise* left Pearl with Carrier Replacement Air Group 10 aboard. On the twenty-sixth, the "Big E" and *Hornet* took on four enemy carriers southeast of Guadalcanal near the Santa Cruz Islands. The *Hornet* fought its last battle that day, and after a long, tragic escort flight, Flatley came back to a battle-damaged *Enterprise*. Very low on fuel and out of ammunition, he ran into yet another group of enemy fighters. Resorting to Thach's weaving tactics, he survived all attacks. Finally back aboard, he repaired to his stateroom immediately aft of other officer staterooms destroyed or damaged on the carrier's second deck. Before writing letters of condolence to the families of his lost pilots, he crafted notes for his action report. Therein Reaper Leader—Flatley's squadron sobriquet—recounted how the beam support tactic had allowed him a successful defense in a most untenable situation. And he wrote that "the four-plane division is the only thing that will work, and I am calling it the Thach Weave."

THACH WEAVE POSTSCRIPTS

Beginning late in World War II and continuing for many years, fighters pilots—some still in uniform, others who had returned to civilian life—told Thach that his weave had saved their lives in combat, a validation he greatly appreciated. Still, some have questioned whether Thach was the innovator of the tactic. Capt. Hugh T. Winters, commanding officer of Air Group 15 on the *Essex*, commented in 1996 that the tactic was a natural defense that did not require great insight and would have occurred to fighter pilots without having formal introduction.[21] But Winters went on to say that it was good that the tactic was named because it focused attention on the maneuver. Winters, however, may not have been acquainted with the specific distances between planes Thach desired (tactical turning radius of the Wildcat) and the necessity for both pilots to constantly look at the other (optimum lookout) and turn at the precise moment an enemy fighter committed to attacking. Thach's formula required a more formal understanding of the maneuver than the British tactic with which Winters was familiar.

Another challenge in 1984 was raised by Rear Adm. John Crommelin, USN (Ret.) in a long letter to Vice Adm. Bernard M. "Smoke" Strean, USN (Ret.). Strean, commanding officer of VF-1 aboard *Yorktown* CV-10 in 1944, had written Crommelin asking for membership nominations to The Early and Pioneer Naval Aviators Association (the "Golden Eagles"). In his six-page, single-spaced typed reply, Crommelin, air officer and then executive officer of *Enterprise* during its critical 1942–43 battles off Guadalcanal, assigned credit for the weave tactic to George W. "Red" Brooks and Gordon E. Firebaugh. According to Crommelin, the two pilots came to him while *Enterprise* was en route to

Guadalcanal in early August 1942 and asked to demonstrate a defensive tactic Crommelin later dubbed the "Enterprise Scissors." The tactic, he noted, was shared with all of VF-6 before combat began on 7 August, and two pilots (Chief Machinists Thomas W. Rhodes and Lee P. Mankin Jr.) told him after an early engagement that they had each shot a Zero off the tail of the other. When Enterprise returned to Pearl in September for repairs after the Battle of the Eastern Solomons, Crommelin stated that he sent Brooks "to teach Jimmy Flatley and VF-10 the 'Enterprise Scissors.'" Finally, Crommelin wrote that Brooks reported back to him "and told me he thought VF-10 had the good news."[22]

Crommelin distributed numerous copies of his lengthy letter to Strean, friends, peers, and researchers—who approached him in following years as it addressed several other incidents from the war years. In the early 1980s, several Hall of Fame programs were initiated and Crommelin further distributed the letter to support several veterans he believed should be inducted. In time, some veterans of the Enterprise had copies of the letter—and conversations with their former shipmate, and by word of mouth or articles, word of the Enterprise Scissors spread. Brooks retired as a captain in 1956 and died before Crommelin's 1984 letter began circulation. However, Gordon Firebaugh, who retired as a captain in 1957, wrote in 1987 that Crommelin's 1984 letter to Strean "contains adverse information." He continued, "I too, have never heard of the 'Enterprise Scissors' until (Commander Howard S.) Packard wrote me a few weeks ago." Firebaugh also wrote there was no question in his mind that John Thach and Jim Flatley were responsible for the weave.[23] Finally on this matter, as noted above, it was Jimmy Flatley who coined the phrase "Thach Weave."[24]

It is entirely possible that the Enterprise Scissors resulted from one of Thach's lectures to fighter pilots at Pearl Harbor in the last week of June. Entries exist in the diaries of several pilots about these lectures, and it is highly probable that VF-6's commanding officer, Lt. Louis Bauer, picked up on the tactic and carried it aboard the Enterprise when it steamed toward Guadalcanal on 15 July 1942.[25]

It is understandable that some pilots who did not fly with or near Thach during the first year of World War II might not have known of his prewar experiment with Butch O'Hare, his continuing dialogue with Jimmy Flatley, or his combat experience at Midway. Younger pilots passing through training during the war had far greater exposure to both Thach and his tactics. Too, the character of aerial combat changed considerably in 1944 and 1945, as well-trained naval aviators in planes equal or superior to the Zero engaged Japanese pilots who were not as well trained as their predecessors. Rather than rotating experienced pilots back to training duties, the Japanese left most experienced fighter pilots in combat until they were lost or the war ended.

After World War II, when Thach was director of training at Pensacola, he was invited to critically review a revision of USF 74 "in view of your experience in developing fighter tactics."[26] Thach's lengthy response touched on several matters, one of the more interesting being his distinction between the word "scissors" and "weave":

> There is one matter of nomenclature that would cause a great deal of confusion. I refer to Paragraph 411, sub-title "The Scissors." The word "scissors" has been, for years, used to describe a maneuver that a single plane works into when attacked by an enemy single plane, each turning toward the other, trying to beat the opponent in the turn to get on his tail. I think it aptly describes that particular maneuver and should not be used to describe any other. The maneuver used in Paragraph 411 and called the "Scissors" has been called the "Weave" and taught as a weave to the many thousands of students who have graduated from the Training Command in the past three or fours years. All of our training films and pamphlets call it a weave and I would hate to see the terminology changed now not because I invented the d—— thing, but because of all the confusion that would result.[27]

During the Vietnam War, the Thach Weave was again reinvented when early dogfight losses against Russian-built MiGs signaled a need for change. Resorting to group tactics, scores significantly improved.[28] This did not surprise Thach, as he had always maintained the team as a whole rather than individual effort resulted in aerial combat success.[29] Teamwork was the key, and the Thach Weave was perhaps its ultimate expression.

7
The Thach Weave Gets Help from Hollywood

As July rolled into August 1942, Lieutenant Commander Thach and his family set up their new quarters at Naval Air Station Jacksonville. The installation had opened for business in January 1941 under the direction of Capt. Charles Mason, with considerable help from his executive officer, Cdr. Joseph James (later "Jocko") Clark and Lt. Jimmy Flatley, who wrote the station manual. Still too new to have many of the amenities of more established naval air stations, such as Pensacola or San Diego, NAS Jacksonville was making satisfactory progress toward meeting the basic needs for which it was established.

Just over three thousand Navy pilots had been trained in 1941, nearly eleven thousand would be trained in 1942, and that number was projected to nearly double in 1943. There was no way NAS Pensacola and the satellite fields there could produce all the new pilots required. Facilities alone could not handle the influx, and even if they could, there was not enough air space in the area to handle all the planes. Pilots graduating from Pensacola knew how to fly, but they were not ready for combat. Before the war a new pilot was absorbed into a squadron and prepared for potential combat by the squadron. Once the war began in December 1941, this quickly proved undesirable and inadequate because there was not sufficient time to prepare rookies for combat. Up and down and throughout naval aviation this was known, but entrance into war sooner than expected found naval aviation unable to fully implement the solution of advanced training.

The primary function of the Naval Air Operation Training Command at NAS Jacksonville was to place new pilots in the cockpits of the planes they would fly in combat. Familiarity with their plane was followed by introduction to the latest tactics utilized by their type of aircraft (fighters, dive-bombers, and torpedo planes). Thach's first official title was staff gunnery officer. The opportunity to serve in a training role was not a disappointment to him as it was to some of the other fighter pilots who had recently rotated back to various billets throughout the expanding training command. Had he not gone into the Navy, Thach would have followed in the footsteps of his parents and become a teacher and coach. At Jacksonville he would not have time to assist with athletics, but he would be a teacher of the first rank.

FILM WRITER, DIRECTOR, AND NARRATOR

Among the many naval aviation matters Thach and Jimmy Flatley discussed in San Diego in late July was the challenge confronting an operational training command. Thach mentioned that there were too few instructors with combat experience to effectively communicate what had been learned to the hundreds of pilots then in training and the thousands to come. Flatley, who had been in San Diego both before and after the Battle of the Coral Sea, commented that he had been talking to Walt Disney about the same matter and that Disney had expressed an interest in helping. Both fighter pilots had knowledge of camera guns used in training and both were strong advocates of its use and value. However, both also knew the limitations of using film in training films because the maneuvers often required more distance than a camera could capture. Flatley's thought that animation might be the answer brought quick affirmation from Thach as well as a promise to pursue the matter, as Flatley was then preparing for transfer to the Pacific with his Grim Reapers.

Thach visited Disney before leaving for Jacksonville and found the creator of Mickey Mouse, Donald Duck, and other cartoon favorites more than willing to help. Initially there was a question in the minds of both how effective the endeavor might be, but there was only one way to find out. Upon arrival in Jacksonville, Thach first had to sell Rear Adm. A. B. Cook on the value of animated films as training aids, not a difficult sell, since Cook knew that several hundred training films either already existed or were planned. Soon Thach enlisted the aid of Capt. Arthur Radford, his former squadron commander from *Saratoga*. In mid-1942, Radford was in Washington at the Bureau of Aeronautics and promised the ten suggested training films would be produced and distributed.

Letters of intent were written and approved, and the necessary contracts were prepared. As it turned out, Warner Brothers contributed by producing live-action portions of the films, while Disney produced the more voluminous animated sections.

Before the end of 1942, Thach realized he actually had two different jobs despite having only one official designation. Although he expected to participate in the development of the films, he did not expect to become the main writer, director, and narrator. And he did not realize how much time he would have to spend in Burbank, California, home of the Disney Studios. Throughout the course of working on the films, he usually spent two weeks in California for every three in Florida. Although the two-versus-three schedule was not planned as such, artists often got to a point in their work, discussions, and debate that only Thach could resolve the matter. When that occurred, a wire was sent and Thach made his way to the flight line and thence to California. He was in the Disney studio so often that he had his own mailbox there.

In Burbank, Walt Disney provided a three-story building and a host of artists to work on the "Jacksonville Project." Erwin Verity, one of Disney's more experienced executives, was appointed as project manager. Verity and Thach made a good team, each cooperating with the other to a high degree.

At the beginning of each of the ten films, Thach had to meet with the more experienced artists to demonstrate with his hands, models, and cutouts what a plane did and how it maneuvered through space. Several of Disney's top animators drew the several critical aircraft maneuvers, while younger, less-experienced artists drew the progression pictures in between so that the movement of images was smooth. Quickly Thach realized he was actually teaching Disney's artists how to fly, but they in turn knew the value of what they were doing, so the give and take was beneficial to all. Relative movement between a plane and target was particularly important but the concept proved to be understandable to the artists despite its degree of difficulty. Even the degree of a plane banking for a turn or dive had to be accurately presented, as it was important for pilots to see precise maneuvers. The film might be similar to a cartoon, but no laughs were wanted or intended. In time, however, Thach became amused at the enthusiasm and dedication of the artists as they adopted the language of fighter pilots and even began to discuss tactics.

Thach the writer had become Thach the director because no one else could ensure that angles were correct, that the nose of a plane was not too high or too low, and that all relative positions were correct throughout the sequence of a particular maneuver. Originally, no one expected Thach to be the narrator, but the narration also had to be perfect to ensure that pilots in training were offered a realistic representation of what they would see in combat. Several people were auditioned, with Thach and Disney sitting as judges, to narrate the films. Neither was satisfied with the auditions, and Disney finally turned to Thach and told him, "You've got to do the narration." Thach protested, but in vain. Disney's final sell was that Thach not only knew the information but also knew what words in the script had to be emphasized. As there had been an instance of misplacing emphasis during the auditions that changed connotation, Thach was left without an effective rejoinder. At Warner Brothers Thach was also asked to portray himself in a segment that called for an instructor to illustrate several salient points to student aviators.

While Disney, Verity, and the building full of fighter-pilot artists worked feverishly, fervently and happily, few projects of special importance wrap up without some problems along the way. Problems with the Jacksonville Project were few and relatively insignificant when viewed in perspective, but there were some bumps with Disney, Warner Brothers, and even superiors at NAS Jacksonville.

The biggest problem was the amount of time Thach spent in California. Rear Admiral Cook understood the need, but he had other projects (discussed later) for which Thach was needed. But it was not until Cook was preparing for reassignment in 1943 that he fully appreciated how hard Thach had worked on the films. Indeed, he offered Thach an apology for thinking he might have been playing around in Hollywood instead of devoting time to work. After seeing the first several films, Cook not only was greatly impressed but also began to realize just how much time Thach had to put in to get results that even people in Washington were lauding. Although Thach assured Cook no apology was necessary, he appreciated the admiral's complimentary parting remarks.

A second bump required Thach to accept Radford's invitation to call him if any problems arose. Disney officials were pleased to have many costs of production come in under budget projections listed in their letters of intent. For the first three films, costs were projected to be $155,780 but instead were completed for $127,565, a difference of $28,215.[1] However, at Warner Brothers one executive objected to a particular sequence he knew would bust the budget. Thach insisted that the scene in question was so significant it had to be included regardless of cost, as it would make the difference between life and death for pilots. The executive's rejoinder was that the Navy would not spend the large sum of money for one scene that might cost as much as an entire film. Thach reminded the executive that he had unlimited authority for the films, but when dialogue failed to resolve the matter he picked up the phone. After Captain Radford answered, Thach talked to him briefly and then handed the phone to the Warner Brothers production manager. The scene was produced, and there were no further bumps.

When completed in late 1943, the ten films averaged about thirty minutes in length. As can be seen in the respective titles, all ten—*Use of the Illuminated Gun Sight, Gunnery Approaches, Snoopers and How to Blast 'Em, Don't Kill Your Friends, Group Tactics against Enemy Bombers, Offensive Tactics against Enemy Fighters, Defensive Tactics against Enemy Fighters, Escort Doctrine, Combat Air Patrol,* and *Conclusion and Summary*—were directed primarily at fighter pilots. Thach was pleased that his immediate superiors and peers were highly complimentary of the final product, which not only presented tactical problems and solutions but also gave pilots a feel for the battles they would enter. One magazine article referred to him as the Navy's Chennault, a compliment both to him and the Army's Brig. Gen. Claire Chennault, who developed fighter tactics for the "Flying Tigers" in China.[2]

Although satisfied that he and his Burbank-Hollywood associates had produced films to the best of their ability, Thach nonetheless knew that the films did not guarantee victory in the air. But it was critical that pilots have a plan when thrust into combat. Even a poor plan was better than nothing. In time, Thach came to appreciate that the combat simulations in his films, and the supplementary booklets he wrote and Disney illustrated, had the value he hoped they would. That appreciation was especially meaningful when throughout the remainder of his life he was approached by naval aviators who told him his films helped save their lives in combat.

STAFF GUNNERY OFFICER

As earlier mentioned, Rear Admiral Cook worried that Thach's trips to Hollywood and Burbank were taking too much time away from his responsibilities at Jacksonville. Particularly, Cook wanted Thach to maintain a liaison with other training bases by presenting lectures. As there were a number of training bases in Florida and surrounding southern states, the plan was workable. But Thach reminded Cook that he was only one person and that once the films were ready, more pilots could be presented the essentials of aerial combat in less time. The argument being sound, Cook relented, but he still wanted Thach to visit the other commands as much as possible. In the time Thach was in Jacksonville, he traveled extensively to other nearby training fields. Films would be important, but there were many other training aids and devices that had to be planned, crafted, and put in place. Most of the devices and aids originated in Washington but Thach had to ensure that the right aids were scheduled for the right places. Training needs determined, he turned attention to ensuring that the training syllabus was being followed. The syllabus was constantly a work in progress as weapons scores by pilots in training dictated changes not only in process but also in time allotted to achieve a desirable level of proficiency before a pilot was qualified to join an operational squadron. One of the more difficult tasks attending Thach's responsibilities was dealing with the patriotic desire of young men who wanted to become naval aviators but had difficulty mastering the necessary skills. Their desire to serve and fly, plus the great need for so many new pilots, occasionally resulted in well-meaning officers at basic training commands granting too many second chances. After several months of witnessing this process, Thach gathered a number of pilot training files and took them up the line to Rear Adm. George Murray, chief of air intermediate training command and commandant of the Air Training Center at Pensacola. There he presented the files and stated that the pilots had been murdered because the extra time granted them to meet minimum standards for their wings translated into their deaths. He knew their deaths were not intentional, but he strongly advocated that when two or more instructors in basic training recommended discontinuation, those recommendations should not be overturned. Not having been aware of the problem due to a lack of communication and the everyday press of other duties, Murray promised change. The two officers had known each other for many years, and mutual respect made discussions much easier and more productive. Happily, Thach did not have to return to Pensacola again with another batch of files, and efforts were made to establish feedback from the fleet to operational training and then to basic/intermediate commands.[3]

SECRETS OF THE ZERO REVEALED

Less than a month after reporting for duty at Jacksonville, Thach was in the Bureau of Aeronautics for a technical interview. The major purpose was to obtain an overall view of how extant aircraft (the F4F-4 Wildcat) and its instruments were meeting the needs of naval aviation. The 26 August 1942 transcript reveals that the opening third of the interview centered on air operations, including tactics, from Pearl Harbor through Midway, the remainder being devoted to specific questions pertaining to aircraft equipment. Among Thach's responses not previously cited were (1) a suggestion that a microphone for the plane's radio be incorporated into the oxygen mask; (2) an appraisal that aircraft radios were not state of the art, battles could be lost because not all of a message could be heard, and that push buttons should replace dials; (3) a recommendation to consolidate electrical switches; and (4) a suggestion that planes be camouflaged for water background. Few of Thach's ideas, recommendations, and suggestions failed to find later implementation. Of all the training aids and devices that were invented, created, or researched during Thach's year and a half at Jacksonville, the most significant was the capture of a Japanese Zero. Sighted on 10 July 1942 on Akutan Island, Alaska, the sleek fighter had taken a bullet on 4 June that severed an oil line. A subsequent forced landing resulted in the plane flipping onto its back leaving the pilot dead. Salvage was conducted in knee-deep water, and despite the difficulty, the damaged plane was taken first to Dutch Harbor and then to San Diego on 12 August. Working around the clock, technicians had the rebuilt Zero ready for preliminary tests that were conducted from 26 September to 15 October.[4] Performance tests with allowance for full armament and maximum internal fuel load revealed the strengths and weaknesses of the fighter Thach, Flatley and other pilots had seen conducting its lethal business. The all metal Zero's wing was integrated with the fuselage (coincidentally a feature of the experimental XF6F Hellcat then making initial flight tests). No leak-proof tanks or armor were in evidence thereby documenting earlier American suppositions, and the fighter weighed only 3,718 pounds empty, some 2,000 pounds less than a Wildcat and 5,500 pounds less than a F6F Hellcat. Internal fuel capacity was 103 gallons in the wings, 38 in the fuselage tank, and 87 gallons in an external tank. The Zero's internal capacity of 141 gallons was actually less than the 144-gallon capacity of Wildcats with fuel bladders and 160 without. But the Zero's light weight and high pitch settings for the propeller gave it considerably greater range than the early Wildcats. Initial examinations of the Zero revealed little that U.S. Navy carrier aviators had not already deduced. The enemy fighter was known to have two 20-mm cannon and two 7.7-mm machine guns. When the Zero's characteristics were published in November 1942, Thach was slightly amused to learn that the 7.7 guns had 500 rounds each. Coincidentally, that was exactly the number he had told interviews at BuAer in August what he considered to be the optimum number of shells for U.S. Navy fighters. The most significant value of the captured Zero was the performance characteristics recorded during test flights. When compared to the F4F-4 Wildcat the Zero was superior in speed and climb at all altitudes above 1,000 feet. It was also superior in service ceiling but could not outrun the Wildcat close to sea level. Of particular interest to Thach and other instructors was the Zero's propensity to have its engine cut out in pushovers, which was apparently the main reason enemy fighters often did not follow Wildcats in a dive. That peculiarity aside, the Zero could dive with the Wildcat. As the new F4U Corsair was nearly ready for combat, the results of tests with the Zero were stacked up against the gull-wing fighter. In level and diving speeds at all altitudes, and in climb at sea level and above twenty thousand feet, the Zero was inferior. As with the Wildcat, a Corsair could resort to a pushover or rolls at high speed if it was surprised. But also like the Wildcat, Corsair pilots were wise not to try to turn with a Zero because it still had a marked advantage with that maneuver. The bottom line was that the Corsair could chose the manner of combat, as it could climb to high altitude, where it would have a distinct advantage or speed away from any undesirable situation. The captured Zero gave Thach and all training commands a precise understanding of what they needed to address. And it would continue to be stressed that until the Corsair and Hellcat war-winners were operational, mutual-defense tactics, including the Thach Weave, would have to be employed.

PUBLIC RELATIONS

Training was Thach's major responsibility while assigned to Jacksonville, but not far behind was public relations. Lectures at training bases were not public relations jaunts, but as time progressed, many lecture visits were dovetailed with speeches and interviews. Over time lectures were integrated with training schedules to have Thach or other instructors speak on a specific subject within a day or two of the pilots taking to the air for exercises. Ideally, a lecture, film, or both would precede aerial exercises. Early on, formal demonstration lectures were produced to standardize and correct information that flight instructors had been passing to students.[5] Given the dynamics of the war and the introduction of new technology, the training syllabus was constantly a work in progress. Visits by the media to Jacksonville were not infrequent and when available Thach was part of the team to meet, greet, accompany and educate representatives of the wire services, print media and radio networks. Such tours usually included a visit to the Celestial Navigation Trainer and to Cecil Field for demonstrations in dive bombing, field carrier landings, and even night flying. Visits to the Naval Air Gunners School was also a regular stop, and when the media arrived in groups the visit ended with a dinner at the Commissioned Officers' Mess. For nearly all such visits, it would have been nearly impossible for the officers or reporters to draw a distinct line between education and recruiting. Although not overtly pronounced, recruiting was on the minds of Thach and others with similar duties into 1944. Many speeches, radio broadcasts, and newspaper and magazine interviews concluded with Thach mentioning the need for thousands of new naval aviators and promising that new recruits "will find a job and a cause which will envelope them completely."[6] Toward the end of his Jacksonville assignment, references to the Battle of Midway diminished, as did emphasis on recruiting. Additional significant battles had been fought throughout the Pacific since June 1942, and by early 1944 the Navy could see a need to cut back on the number of new pilots entering the training commands. The media, however, never ran out of a need for information. Most of Thach's experience with the media was positive and enjoyable for him, but there were some exceptions. High on his list of pleasant experiences was an invitation to the Author's Club in Hollywood while he was working with Disney and Warner Brothers and another was an invitation to a ship launching. The public learned of *Lexington*'s loss after the Battle of Midway and because "Lady Lex" had fought so well off New Guinea and in the Coral Sea, the hull of an *Essex*-class carrier scheduled to be named *Cabot* was renamed for the lost CV-2. As Thach had been a significant part of CV-2's combat history, he was invited to *Lexington* CV-16's christening ceremony at Quincy, Massachusetts, on 26 September 1942. Seeing a new *Lexington* in the water at the Fore River Shipyard was enough to make the day memorable, but a second ceremony later in the day made it even more so. With many of the same newsmen who had just covered *Lexington*'s launch present, Thach was awarded two Navy Crosses and the Distinguished Service Medal. One Navy Cross was enough to make news. The award of two Navy Crosses in the same ceremony was bigger news. And the award of the Distinguished Service Cross for a middle-grade officer was nearly unheard of. In the pyramid of medals, Thach received the second and third highest, only the Medal of Honor ranking higher. The first Navy Cross was for the 20 February

1942 battle off Rabaul, and the second was for Midway. The Distinguished Service Medal recognized his "exceptionally meritorious service" for leadership and training of his squadrons in addition to "fighter tactics developed by Lieutenant Commander Thach." Additionally, a Letter of Commendation from Admiral Nimitz was presented for leading the attack on Salamaua and Lae. The downside of relations with the media was most apparent when in September 1942, correspondents in Washington asked him detailed questions on Midway. Thach stated for them, as he had previously to others, that Army B-17s had done no damage to any of the enemy carriers on 4 June 1942. Explaining in detail how difficult it would be for a high-level horizontal bomber to hit a ship, Thach noted that a bombardier aimed where a ship would be if it maintained a straight course. Continuing, he said that a ship captain knew where a plane would be when its bombs were released and that a CO would wait until the plane got to that point and then make a sharp turn away from its base course. Given the lengthy amount of time for the bombs to fall, the chances of a hit were quite remote. Later Thach realized it might have been better for him to stop speaking at that time. But he went on to say that to his knowledge, no high-level bombers had sunk a ship in either the Atlantic or Pacific. Experienced reporters then had Thach in a trap. They asked if he was contradicting the widely disseminated story from December 1941 that a B-17 pilot, Capt. Colin P. Kelly Jr., had sunk the Japanese battleship *Haruna*. The common understanding from the early days of the war and for many years thereafter was that Kelly sank a battleship and was posthumously awarded the Medal of Honor. Without challenge from Washington, Kelly had been presented by the media as a national hero. Indeed, he had been recommended for the Medal of Honor but instead was awarded the Distinguished Service Cross (the Army equivalent of the Navy Cross). Later evidence revealed that Kelly had attacked a cruiser and that his heroism was remaining at the controls of the burning plane long enough to allow his crew to bail out before it exploded and crashed. In fact, *Haruna* survived until late in the war, when half-sunk at Kobe (scrapped in 1946), and the Army knew in 1942 from intelligence reports that it was afloat. In September 1942, Thach did not know the exact details of Kelly's heroism, but he did not doubt he had done something to deserve plaudits. But he also knew that high-level horizontal bombers were not viable ship killers. Thach made the best save he could under the circumstances by acknowledging that by giving his life for his country, Kelly had done more than he had. After further compliments to Kelly's heroism, Thach told the group that he was familiar with enemy ship recognition and saw

Haruna at Midway (most likely he saw another, *Haruna*-class ship). But despite his effort to not tarnish Kelly's image, *Washington Post* headlines blared that Thach had contradicted the secretary of war. Other newspapers chose to headline and emphasize some of his other thoughts such as his blueprint for victory (many carriers), but an article in *Collier's* picked up on the same negative theme as the *Washington Post*. Thach knew that the Army's erroneous damage reports from early in the war would someday come to light, but he was not happy to be the unfortunate person standing in the spotlight when it came on. Close friends joked to him that they expected his next assignment to be Siberia. Unhappy as he was with the B-17 episode, it nonetheless placed him in more demand from the media. An article he prepared for *Collier's*, "The Red Rain of Battle: The Story of Fighter Squadron Three," brought an invitation from publisher Simon and Schuster to discuss possible expansion to a book if he was "not already committed elsewhere."[7] Later, other members of the media warmed to Thach, Roland "Jerry" Gask of *Newsweek* writing on 28 May 1943, "Thach led with his chin. But his facts were right. Here's to Thach!"[8]

WRAPPING UP AT JACKSONVILLE

On 9 July 1943, Thach was promoted to commander, and not long after that there were refinements to his job title. The initial title, staff gunnery officer, did not fully cover his responsibilities for guiding and directing the operational training program for both pilots and aircrewmen. Following Rear Admiral Cook as chief of naval air operational training was Rear Adm. Andrew C. McFall, an officer as appreciative of Thach's efforts and ability as was Cook when he left. By late 1943, Thach's title was training officer, and the heart of his job was still teaching pilots and aircrewmen how to shoot. Although the European theater was the primary concern in Washington, in Jacksonville most thought was directed to the Pacific. In most of Thach's speeches during the period, mention of Germany or its planes usually got no more than a phrase in one sentence. By a wide margin graduates from Operational Training Command went to the Pacific.

One of the more memorable of many ceremonies Thach participated in while at Jacksonville was on 14 March 1944. That day marked the first formal graduation of naval aircrewmen. Aircrewmen, often informally referred to as rear-seat gunners or just gunners, were not new to naval aviation but a formal graduation ceremony with the award of distinctive aircrew wings was. Already trained at a technical center as an aviation specialist in radios or ordnance, or as an aviation machinists mate, a prospective gunner already had skills. At the Naval Air Gunnery School he learned the use of free guns, both .30 and .50 caliber, and tactical training in the defense of their aircraft. With pilots they formed a mutually dependent and lethal team, each depending on the other for safety and destruction of the enemy.

Among the speakers on 14 March were Rear Admiral McFall and Thach, who had an opportunity to introduce entertainer Bob Hope during a radio broadcast for the National Broadcasting Company. Introducing Hope as "a one-man USO," Thach went on to say that while Hope was in North Africa "the Germans bombed the base where he performed for the first time in six weeks … [so] we have fighter planes standing by just in case the enemy knows he is at Jacksonville today." When the program was planned, Thach had not expected to participate, and McFall would have introduced both Bob Hope and scheduled speaker, Vice Adm. John S. "Slew" McCain, then deputy chief of naval operations for air. As it turned out, McCain was a last-minute no-show for the noon program, so McFall read the speech and Thach, a veteran by then of numerous radio broadcasts, was drafted to assume the ceremony duties McFall vacated.[9]

The absence of McCain from Jacksonville was only momentary. When he arrived, he had a brief conversation with Thach that proved every bit as memorable as introducing Bob Hope and speaking during the aircrewmen's first formal graduation.

8

Battle of the Philippine Sea

By the summer of 1943, several of Commander Thach's early training films had been distributed. All film production demanding his time was complete, and he was able to devote time exclusively to overseeing operational training.

In March 1944, Thach was very happy in his training billet, given how much he enjoyed teaching. As he was approaching the end of nearly two years in training command, he wondered if he was going to be ordered elsewhere. A visit to Jacksonville that month by Vice Adm. John McCain resolved the matter when McCain pulled Thach aside and offered him a job. McCain was then serving in Washington as deputy chief of naval operations for air, the first flag officer to serve in that capacity. Immediately before, he had been chief of the Bureau of Aeronautics (October 1942–August 1943) and had served as commander of all Navy land-based aircraft in the South Pacific during the first two months of the Guadalcanal campaign. In the spring of 1944, he knew he was headed back to the Pacific and he needed a staff.

Although flattered and delighted by the offer, Thach had to ask why McCain was offering the job of operations officer to him since the two officers were not well acquainted. Although McCain knew Thach's reputation in naval aviation as a successful combat pilot, tactical innovator, and training officer, his reply focused on another of Thach's traits: Thach was known not to be a yes man. McCain did not want a staff full of yes men. With Rear Admiral McFall's blessing already secured, Thach packed his bags.

McCain ordered Thach to temporary duty with Vice Adm. Marc Mitscher, then serving as commander, Fast Carrier Task Force Pacific (Task Force 58). McCain instructed Thach to get to the Pacific immediately and begin assembling a staff in addition to acclimating himself with current operational procedures. Given the highest priority for travel, Thach flew to Majuro, a new fleet anchorage in the Marshall Islands some two thousand miles from Pearl Harbor that had been captured only two months earlier. There he boarded Mitscher's flagship, *Lexington* (CV-16), the Blue Ghost, only hours before it steamed for combat on 6 June.

A DIFFERENT WAR

In the nearly two years since Thach left combat, the character of the war had changed drastically. In June 1942, the U.S. Navy had averted disaster at Midway and was still on the defensive. The Navy had only four aircraft carriers in the Pacific. Four months later there would be only one, with *Wasp* succumbing to submarine torpedoes on 15 September and *Hornet* to carrier bombs and torpedoes at Santa Cruz on 26 October. The *Saratoga* was torpedoed again on 31 August and required repairs that kept it away from the Solomons until early December. The 24 August 1942 Battle of the Eastern Solomons cost the enemy one light carrier and kept the Japanese Navy from assisting troops on Guadalcanal. Although a tactical loss, the Battle of Santa Cruz was a strategic victory and the last of the four carrier-versus-carrier battles up to the spring of 1944.

Although he knew from official sources and the media that the Navy was growing stronger with the passing of each month, Thach was nonetheless greatly impressed with what he saw at Majuro. Anchored there were more carriers than he had ever seen together, among them the new *Lexington*. When he last saw it at Quincy, Massachusetts, in February 1943, only the hull was complete. Now its island was not only in place but it was glistening with electronics and guns. In place of his old mount, the Wildcat, was the now combat-proven F6F-3 Hellcat fighter, TBF Avengers in place of the old TBD Devastators, and SB2C Helldivers along with a few SBDs that would soon be phased out. Also within his view were several new fast battleships, new cruisers, and destroyers. By the spring of 1944, the U.S. Navy had quantitative and qualitative superiority over the Japanese in both surface ships and aircraft.

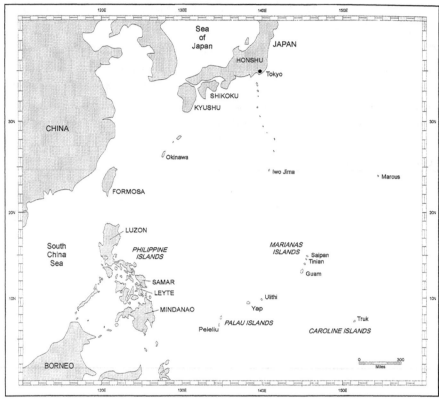

THE WESTERN PACIFIC

While he had not been there to see it, Thach knew that in the European theater the carriers *Wasp*—before it was sunk—the *Ranger*, and new escort carriers had supported the invasion of North Africa in November 1942. And new escort carriers had proven their great value operating with destroyers and new destroyer escorts in hunter-killer groups. Escort carriers in the Pacific also countered Japanese submarines, but many were utilized for close air support and fleet defense for amphibious landings that had already occurred in the Marshall and Gilbert Islands. With a large and still-growing Navy on the offensive in the Pacific, Atlantic, and Mediterranean, it was a much different war than Thach had fought. Also different in June 1944 was the attitude of all. The United States and the Allies were going to win this war on all fronts. It was only a matter of when.

PRELUDE TO THE MARIANAS

The protracted struggle for Guadalcanal in the southern Solomon Islands ended in February 1943, when the Japanese withdrew, and in March 1944, the Solomons campaign itself was on the verge of successful completion. Raids on the Gilbert and Marshall Islands in late 1943 led to invasions and final occupation in February 1944. Just prior to Thach's arrival, the dozen fast carriers and escorting units of Task Force 58 moved westward toward the Palau Islands before turning south to New Guinea. The objective was to support the invasion of Hollandia off central New Guinea (21–24 April) before returning to Truk—Japan's Pearl Harbor—in the Caroline Islands. Located almost equidistant from the Solomons, Marshalls, Marianas, and Palau Islands, Truk had been assaulted in February and April's attack was intended to neutralize it for good. On 29 April, carrier planes again swept over the once feared enemy base and by the next day it had been relegated to the same neutralized status as Thach's old nemesis, Rabaul. The Zero was still a menace and improved fighters were being produced, but enemy pilot performance was noticeably inferior to what it had been in the early months of the war. With a shortage of well-trained pilots flying from carriers or land bases, the enemy was losing control of the air and, consequently, the bastions around their outer defensive perimeter.

Reporting to Mitscher aboard *Lexington* on 5 June, Thach understood why McCain had ensured he was given such unique top priority in gaining passage to Majuro. The following morning the task force was steaming out of the anchorage for Operation Forager to prepare the Marianas for invasion and occupation. An official title was not required for Thach as he was aboard as an observer, but he found himself listed as a special assistant to Mitcher's chief of staff, Capt. Arleigh A. Burke, an officer he knew only by reputation.

The timing of Thach's arrival could not have been better. Burke, who had earned a well-deserved combat reputation as an outstanding destroyer flotilla leader, had come aboard *Lexington* on 26 March, unhappy to be assigned Mitscher's chief of staff. Mitscher, whose leadership of the fast carriers since January had earned him a similar combat reputation, was equally upset to have Burke forced on him. Admiral King's decision to have non-aviators as chief of staff for aviator commanders (and vice versa) did not initially set well with anyone in the Pacific. From 26 March into early April, the relationship between Mitscher and Burke was formal and frosty. During the Hollandia-Truk operation from 13 April to 30 April, it improved to functional. By the time Thach came aboard on 5 June, it had further improved to respectful, and by early August, Mitscher was recommending Burke for promotion to rear admiral. The professional and personal relationship between the two warmed to such a degree that Mitscher requested Burke for his staff when he took over the Eighth Fleet in 1946. But on 5 June 1944, the two officers had progressed only to the functional-respect relationship.

For Thach a major advantage of the still budding professional relationship between Mitscher and Burke was that Burke wanted to learn whatever he could from aviator Thach. The two enjoyed a good relationship from the beginning especially while comparing similarities of surface and aerial combat. The two agreed that timing in the air was as important as on the sea, and that combat in both required thinking ahead. Of course Burke had spent time with Cdr. William J. "Gus" Widhelm, the staff's operations officer and renowned combat pilot. He also learned what he could from longtime aviator friend Capt. Truman Hedding, Mitscher's former chief of staff who remained on briefly as deputy chief of staff. But Mitscher still had a habit of bypassing Burke to speak with Widhelm, and Hedding when he was not away on temporary duty, on aviation matters. Burke realized Mitscher had a special relationship with Widhelm dating back to their days together on the old *Hornet*. Widhelm was every bit the fighter Thach was but was not overly happy to be serving in a staff job. Although effective as an operations officer and appreciated by fellow staff members, in August he would be detached for duty as Training Officer at Quonset Point, Rhode Island.

Burke assigned special projects to Thach, mainly plans for the upcoming operation. As important were the continuing informal conversations through which Burke continued his rapid education of carrier operations. Beginning in June 1944, the two men developed a personal and professional relationship that eventuated in Thach influencing Widhelm's replacement as operations officer. Later Thach would work with Burke during the 1949 unification crisis, and in 1958, Burke, then chief of naval operations, assigned Thach a role of particular significance.

At sea on 6 June, Thach was as elated as everyone else in the task force, and many others around the world, when word of the invasion of Normandy was announced. On that day, massive naval armadas were moving toward the Axis powers in both Europe and Asia. Victory was not assured on 6 June 1944, but its promise was genuine.

McCain had specifically instructed Thach to look for mistakes that might occur on Mitscher's staff so his staff would not duplicate them. From time to time Thach pulled one of his little black books from his pocket to make an entry on some procedure or action that could be improved. Dividing his time between the flag bridge and combat information center (CIC) on *Lexington*, Thach completed his special assignments, talked with Widhelm, the occasionally present Hedding, and Burke, and observed the staff. Moving from the bridge to flag plot and CIC was no problem as all were within steps of the other. Not long after CIC was moved from the island of *Essex*-class carriers to the gallery deck, thereby forcing an operations officer to choose between the command center (flag plot) which remained in the island or information center (CIC) immediately below the flight deck.

BATTLE OF THE PHILIPPINE SEA

On 11 June, planes from Task Force 58 struck Saipan, Tinian, Guam, and Rota, yet for the moment the enemy did not know whether this attack was a raid or invasion. By 13 June, it was apparent to them that invasion was imminent. Saipan and Tinian were not only fortified but also had the support of the local populations, whereas Guam to the south had been an American protectorate at the beginning of the Pacific war. Although recognizing the possibility of assault on the Marianas, the Japanese did not expect it so soon. More likely, the enemy thought, the Allies would move up from the southwest and attempt to jump from New Guinea to the Philippines. Still, plans were ready to meet such a thrust into the Marianas. Loss of these islands was not acceptable as long-range bombers could reach Japan's home islands.

Japanese plans to meet a push into the Marianas had merit on paper. Aircraft would be flown into the islands to supplement those already there and carriers would engage an American task force for the first time in nearly two years. Particularly helpful was the fact that the predominately easterly winds in the area favored Japanese carriers as they moved toward the Marianas and Task Force 58. Conversely, American carriers would have to turn away from the Japanese carriers both to launch and recover planes (just as they had at Midway). This unfortunate reality not only precluded Task Force 58 from closing the distance between the two fleets but also markedly slowed the speed of launch and recovery. With the advantages of an easterly wind and aircraft with longer range, the Japanese could also launch from outside the operational radius of American carrier planes, land their aircraft on Guam to refuel and rearm, fight again, and return to their carriers later.

On the thirteenth Fifth Fleet commander Spruance and his subordinate fast carrier commander, Mitscher, had enough reports from submarines to know that enemy carriers and heavy surface units were heading east. They also knew planes were being staged from Japan to the Marianas and two task groups were detached on the fourteenth to raid their fueling bases in the Bonin and Volcano Islands. The major problem for Spruance, however, was that he did not know for sure whether the enemy was approaching as two separate threats or one. Was the enemy coming from the west or southwest, or both? The propensity for the Japanese to attack in front and on the flank was well appreciated in the minds of American commanders and historically it had served the Japanese well. The potential and fear of an end run permeated Spruance's mind as Thach and others would soon discover.

Four days of air attacks concentrated on specific targets ashore and at sea to ensure no Japanese left the islands and no more came in. On the thirteenth, fast battleships rather ineffectively weighed in and the next day old battleships, including four earlier sunk or damaged on the day of infamy, moved toward the beaches with cruisers and destroyers to more efficiently soften up landing beaches. On the fifteenth, the Marines stormed ashore on Siapan, and Thach was particularly interested to learn of the improvement in close air support and call fire. Consensus was that it was better than at the Gilberts and Marshalls, but there was still room for improvement. And the Marines made known the Corps' desire for control of air support and its own pilots, a request that Thach not only noted but vowed to champion when he was in a position to do so. Knowing he would have McCain's ear very soon, he would broach the matter then. What he could not know then was that this matter would soon become paramount to him against the Japanese and in another war six years later.

At dusk on the day of invasion Thach secured notebooks in his pocket and strapped on a helmet like everyone else on the bridge except Mitscher who would not trade his long-billed cap for anything heavier. Taking advantage of the diminishing light, enemy twin-engine torpedo planes swept in to attack. Watching from an out-of-the-way location on the carrier's starboard bridge wing, Thach stood beside a *Time-Life* correspondent and photographer, J. R. "Ed" Eyerman, whose role of observer was the same as his. Standard procedure was for a ship or formation to turn into a torpedo attack to comb the tracks of the lethal underwater projectiles. Torpedoes dropped and the *Lexington* turned sharply toward the drop point, resulting in an enemy plane bearing straight for the carrier's flight deck.[1] Antiaircraft guns forward, including a 40-mm quad within feet of where Thach was standing, opened up on the rapidly closing intruder and pounded it. Everyone held his breath as the torpedo-bomber flew down the deck, its left wing barely missing the island and the plane just missing fueled and armed aircraft parked aft. A roaring fire from the plane's starboard engine silhouetted the two pilots as it shot past the flag bridge, and just after clearing *Lexington*'s stern, the wing burned off and the plane crashed into the ship's wake. The threat of the enemy plane crashing into the island where he stood or into the aircraft on the carrier's flight deck past, Thach hung his head over the side of the wing bridge to watch a torpedo slide harmlessly by some forty feet away from the hull.

The torpedo threat over, Thach turned and congratulated Eyerman for taking what would have to be a classic picture of a Japanese plane over an American carrier deck. Eyerman, however, stated that the small camera around his neck was too heavy to lift and he had missed his chance at a photograph. That explanation was too remarkable to forget, so into a little black book it went, along with a testimonial to the effectiveness of the post-Midway 40-mm battery.[2]

The attack on the task force late on 15 June was not the only matter of considerable interest in flag plot. Additional submarine reports and other intelligence confirmed the enemy was still heading east and on the sixteenth a second large enemy surface force was known to be heading northeast toward the Marianas. Additional sighting reports came in on the seventeenth and eighteenth, but there was no agreement between Spruance's staff and Mitscher's staff as to their interpretation. An enemy radio transmission was picked up, but there was no agreement whether it was actually from the approaching fleet or a destroyer sent to that location to intentionally mislead.

The salient decisions how to fight the impending battle were made on the night of 18–19 June, but only after a debate that lingers over half a century later. Convinced that the location of the Japanese Mobile Fleet was known, Mitscher and his staff drew up a plan to steam westward late that night in order to begin launching at 0500 to attack the Japanese carriers. Whether the enemy would be caught before or during his own launch was not the important consideration it had been at Midway. The idea was to get in a strike and hopefully damage enemy carrier flight decks, but the greatest advantage was that Task Force 58 would then be heading east into the wind to recover and launch just as the Japanese would be doing. The end result would be an eastward running battle allowing Task Force 58 to not only be within striking distance of the enemy and quickly launch and recover but also be heading in the direction of any potential Japanese end run. Just after midnight, however, Spruance rejected the plan, and the air within and above the Blue Ghost turned blue from profuse profanities.

Despite his unofficial role, Thach joined Burke and other members of the staff to draft multiple messages to explain the error in Spruance's thinking. Early drafts were educational and diplomatic; later drafts approached mutiny. None were sent. As the early morning wore on attention shifted to planning for battle. Thach suggested that all planes except fighters be struck to the hangar deck so the flight decks of all fifteen carriers in Task Force 58 could be devoted to the defensive role Spruance's decision mandated. The idea to get all fighters into the air for defensive purposes was quickly agreed upon, but Thach's idea for the Avengers, Dauntlesses, and Helldivers was rejected. Rather than stow them in hangar decks, they were to be launched and ordered to orbit away from the carriers. Later in the day, after the battle began, some were ordered to crater runways on Guam to delay the shuttle process between enemy carriers and land bases. It worked.

Nearly a hundred miles west of Guam, southernmost of the three major islands to be taken, Task Force 58 awaited the appearance of enemy planes on the morning of 19 June. The sky was clear and visibility perfect. Emotion and energy expended throughout the long night, Thach and Mitscher's staff now looked to Lt. Joseph R. "Joe" Eggert to orchestrate defense of the task force. Eggert, who depended on tea rather than coffee to quench thirst and keep him alert, coordinated the work of other task group FDOs.[3]

For this battle, Thach had no direct responsibilities other than to observe. Looking across *Lexington*'s task group some four miles wide, he felt some pride in being an early advocate of the formation that placed carriers inside an outer ring of battleships, cruisers, and destroyers. Fighters from each of the four carriers of his task group were above to protect all four. Other task groups with their fighters above stretched to the horizon. This vast improvement of task group and task force organization would be difficult to penetrate. By the end of the day, his appraisal proved correct, as only a dozen or so of four hundred enemy planes managed to get close enough to drop ordnance. No Task Force 58 ships were sunk, none were seriously damaged, and Thach saw only one of the few Japanese planes that did appear over the force and it did not survive. It was

Battle of the Philippine Sea

a matter of both professional and personal pride for Thach to see how well the Navy was profiting from the mistakes of 1942.

The qualitative decline of Japanese pilots resulting from insufficient training and combat experience was well documented during the morning and early afternoon of 19 June 1944. Even Eggert briefly questioned within what radar scopes were showing. But what he could hear from pilots and what he could see on the large plotting board before him forced him to believe that enemy planes were falling in great numbers. Occasionally doubt crept back in when he heard antiaircraft gunfire, but very quickly the guns were silent again.

By sunset on 19 June, the Marianas Turkey Shoot was over and twelve of Mitscher's fifteen carriers were released to pursue the enemy carriers. Best intelligence from reports available indicated the Japanese had committed nine carriers. Not known was that only seven remained on the surface. The *Albacore* (SS-218) had placed only one torpedo in *Taiho*, Japan's newest large carrier, and *Cavalla* (SS-244) put three into *Shokaku*, veteran of the Coral Sea and Solomons. Poor damage control on both carriers led to internal explosions that put both under the waves before dark.

The Japanese Mobile Fleet was finally located late (1540) on the afternoon of 20 June. Given the distance to the enemy and the time of day, everyone in flag plot knew the planes would have to return after dark. This was a matter of particular concern because many pilots had not landed on carriers at night. Having just reported from training command Thach knew many pilots were not night qualified because of the pressure to produce as many new pilots as possible. Still, he noted that all a pilot needed to see was the illuminated paddles in the hands of the LSO, not the deck.

Just over two hundred Avengers, SBDs, Helldivers, and Hellcats lifted off and headed toward the northwest to attack the retreating Mobile Fleet. Turning east again to launch, the distance between the two forces widened. Concern for arriving back in the dark was competing with fears that some planes might not get back at all. But for about thirty minutes, concerns and fears were momentarily displaced by reports as the task force's planes attacked enemy ships. Anticipation that aerial opposition would be weak was confirmed but antiaircraft fire was intense and fatally accurate for nearly two dozen attackers. Particularly regrettable was the loss of one of the Avenger pilots from *Belleau Wood* (CVL-24), which helped torpedo *Hiyo*, the only enemy carrier lost in the battle as a result of U.S. carrier aircraft. The other enemy ships sunk were two oilers, and the carriers *Zuikaku*, *Junyo*, and *Chiyoda* were damaged. Four months later, *Zuikaku* and *Chiyoda* were at sea again to challenge Mitscher's carriers.

The sun set shortly after 1900 and by 1930 the Philippine Sea was totally black. Shortly after 2000 friendly planes were close enough to the task force to land. What then followed was simultaneously heroic and tragic, exhilarating and frustrating, suspenseful and relieving. Justifiably anxious, pilots were initially exuberant to see that Task Force 58 was ablaze at night with searchlights from many ships pointed straight up in the dark night with others played against the carrier hulls. Heading into landing patterns, however, the lights proved too much of a good thing, as some pilots were blinded by the well-intentioned beams.[4] As it was quickly apparent that pilots could not determine their own carrier homes, a message went out for pilots to land on whatever deck was clear. Few planes landed aboard the carrier from which they earlier had launched, but pilots counted their blessings on finding an open deck. Many did not. Over the next two hours, some attempted to land on cruisers and destroyers, and some ended up in the water as the combination of fuel exhaustion, fatigue, deck crashes, and an overabundance of illumination made a carrier landing impossible. Thach later recalled, as did a number of veterans of the battle, that although not chaotic, "the worst thing about the whole day's operations was trying to dodge your own airplanes in the dark."

Throughout the recovery process Thach continued to alternate between the bridge and CIC, although most of his time was on the bridge. Burke constantly sought Thach's advice, as he had since the former fighter pilot reported aboard. One comment that probably stayed with Burke was Thach's comment that in the future, flight decks should be illuminated at night just like airport runways. The logic was that in the future any enemy would have radar and already know a carrier's position, so the carrier might as well ensure getting its planes safely down. But Burke had much more to do than Thach on the evening of 20 June, thereby leaving the prospective operations officer time to suffer with the thought "we didn't have to have this happen if we'd done the right thing last night."

Although the enemy fleet was sighted on 21 June, it was too distant for Task Force 58 to assault, and the Battle of the Philippine Sea ended. Combat and operational aircraft losses for TF 58 were approximately 20 percent, but only 10 percent of the pilots were lost. No ships were lost or seriously damaged, and prospects for the Japanese Navy to immediately further contest U.S. occu-

Battle of the Philippine Sea

pation of the Marianas were bleak. Saipan was secured on 9 July, Tinian on 1 August, and organized resistance on Guam terminated on 10 August.

Within days after the Battle of the Philippine Sea, *Lexington* briefly returned to Eniwetok in the Marshalls before heading back into combat. Thach left the carrier to board a plane to Pearl Harbor. There he would conduct interviews and select officers and men for McCain's staff. His orientation on the Blue Ghost could not have been more instructive. He had seen how the newer ships and planes performed in combat, and he was well acquainted with the talents and potential limitations of the senior officers he would soon be working for and with. With ideas in his mind and recorded in his pocket notebooks, he was ready to trade his observer role for the official responsibilities of an operations officer.

9

Operations Officer: Destination Leyte

As Commander Thach began packing for his Pacific assignment, a two-platoon system of fleet commanders was approved in early May 1944, thus allowing one of the two fleet commanders to plan a subsequent operation while the other was at sea conducting a current operation. Beginning in August and for at least a year, the only changes would be the two fleet commanders and staff while the ships remained the same. By late 1945 there would be enough large fast carriers available to allow both commanders to actually command two different fleets if the war was still in progress at that time. To differentiate between the two Pacific commands, it would be Third Fleet when Admiral Halsey was on the bridge and Fifth Fleet when Admiral Spruance commanded.

The organization of the fast carriers also changed. On 5 August 1944, commander, Fast Carrier Force Pacific Fleet was changed to commander, First Fast Carrier Task Force Pacific (Vice Admiral Mitscher) and commander, Second Fast Carrier Task Force Pacific (Vice Admiral McCain). When McCain learned he would soon command the fast carriers, he made several understandable assumptions that did not eventuate. First, he initially believed he would relieve task force commander Mitscher when Spruance turned the fleet over to Halsey. Not having been at sea as long as Spruance, Mitscher, four years junior to McCain, declined relief, much to McCain's surprise and consternation. Second, when McCain arrived at Pearl on 9 August, he expected to inherit at least some personnel for his staff. The following day he found out that was

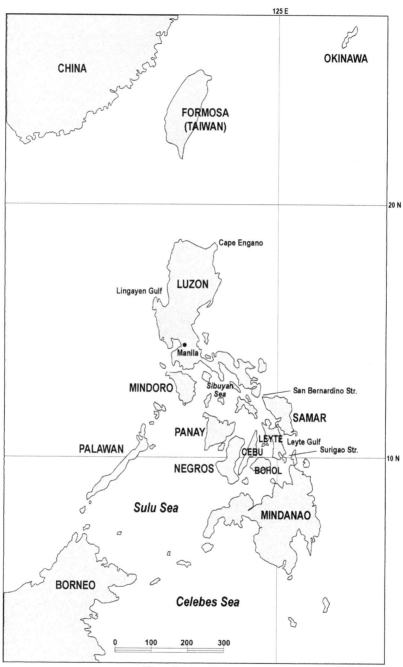

The Philippine Area

not going to be the case. He instead would go to sea as a task group commander to gain experience ("makee-learn"), and like all task group commanders he would be subordinate to the task force commander, Mitscher.

BUILDING A STAFF

There was nothing Thach could do to help McCain convince Mitscher to go on leave or obtain some of his staff and key personnel from Jocko Clark's task group staff. But Thach had already begun the process of finding officers for McCain's new staff and had helped set in motion selection for an opening on Mitscher's staff. Before Thach left *Lexington* in July 1944, he discussed one personnel matter with Burke that would benefit both. With Gus Widhelm leaving, Thach "strongly recommended" that Jimmy Flatley take the job of operations officer. "I . . . told him [Burke] how good Jimmy Flatley was," Thach recalled. "Also, Jimmy and I though a lot alike, and it [would be] easier to relieve when we rotated." Mitscher needed little prodding in regard to Flatley, as he had written a very favorable fitness report for him while both were serving at Fleet Air West Coast in 1943. Burke knew Flatley's reputation, and when they began working together in September, a strong personal and professional association quickly evolved. Recalling the 1944–45 operations, Thach said of Flatley, "We worked very well together, very closely, and whenever he discovered something that he thought would be useful to us, he passed it to me immediately . . . and I did the same thing."[1]

Jimmy Flatley did indeed become a valuable asset, not only to Thach but also to Mitscher and Burke. But in July and August, Thach had to find other officers who had Flatley-like talents if not his experience. The new staff required nearly two dozen officers and over a hundred enlisted men to round out 130 to 140 billets. Feeling very much like a coach again, he began to search lists of personnel who were available and other lists of those not available but who might be enticed. Just like a coach, he was looking for talent whether it came from people he knew or did not know.

Fortunately in July, August, and later, Thach found some talented officers he did know. For his own assistant operations officer, he selected Lt. Cdr. Gordon Cady, but Cady died on 30 August in an operational accident and was replaced in November by Lt. Cdr. Bill Leonard, Thach's VF-3 executive officer during the Battle of Midway. Leonard reported from Fleet Air West Coast where he had served with Mitscher and Flatley. While at San Diego he had flown the captured Zero, an experience that not only enhanced his value as assistant operations officer but also as an air ordnance expert. In some respects in late 1944, Leonard's recent experience in the air with newer aircraft and ordnance placed him in position to acquaint Thach with those developments. Another old VF-3 fighter pilot whose administrative skills would eventually propel him to four-star rank, Cdr. Noel Gayler, joined McCain's staff before joining Admiral Towers's staff in late 1945, when Towers assumed command of the Second Fast Carrier Force.

Like Leonard and Gayler, some of the officers who came to McCain's staff brought solid credentials. Lt. Cdr. J. H. Hean, tactical navigation and gunnery officer, had a master's degree in science and had been in the ordnance test business, and Lt. Cdr. John Tatom had a strong background in aerology. Other officers known to Thach were equally capable, and their appointments to the staff quickly filled the open slots. One of the officers Thach and McCain hoped to bring aboard for permanent staff duty was Lt. Charles D. Ridgway III, USNR, a fighter director officer who had demonstrated exemplary proficiency during the Battle of the Philippine Sea and earlier actions.[2] A force fighter director not only had to make intercepts but also had to control patterns and plans and determine augmentation of fighters to meet sudden changes in the air. Happily, Lt. Cdr. J. N. MacInnes was available for this important responsibility. Experienced enlisted men were at a premium and were as scarce as good officers. Fortunately for Thach, his primary purpose was to find officers. Whereas he was tasked to find a communications officer, the thirty enlisted men to work for that officer in decoding were assigned in blocks soon after graduation from their specialized training school.

Among the officers Thach selected whom he already knew was Lt. William T. Longstreth who had experience as a ship's air intelligence officer. Before the war was over Longstreth proved to be one of the most outstanding authorities on Japanese ships and aircraft. He could recognize and identify by name enemy combat vessels and merchant ships in photographs. Subtle modifications on a merchant ship might cause him to look at a photograph a little longer than usual, but soon he would see the changes and know exactly which ship he was viewing. Longstreth was usually the first staff member pilots talked to upon returning from a strike. An excellent interrogator, the six-foot four-inch Longstreth was as successful in piecing together information from pilots as he was in identifying their targets both ashore and at sea.

Another acquaintance from earlier in the war was Lt. Cdr. Don B. Thorburn, whom Thach met at the Disney Studios in 1942. Although not trained in aerology Thorburn had done the writing for films to help train aviators understand the basics of weather and how to read a map. That writing plus his prewar experience in advertising were the reasons Thach invited Thorburn to be the staff's public relations officer. As war correspondents were aboard for nearly every operation, someone had to coordinate their schedule to meet both the needs of the Navy and the media.

There were several important slots for which there were no applicants. When Lt. Alfred M. Grafmueller came to apply, he did not know exactly where he might fit, but he asked to be considered. During the interview, Thach learned that Grafmueller had operated a successful gift card business in New York before the war. Somewhat to Grafmueller's surprise, Thach offered him a job as logistics planner. Asking how his card business qualified him for the job, Thach responded that anyone who could start a business and operate it demonstrated a talent for getting things done. Logistics planning, he explained, would require arrangement for replenishment of food, pilots, planes, ammunition, and everything else needed to keep men and ships functional. Sensing that Grafmueller wished to go to sea in a significant billet, Thach thought the young man would excel if given an opportunity. He got the opportunity and proved his worth.

Thach had jokingly told McCain that he was also willing to choose the chief of staff, Thach's immediate superior, and he advised McCain that he would like to have Capt. Herbert Duckworth. Both knew that Duckworth was ineligible, as he was an aviator and the chief of staff had to be a non-aviator, so McCain tabbed Rear Adm. Wilder D. Baker for the slot. Although a surface officer, Baker did have the distinction of being the only non-aviator to command a task group from 29 June to 7 July, when he relieved medically incapacitated Rear Adm. Keen Harrill as commanding officer of Task Group 58.4. Earlier Baker distinguished himself as one of the first officers to critically study the antisubmarine effort in the north Atlantic. An outstanding leader in both destroyers and cruisers, he was well qualified for his duties as McCain's chief of staff and he worked effectively with Thach.

Thach was pleased with his staff even before they left Pearl Harbor for operations. Most of those earlier known to him had already proven themselves, and the others had expressed a keen desire to serve in combat. Like any Naval Academy graduate, Thach knew regular officers wanted combat either as a matter of duty or to enhance promotion potential. Regardless of motivation, he expected to be able to count on them in difficult circumstances. That the several reserve officers who had applied to be members of his staff—including Longstreth, Thorburn, and Grafmueller, as well as Lt. Cdr. J. B. L. Reeves, Lt. Cdr. Charles Martin, Lt. Charles A. Sisson, and Lt. John Moss—were so anxious to serve in the heat of battle was especially uplifting. Their only motivation was to serve, and that alone caused them to step forward. After the war, they returned to their respective civilian pursuits, but Thach respected them every bit as much as he did his fellow professionals.

RESPONSIBILITIES AS AN OPERATIONS OFFICER

Vice Admiral McCain and Vice Admiral Mitscher depended heavily on their respective operations officers. Other members of the staff were no less important as each had a role to play as a member of the team. Poor performance by any member of the team deleteriously affected all. Still, the operations officer was critically important because he and the chief of staff were the first two officers the admiral turned to for advice when an important decision had to be made, and especially if it needed to be made quickly. Rear Admiral Baker fit well as McCain's chief of staff in part because as a non-aviator he did not thrust himself in between McCain and Thach when aviation questions required resolution.

McCain well knew that he had to have total confidence in Thach. His life and the lives of everyone aboard and in the task group—and later task force—depended on the sound thinking of his operations officer. McCain was not alone in believing that fighter pilots were ideally suited for the critical job of operations officer. Hugh Winters, CO of Fighter Squadron 19 before assuming command of Air Group 19 (CAG-19) at Leyte Gulf, knew that from firsthand experience. Years later he stated that fighter pilots not only had the responsibility of escorting attack aircraft to and from their targets but also had the role of attacking antiaircraft batteries to lessen the danger for bombers that did not have the same degree of maneuverability.[3] The combat experience a fighter pilot brought to the job also enhanced capability to improvise as combat often presented unforeseen challenges.

No matter how good a fighter pilot may have been in combat and how much experience he had, it all went for naught if he could not effectively interact with other members of the staff, those both above and below his rank. The ideas of others had to be given consideration, not only to meet the demands of war and everyday needs at sea but also to maintain morale. In this respect, Thach was especially gifted. When Bill Leonard, Thach's assistant operations officer who had just worked directly for Jimmy Flatley, was asked to compare the working relationship with the two operations officers, he responded that "he could not choose between two princes."[4]

Both McCain and Mitscher by late 1944 were relatively unfamiliar with recent rapid advances in aircraft, ordnance, and tactics. Consequently, both relied heavily on their operations officers. Technology was advancing so rapidly by late 1944 that even Thach and Flatley were sometimes pressed to deal with it. One example was the advent of the new 11-inch Tiny Tim rockets. At Pearl in December 1944, helping write plans for the Iwo Jima and Okinawa campaigns, Flatley wrote Thach to advise him that the large rockets required too much storage space aboard carriers, carried too little explosive, and that accuracy left much to be desired.[5] As time progressed during the last year of the war it became apparent to both Thach and Flatley that McCain was more willing to adopt newer ideas than Mitscher. Ideas Thach presented to McCain got a quick yes (mostly) or no, while Flatley, often with Burke in support, had to lobby longer to influence Mitscher. The use of night fighters and Thach's ideas for keeping enemy planes grounded did not initially find Mitscher in agreement. Wisely, McCain allowed Thach to contact Flatley and for him then to handle these debates with Burke and Mitscher.

Being the operations officer for a task group and for the entire task force was similar in many ways. The major difference was in the volume of work. The Joint Chiefs of Staff and CNO (Admiral King) in Washington, D.C., handled the strategic direction of the war. The next level was commander, Pacific Fleet (Admiral Nimitz), at Pearl Harbor until he moved his headquarters to Guam in late January 1945. From Nimitz, command was exercised by the fleet commander (either Spruance or Halsey), then the task force (either McCain or Mitscher), and finally the task group. Each command had its own operations officers, but the task force and task group operations officers were the ones most directly associated with the actual fighting. With their flags usually on a battleship or cruiser, the fleet commanders did not have to look directly into the faces of the pilots and aircrew who were perhaps about to see their last day on earth. Task force and task group commanders and their staffs did. Fleet commanders cared equally for the lives of all, but the point is made here only to note how the decisions of task force and task group commanders were closer to the point of contact with the enemy. And usually the last staff member on the bridge that strike leaders talked to was the operations officer.

When functioning as a task force operations officer, plans for implementation had to be written. After gathering specific written assignments from the staff, the separate sections were edited and submitted for approval by the operations officer, chief of staff and admiral. Mitscher was not given to reading the final versions and expected Burke to give him a verbal synopsis. McCain was given to being more involved as the plans were being written and therefore required little or no final editing before approval. The admirals' blessing rendered, the plans were then distributed to air groups for implementation. At the task group level, Thach also had to write plans but they were shorter than those prepared for the entire task force. His major objective in the task group role was to get everything done early in the evening so pilots could get some idea of what could be expected the following day and then get some sleep. Before launching from the carriers, the pilots had a final briefing and these went faster when most of the essentials were known the previous evening.

Usually neither Thach nor Flatley attended air group or squadron meetings. When they did, it was at the invitation of the air group or squadron. However, meetings between Thach (or Flatley) and air group commanders were frequent. As air group commanders had to deal with Thach when McCain had the task force, and Flatley when Mitscher was on the bridge, these meetings had a different tone. While both operations officers were willing to listen and were solicitous of comments from the CAGs, there was a distinct variance in philosophy. Thach wanted to bring all pilots up to his level, while Flatley was inclined to plan operations directed more to what the ordinary pilot could do.[6] The give and take in these meetings was normal procedure as the 1944 air group commanders had by then seen as much if not more combat than Thach and Flatley. There was, therefore, mutual respect. Despite the dangers confronting the air group commanders in 1944, they understood that they had Hellcats and many friends with them in combat instead of Wildcats and the few that flew with their two operations officers in 1942.

It was extremely rare for Mitscher to attend the meetings with air group commanders. His visit was usually only for a few comments at the beginning of a major operation. McCain was more inclined to visit and occasionally did, but it was the operations officer who "really ran the show." As pilots returned from strikes, Thach often escorted a strike leader and/or other pilots to McCain as the admiral thrived on the direct association with the pilots. Too, Thach wanted his boss to hear firsthand what was happening to help the two of them in future planning. Conversely it was seldom that pilots were escorted to Mitscher. If they were it was because they had done something extraordinarily good or excruciatingly bad. Despite their different personalities both McCain and Mitscher were favorites because the pilots knew both would do anything in their power to support them, especially rescue operations.

Before a task group operation Thach might need only thirty minutes to complete planning. But for a task force operation he usually needed three days to help write and coordinate plans, the size and scope of the operation dictating the degree of necessary planning. Intelligence was included in discussions so everyone would know what they might encounter both in planning and operations. High on the list of concerns was the pattern of attack the enemy might use and how the combat air patrol could respond. Ordnance was always a popular subject as were enemy efforts to conceal their aircraft and gun emplacements. Discussions of ordnance and concealment were usually intertwined as some ordnance, like cluster bombs, worked better on hidden aircraft, as a direct hit was not necessary to destroy the target. Napalm was new to the fleet in 1944, and its potential (and problems) was not fully appreciated by Thach until after he could see photographic evidence and absorb reports from the front. Weather was a special priority for Thach, as the first data he checked every morning was the forecast. On one occasion he had to stop a launch of one thousand planes due to a last-minute evaluation of the weather.

To stop the launch of one plane or a thousand required Thach to be on the bridge immediately before all launches. A sudden change in the weather or intelligence indicating a change in enemy actions required him to be where he could respond to such matters. Usually rising at 0315 from the bunk in his stateroom amidships on the second deck, he first went to his office three levels up on the gallery deck to briefly check for any information that had appeared since he had turned in just before midnight. From his office he went directly up two more levels to the flag plot, the bridge being only steps away. After the first strike of the day was off, his routine was to go down to the wardroom for breakfast and then back up to his office to check reports. From his office it was up back to flag plot and the bridge to check strike progress before descending once again to his stateroom for a quick nap.

When the first strike was back to the force or group, Thach was on the bridge to await strike leaders' reports and look at photographs taken during the strike. Without waiting for the photos to dry, decisions were made on whether

Operations Officer: Destination Leyte

to make changes in the plan for the second strike. Very often the wet photographs revealed hidden and camouflaged enemy planes that the pilots had not seen. Attack issues resolved, Thach then had to decide whether or not to augment the combat air patrol. That decision made, it was back to the wardroom for another few bites, the stateroom for a few more winks, and then more of the same routine.

Preparing for strikes was not routine. The fleet's general objectives took considerable time to translate into plans at the task force level. From early September 1944 into late November, Thach did not have to write the lengthier, more in-depth task force plans. As a task group operations officer, his main task was to set the schedule for the usual four or five carriers within his group, 38.1. But in addition to the schedule there were other decisions to make. Targets had to be designated and pilots were usually assigned to strike certain areas they had already attacked as their familiarity with a target promised better results. The number of planes needed and the ordnance loads required had to be determined. Searches had to be organized, and a check was made to ensure that sufficient planes were aboard, that all aircraft parts and fuel supplies were available, and that known shortages had been identified and replenishment requested.

Dealing with aircraft alone was a significant challenge, but an operations officer also had to plan for all ships in the formation. As would be expected, Rear Admiral Baker was particularly helpful in assuring that Thach and the staff had determined the proper formation for all ships and refueling. Refueling was almost a daily chore, the short legs of destroyers particularly demanding. When operating as task force operations officer, Thach seldom saw a day go by that a ship did not join or depart. The location of friendly submarines had to be updated to ensure they were not attacked and that they were close enough to operations to assist in rescue or other tasks. Point option—where ships would be at a future hour—was established and communicated to all ships and a duty task group had to be assigned. Destroyer scouting lines had to be set, fighter director stations assigned, mail deliveries scheduled, and the operations of other Allied forces had to be assessed. When all these and other requirements were addressed, a daily report had to be prepared to pass on up the line.

Few jobs for middle-grade officers were as difficult and demanding as the responsibilities of a task group and task force operations officer. Certainly, few other duty assignments held the lives of so many men and the welfare of so many ships and planes in their hands.

TO THE PHILIPPINES

On 18 August, McCain relieved Rear Admiral Clark as the commanding officer of Task Group 58.1, but Clark, along with several key staff members, remained for nearly two weeks to assist in orienting McCain, Thach, and others to their new jobs. Selecting *Wasp* (CV-18) as his flagship, McCain and his staff hastened to prepare themselves for the next sortie. On 26 August, Spruance transferred command of the fleet to Halsey, and McCain's designation changed to Task Group 38.1. Two days later, Task Force 38 steamed from Eniwetok heading southwest to conquer or neutralize island bases required for the Allied advance toward the Philippines and to destroy all the enemy combat and merchant ships that could be found along the way.

McCain's Task Group 38.1 consisted of two *Essex*-class carriers, *Wasp* and *Hornet* (CV-12), plus three *Independence*-class carriers, *Belleau Wood*, *Cowpens* (CVL-25), and *Monterey* (CVL-26). Knowing that resistance would become more intense as the fast carriers neared the enemy's inner defense perimeter, Admiral King had on 31 July authorized new aircraft complements for the large *Essex*-class carriers. Instead of thirty-six bombers and thirty-six fighters, the big carriers were to operate only twenty-four bombers and fifty-four fighters. Allocation for the TBF/TBM Avengers remained at eighteen, as the torpedo was still the best weapon to sink a ship. The Avengers could carry bombs for appropriate targets ashore and Hellcat fighters would be expected to carry 500-pound bombs when needed to make up for the lesser number of SB2C Helldivers. The SBD Dauntless fought its last battle in June at the Battle of the Philippine Sea and performed well, but the character of battle was about to change and more fighters were the answer.

To achieve the strategic objective of taking the Philippines, Halsey's Third Fleet struck targets as far north as Chichi Jima and Iwo Jima, as far to the east as Wake Island, and as far south as the Palau Islands. From 6 to 9 September, the task force concentrated on the Palau Islands, particularly Peleliu, Ulithi, and Yap. On the ninth and tenth, planes were over Mindanao, the southernmost main island in the Philippines, followed on the eleventh and twelfth with attacks on the central Philippines. Surprisingly little resistance was encountered over Mindanao, leading Halsey to recommend on the thirteenth that invasions planned for that southern island be canceled in favor of moving on to Leyte in the central Philippines. On the fourteenth, the recommendation was approved, leading Halsey and Mitscher to begin planning for landings at Leyte on 20 October. An assault on Luzon, northernmost of the main Philippine Islands, would follow about two months later.

Invaded on 15 September, with McCain's 38.1 fifty miles offshore for air strikes, opposition on Morotai—approximately halfway between New Guinea and the Philippines—was relatively weak. But Peleliu, secured on 22 September with resistance lasting into late November, was a killing ground equal to the anguish encountered earlier at Tarawa and, later, at Iwo Jima and on Okinawa. Marines were encountering major problems with the network of caves, many with artillery that the Japanese fired and then quickly pulled back inside. Unable to eliminate some caves that were taking a fierce toll of lives, the Corps called for close air support. Thach talked with McCain and asked that instead of sending the eight fighters the Marines requested for support on Bloody Nose Hill, the task group send forty-eight to pulverize the area. McCain agreed, and tons of rockets, bombs, and .50-caliber shells poured into the cave-pocked hillside. As a token of their appreciation, the Marines later sent Thach a Japanese flag they captured when the hill was overrun.

On 23 September, the excellent anchorage at Ulithi was occupied. Located in the northern Palaus to the southwest of Guam, it provided yet another base for supplies and rest. Rest, however, was about to become an unfamiliar concept as the At Sea Logistics Service Group comprised of fleet oilers, escort carriers, destroyers, and destroyer escorts was being formed to carry fuel, supplies, and replacement planes to the carriers. With fueling rendezvous set at the maximum range of enemy land-based aircraft, the possibility of interference was minimal.

While pilots and sailors felt the pressure of constant operations, the pressure on the enemy was even more intense. And when the task force raided Manila for three days, 21–24 September, all Japanese knew that a decisive battle for the Philippines would soon be fought. Enemy leaders knew they had no chance for victory in the Pacific if they lost the Philippines. Without control of the skies and seas in and around the islands, crucial raw materials, especially oil, from Southeast Asia, could not be transported to the Japanese home islands. Consequently, major Army and Navy assets had to be committed to the defense of the Philippines.

Pleased with the large numbers of Japanese planes destroyed in the air and on the ground and enemy ships sunk during the month of September, McCain's task group on the twenty-ninth entered Seadler Harbor at Manus in the Admiralty Islands just northeast of New Guinea. Time there was short, however, as the carriers were back at sea on 4 October, steaming northwest to begin attacks designed to isolate the Leyte landing area. To seal the area, the carriers refueled on the eighth, struck Okinawa well to the north on the tenth and then worked south to Luzon and Formosa.

Unknown to Halsey the Japanese had concentrated air power on Formosa and was ready when task force planes attacked from 12 to 14 October. Based on photographs brought back by pilots, Longstreth advised Thach that enough enemy planes had survived the first attacks that there would be a strong counterattack. In that assessment he was correct as enemy planes damaged two cruisers, *Canberra* (CV-70) on the thirteenth and *Houston* (CL-81) on the fourteenth. Thach was little surprised by the change in tactics the torpedo planes used in attacking the two ships. Approaching very low on the water under radar, they did not lose time by circling for optimum position. Instead they came straight in. The new tactic did not result in many hits, but any hit by a torpedo usually had devastating effect. Orders from Halsey were for McCain's group to remain with the cruisers while the other task groups departed for more strikes. With 38.1 nearly immobile, Thach's response was to suspend air strikes in favor of keeping all five carrier decks free to operate only fighters just as he suggested during the June Battle of the Philippine Sea.

Halsey also decided to use the damaged cruisers as bait in the hope that enemy ships might sortie to sink the cripples. That plan almost worked but the enemy soon realized it was a trap and did not commit surface ships to the operation. Planes, however, were committed and Thach, his radar operators, and fighter director had their hands full repelling the frequent attacks. Rather than sending all their planes in one or two big raids, the enemy chose a series of attacks with smaller formations. Using aluminum chaff to confuse radar sets, the enemy approached at both high and low altitude and was rewarded on the sixteenth by placing a second torpedo into *Houston*. The cruisers were too tempting to ignore, and the result was that the Japanese planes were basically providing targets for the task group Hellcats and ship's antiaircraft gunners. Using the principle of concentration, Thach had enough fighters in the air for each attack with sufficient reserve to intercept others. For his role in placing the right amount of fighters at the right altitudes during the defense of the cruisers, Thach was awarded the Silver Star.

By the evening of 16 October, the Japanese were convinced they had inflicted serious damage on the U.S. Navy and even thought ensuing invasion plans might be postponed. Within days, however, it became apparent that they had not won tactically or logistically. That *Canberra* and *Houston* were successfully towed to safety did not significantly enter into the enemy's calculus. The more important math was that they had lost over five hundred planes during the four-day battle, including a significant number scheduled to go aboard carriers later in the month. The loss of so many carrier air group pilots forced enemy planners back to their charts to draw up a new defense plan, one that would have to take into account that their carriers were suddenly only empty hulls. And by the seventeenth when preliminary operations on Leyte were observed prior to the main landings on the twentieth, it was readily obvious that the air battle off Formosa could not be counted as a strategic victory.

THE BATTLE OF LEYTE GULF

Japan's hastily drawn new plan for the defense of Leyte had merit despite the short amount of time that could be devoted to it. First, a Northern Force of four decoy carriers with only two-thirds of their normal complement aboard was to entice Halsey's task force away from the Leyte landing beaches. Second, a Center Force comprised of battleships, cruisers, and destroyers would approach the landing beaches through the Sibuyan Sea and San Bernardino Strait. And third, a Southern Force of battleships, cruisers, and destroyers was to steam toward Leyte through Surigao Strait. If the plan worked as designed, the carriers of the task force would leave Leyte and follow the decoy carrier force north. Then, the Center Force with superbattleships *Yamato* and *Musashi*, would approach the Leyte beaches, as would the Southern Force. Even with Halsey's carriers too far away to counter the Japanese approach, the U.S. Navy would still have six of its old battleships and the planes aboard escort carriers. Given all the circumstances of the moment and without other viable options, it was as good a plan as could be devised.

Had Halsey, Mitscher, Flatley, McCain, and Thach known in mid-October that they would have an opportunity to chase enemy carriers, all would have rejoiced at the prospect. Admirals King and Nimitz had directly expressed a desire for the destruction of the enemy's carriers in the forthcoming battle. Certainly the commanders of the Third Fleet, Task Force 38 and Task Group 38.1, and all air groups were not happy about the prospect of being tied to the area near the invasion to support the Army. That was something that the escort carriers could and should handle along with surface ships while the fast carriers relied on their mobility to strike distant enemy airfields and ships. With aviator Halsey present, Mitscher would most likely not be delegated tactical command of his carriers. Before the battle, however, there was no undue worry as general understanding was that if enemy carriers appeared there would be no debate similar to the ones off the Marianas in June, particularly as Halsey had no amphibious responsibilities.

On 18 October, 38.1 launched strikes against installations on Luzon and then attacked Manila on the nineteenth. The attacks to the north were successful in decreasing the enemy's aerial threat, and on 20 October, Task Force 38 began concentrating on the Leyte beaches to cover the Army's landings. Aerial opposition was light and many enemy planes were burned in their revetments, no small accomplishment given the heavy rains during the period of the invasion. After two days of attacks, carrier air groups needed rest and the carriers needed their ordnance replenished. Consequently, on the twenty-second the five carriers of McCain's 38.1 were ordered almost due east to Ulithi, with *Bunker Hill* (CV-17) following the next day. This reduced the number of fast carriers available to Halsey and Mitscher from seventeen to eleven.

On the early morning of 23 October, submarines *Darter* (SS-227) and *Dace* (SS-247) discovered the Japanese Center Force in the Sibuyan Sea and sank two heavy cruisers and seriously damaged a third. Given the enemy's propensity to attack with divided forces, Halsey recognized there was a possibility that the enemy might approach Leyte from the south. Late on the twenty-third, Jimmy Flatley plotted the Center Force and the presumed Southern Force on his maps. Although the Northern Force had yet to be seen, Flatley also placed it on his map. Mitscher and Burke agreed with Flatley that the Japanese would not bring surface forces to Leyte without also committing their carriers that intelligence placed to the north in Japan. The developing enemy plan was becoming apparent early on the morning of the twenty-fourth, and Flatley's calculations projected the Center and Southern forces off Leyte on the morning of the twenty-fifth.

After Flatley and Burke marked maps and charts on flagship *Lexington* with Mitscher uncharacteristically glancing over their shoulders, the admiral was sufficiently convinced that battle was imminent. Indeed nine enemy battleships, one large carrier, three light carriers, over a dozen cruisers, and many destroyers were steaming toward Leyte. On the morning of the twenty-fourth, however, Mitscher's carriers were widely separated. Northernmost of the four task groups was Sherman's 38.3—Mitscher's flagship therein—then east of

Operations Officer: Destination Leyte

Luzon. One hundred fifty miles to the south off San Bernardino Strait was Rear Adm. Gerald Bogan's 38.2, while another 150 miles south Rear Adm. Ralph E. Davison's 38.4 steamed east of Leyte.

The call to action came from Halsey soon after dawn on the morning of 24 October, when the Southern Force was spotted and reported. Even the call for McCain's 38.1 to return was broadcast directly from Halsey rather than through Mitscher. When McCain got the news to abort his voyage to Ulithi, he was six hundred miles away and low on fuel. Consequently, a fueling rendezvous was set for early on the morning of the twenty-fifth as he headed northwest. At the point of refueling, 38.1 would be some four hundred miles east of Samar.[8]

On the twenty-fourth, good news alternated with bad. In the afternoon hours a report came that one of the two enemy superbattleships (*Musashi*) was severely damaged—it sank that night—along with other battleships and cruisers. The bad news was that *Princeton* (CVL-23) had been hit. Later it was learned that the light carrier had taken only one bomb hit, but fires found fueled and armed planes on the hangar deck. When the ordnance and fuel cooked off, *Princeton* suffered the same fate as Japanese carriers at Midway. The burning carrier could be seen from *Lexington* and that view convinced the flagship's CO and air officer not to reopen magazines to arm AG-19's planes with torpedoes and armor-piercing bombs despite the immediate need for that exact ordnance.[9] And the loss of *Princeton* only affirmed Halsey's belief that enemy aerial opposition had to be eliminated.

Good news arrived late in the afternoon of the twenty-fourth when search planes spotted the Northern Force, composed of surface units two hundred miles in front of a carrier force, all heading toward Leyte. But Thach and McCain did not have enough information to see everything that was going on as Halsey was sending only the information needed for implementation. And so many messages were filling the air that McCain and Thach were hearing much more than they could send. Early the next morning the communications problem only got worse.

Late on the twenty-fourth, Halsey decided to take all eleven available carriers of the task force and all six battleships north to crush the Northern Force of four enemy carriers that was finally seen at about 1700 some two hundred miles north of the task force. Reports seemed to indicate that the Center Force had been so badly hurt that it was no longer a threat, particularly as it had reversed course to the west at about 1400. Later that day, however, it did reverse course and was detected by planes from the night carrier *Independence* (CVL-22). With that information in hand, around midnight Jimmy Flatley and Arleigh Burke were convinced that the Northern Force was a decoy and they woke Mitscher to beg him to signal Halsey. Mitscher asked if Halsey had the same information and when informed that he did, he commented that the fleet commander would welcome his advice only if he asked for it. With that, Mitscher rolled over in his bunk. Unquestionably, he was as exasperated as Flatley and Burke and well understood that the landing beaches might be exposed. But he knew Halsey well enough not to challenge his decision. Exhausted, he tried to get what sleep he could in order to meet the next day's strain of battle. On the *Wasp*, both McCain and Thach slept a little better because they did not know all that was developing, changing, and being misunderstood aboard *Lexington*.

Halsey was very clear in his own mind that his primary mission was the destruction of enemy carriers. He was confident that Adm. Thomas C. Kinkaid, commander, Seventh Fleet and Central Philippines Attack Force, had a sufficient number of escort carriers, battleships, cruisers, and destroyers to cover the landings. Just before 0800 on the morning of 25 October, planes lifted off carriers of the three task groups to intercept the four enemy carriers off Cape Engano. Over the next seven hours they pounded all four and left them sinking. Considerable pleasure was taken by all when news was received that *Zuikaku*, last of the six carriers to strike Pearl Harbor, rolled over and sank just after 1400. Smiles also appeared with news that Rear Adm. Jesse B. Oldendorf's battleships and escorts had sunk enemy battleships *Fuso* and *Yamashiro* before dawn in Surigao Strait, the passage between Leyte and Mindanao and the southern entrance into Leyte Gulf.

Considerably less jubilation was evident when the Center Force appeared off the Leyte beaches shortly after dawn at 0645. In what is known as the Battle off Samar, Center Force, with four battleships, including *Yamato*, and six heavy cruisers, surprised the six escort carriers of Taffy Three, northernmost of the three escort carrier groups, and attending destroyers and destroyer escorts. Heroically, seven destroyers and destroyer escorts charged the Japanese battleships, cruisers, and escorts. The uneven battle cost the U.S. Navy the escort carrier *Gambier Bay* (CVE-73), two destroyers, and one destroyer escort. Running with the wind, other escort carriers caught in the gunfire could not refuel and rearm their Avengers or fighters. Still, the defenders got in some damaging hits on three cruisers. That damage and a belief that opposition from both ships and planes would strengthen as he approached the landing beaches, Vice Adm. Takeo Kurita ordered Center Force to withdraw just after 0900.

Operations Officer: Destination Leyte

Early on the morning of the twenty-fifth, Halsey radioed McCain to cease refueling and head back to help repel the attack of Center Force off Leyte's beaches. At that time McCain was approximately four hundred miles from San Bernardino Strait to the west and the same distance from the enemy carriers to the northwest. Aboard *Wasp*, Thach was confronted with several challenges at once. Scout aircraft had already been launched in the expectation of finding the four carriers with the Northern Force. That expectation also impelled Thach to order all five carriers of his task group to arm Avengers with torpedoes, the best possible weapon to sink a ship. Consequently, when Halsey's message was received and a multitude of other messages began flooding in from the ships under attack off Samar, Thach had to make some difficult decisions. First, he knew that the distance from 38.1 to Samar was too far for Avengers to carry torpedoes. Very reluctantly he ordered that torpedoes be removed and bombs installed. Experienced carrier captains in the group did not challenge the order because they fully understood the combat radius logistics that influenced Thach's decision.

While the change in ordnance was occurring, Thach turned to his second challenge of getting the SB2C scouts back aboard. With prevailing winds from the northeast, Thach ordinarily would have turned the entire task force in that direction. However, the exigencies of combat called for something more creative. Thach's directive was so unique that radio messages were broadcast to the returning scouts explaining the maneuver and that they had to already be in a landing pattern when their carrier was ready to take them.

Given that the scout planes had launched and would be returning at different intervals, the entire task group would have had to steam away from Samar and Leyte for most of the morning. Rather than lose all that time, Thach ordered each of the five carriers to wait until six or more of their Helldivers were overhead and then run ahead of the formation at full speed (twenty-nine knots for the three light carriers and thirty-two for *Wasp* and *Hornet*). One by one, each of the five carriers spurted out ahead of the task group as its Helldivers returned and circled. Then each carrier would turn sharply into the wind, take its planes aboard, turn quickly again back to the west toward the task group—and Leyte—then race back to its assigned location in the formation. For the younger sailors it was amazing to see a carrier charging through the task group formation heading in the opposite direction.

Though pleased with the carrier captains and pilots' execution of his improvised plan to land planes while still moving the task group toward Leyte at high speed, Thach was exasperated that communicators could not get a message through to determine whether or not U.S. forces still held Tacloban airfield near the landing beaches. Who held Tacloban was of critical importance for his instructions to pilots, who might not have fuel to return to the carriers after attacking enemy ships. Despite best efforts, the constant stream of messages from *Wasp* received no reply before the task group was ready to launch. Neither McCain nor Thach knew that Kurita had already broken off his attack when 38.1 was ready to launch. Regardless, the enemy ships were at sea and they had to be attacked.

Just before and during the noon hour, Task Group 38.1 began launching aircraft some 330 miles from Samar. The desired range even with bombs in place of torpedoes on the Avengers was 250 miles. Consequently, Thach recommended and McCain approved the task group continue to plow ahead at close to thirty knots to lessen the distance for the pilots' return flight. After noon the planes from 38.1 did find Kurita's Center Force retiring west through San Bernardino Strait and attacked. Damage was not significant, in part due to the lack of torpedoes for use against battleships and heavy cruisers, and—as Thach expected—some of the pilots did have to land at Tacloban. Still, the battle concluded with Thach and McCain believing that their strike influenced Center Force to prematurely disengage off Samar. Postwar comments by Japanese officers confirmed that Kurita thought incoming air strikes would overwhelm his force because he had no fighter cover. Indeed, planes from the other two escort groups, Taffy One and Taffy Two, were sending planes north to assist Taffy Three's carriers. And Kurita thought the U.S. Navy destroyers and destroyer escorts would not have put up such a terrific fight if they did not expect immediate help from Oldendorf's battleships and cruisers that were known to be in the area. Not until after the war did the Japanese discover that those battleships were actually heading away from the Leyte beaches in pursuit of Southern Force's remnants when Kurita arrived at 0645. In sum, they were too far away and too slow to have effectively intervened.

Only the one strike from 38.1 was launched against Kurita's Center Force on 25 October, but on the following day three were launched. Results were better on the twenty-sixth as the light cruiser *Noshiro* was sunk and the heavy cruiser *Kumano* damaged. It was not the score the task force desired, but at least the landing beaches were secure and Japan was short four aircraft carriers, three battleships, six heavy cruisers, and three light cruisers. On 27–28 October, 38.1 operated off the Philippines and then left for Ulithi; this time the voyage was not interrupted. But Thach was presented with a new problem—a problem that would prey on his mind day and night. Just before World War II he had stayed up late at night moving matchsticks around his kitchen table, looking for a solution to a seemingly impossible problem while his wife urged him to get some sleep. Now he had another, equally important problem keeping him up late at night. The only difference this time was that it was McCain who urged him to get some sleep.

10

Operations Officer: Destination Tokyo Bay

The separate actions off Samar, Cape Engano, in the Sibuyan Sea, and in Surigao Strait were together known as the Battle of Leyte Gulf. But on 25 October 1944, there was little time to feel good about the great victory then being won. Before victory was achieved with Kurita's midmorning withdrawal off Samar and the sinking of four carriers off Cape Engano that afternoon, the Japanese introduced aerial suicide attacks. The minor damage inflicted on the *Franklin* (CV-13) on 13 October was not then recognized as a predetermined desire to crash into Allied ships, and it might not have been.[1] But on the twenty-fifth, the desire was made unquestionably clear. Just before 0800, the escort carrier *Santee* (CVE-29) became the first victim of a suicide hit, with 16 of its crew killed. *Suwannee* (CVE-27) was struck shortly after, resulting in 270 killed or seriously wounded and a severely damaged flight deck. At about 1100, *Kitkun Bay* (CVE-71) and *Kalinin Bay* (CVE-68) absorbed suicide hits, but with little damage. *St. Lo* (CVE-63), however, did not escape and sank in less than an hour with over a hundred lost.

INTRODUCTION TO THE "SPECIAL ATTACK BOYS"

After the Formosa battle less than two weeks earlier, some Japanese naval officers knew that conventional tactics in the air were doomed to failure as a result of inadequate pilot training. Inspired in part by typhoons that destroyed invasion fleets from the Asian mainland over six hundred years earlier, the kamikaze attacks were desperate attempts to reverse the fortunes of war. Without command of the air, battleships and cruisers were exceedingly vulnerable and merchant ships had no hope for survival. As pilot quality declined dramatically, the previously informal practice of ramming an Allied plane or crashing a plane into a ship after that plane's pilot was wounded became a formal plan to guide undamaged, bomb-laden planes into U.S. ships. Approved only days before the Leyte invasion, the original goal was to damage flight decks in order to neutralize U.S. Navy carriers for one week. That accomplished, Kurita's Center Force, devoid of air cover, could reach Leyte's beaches.[2] Although the initial goal of the kamikazes was not achieved, their success on the twenty-fifth alerted both sides to their continued use. As Bill Leonard recounted, once the kamikaze attacks began, days aboard the flagship "were busy but scary."[3]

Late on 25 October, Thach realized his tactical planning would be affected by "the special attack boys," as he initially termed the suicide pilots. Later he saw the kamikaze as a weapon far ahead of its time, similar to the German introduction of a radio-controlled glide bomb first used against the U.S. Navy in the Mediterranean in September 1943. But on 25 October, he did not fully grasp the degree to which the new threat would affect him or his planning. Planning offensive operations was already sufficiently complicated, and countering the new threat required time that was not readily available. Additionally, morale and confidence among officers and men had to be addressed. Initially the psyche of the American sailor approached a state of anomie—a loss of norms, a loss of knowledge of one's place and purpose in the Pacific war. The rules of war had suddenly changed. Continuing kamikaze attacks in the following weeks brought the problem squarely upon Thach's shoulders. As soon as possible, he had to devise a new Thach Weave for the "special attack boys."

Associated with the opening of the kamikaze phase of the war was the appearance of some previously unseen enemy planes. Appearance of at least one such plane, the Kawanishi N1K1 Shiden (Violent Lightning), known as George, caused some additional apprehension as it was capable of flying at altitudes Hellcats, loaded with fuel and ammunition, could not reach. Circling at forty thousand feet, the enemy plane was an excellent scout, reminding Thach of a tactic Genghis Kahn utilized. The infamous Mongolian warrior often sent a scout on a fast horse to briefly study an objective and then speedily disappear. Soon after followed a devastating attack on the objective. In the late fall of 1944, Thach could not help but wonder what mayhem would soon follow the appearance of the fast scouts. Additionally troublesome was the Nakajima Myrt, a high-altitude reconnaissance plane with exceptional range and the speed (352 mph) to outrun a Hellcat. Although worrisome, Thach soon realized these new planes were not in 1944 the problem they would have been earlier, as both sides already knew the location of the task force.

SECURING THE PHILIPPINES

By 30 October, the *Wasp*, with Task Group 38.1, was at Ulithi, and transfer of command of the task force was passed from Mitscher to McCain. Rear Adm. Alfred E. Montgomery assumed command of Task Group 38.1. Although most sources that address McCain and Thach's whereabouts from 30 October to 5 November 1944 place them aboard *Lexington*, some compelling evidence places them on the *Wasp*. While the identity of McCain's flagship for that week would seem an insignificant matter and understandably overlooked, it will be seen that it was important.

On 1 November, Thach was the operations officer for the entire task force rather than for just one of its task groups. For the next three months instead of worrying about four or five carriers, he had the responsibility for all—in late 1944 anywhere from thirteen to seventeen. In the several days prior to 1 November, his problems were being magnified. Even as he was heading away from Leyte on 28 October, the enemy began another series of air attacks. Although their grand naval operations had failed on 24–25 October, the Japanese did not concede Leyte to the invader and thousands of troops from nearby islands began streaming toward the western side of the island in vessels large and small. Japanese planes from all points were ordered to the area, primarily Luzon, to continue attacks on the Leyte beachhead and Tacloban airfield—still and for many days the only Allied-held land base on Leyte. With both the fast carriers and escort carriers requiring replenishment, several groups in addition to Task Group 38.1 had been ordered to withdraw. However, strong enemy air strikes before and by the time Thach reached Ulithi forced a recall of some carrier groups because the Army Air Force had been unable to develop airstrips and logistics as a result of the seasonably heavy rains.

The considerable inconvenience of not being able to have all carriers return to Ulithi for rest, replenishment, and transfer of air groups was aggravation enough. Making things worse were the continuing kamikaze attacks.

On 29 October, *Intrepid* (CV-11) was slightly damaged, but it was able to remain with the fleet. On the thirtieth, however, both the *Franklin* (CV-13) and *Belleau Wood* were hit, inflicting heavy loss of life and causing both carriers to leave the area for repairs. Hastening back to the fight on 3 November, Task Force 38 prepared to launch raids on enemy air bases beginning on the morning of the fifth. Named Operation Pulverize in Halsey's message to the task force (less Task Group 38.4), he directed fueling be complete on 3 November and commence a high-speed run on the fourth to be in position to launch against airfields and shipping on and near Luzon on the fifth and sixth. Although the primary mission was destruction of planes, airfield installations, and shipping, photo coverage of the Lingayen Gulf area was also to be obtained. The message concluded with precautions for defense against "suicide dive bombers" by "alert radar operation, adequate CAP effectively positioned and stacked, efficient fighter direction and as [a] last resort good gunnery with high speed radical maneuvers."4

Gen. Douglas MacArthur preferred to have the carriers assume a defensive posture off Leyte to protect his troops, but Nimitz, Halsey, McCain, and Thach knew that defending MacArthur's troops meant going on the offensive and maximizing the power of the carriers. Thach fully understood that a mobile force not using that mobility essentially made it land-based air because the enemy would know where it could strike well in advance. The worst thing that could be done to a mobile force was to render it immobile.

Looking ahead, Halsey knew early analysis of the photos obtained in and around Lingayen Gulf would pay handsome dividends to planners at all levels. Strikes on the fifth and sixth were highly successful with most of the four-hundred-plus Japanese planes destroyed and torn apart on the ground. These attacks unquestionably reduced the pressure on Leyte but did not totally return control of the air to Allied forces.

On 5 November 1944, the location of McCain, Thach, and their flagship took on the significance noted above. Apparently they were not on *Lexington*, the flagship recently used by Mitscher and his staff. Instead, they were still aboard *Wasp*, the carrier that had been McCain's flagship when he was commanding Task Group 38.1. The assumption was that McCain and his staff trooped aboard Mitscher's flagship, and that assumption has been repeatedly printed in books and articles down through the years. Evidence to the contrary appears from several sources. Writing to Admiral Nimitz on 6 October 1944, Admiral Halsey stated that he had the impression "received from you verbally that

McCain was to relieve Mitscher at the time I was satisfied it would be appropriate. Because of movements of Task Groups, I have not seen McCain since taking over command. It is obvious to me, that the Carrier Task Force Commander should be in the same group as the Fleet Commander. It is my intention to move McCain to T.G. 38.2. This may be accomplished in one of two ways. Transferring *Wasp* to this group or having McCain transfer to *Hancock* (CV-19). Unless I get word to the contrary from you, shall let McCain decide on the method."[5]

On 5 November, Mitscher's former flagship *Lexington* was struck by a kamikaze. A crewmember on an escorting ship several hundred yards to port focused a movie camera on the diving enemy plane and followed it all the way down. Although frequently seen in documentary films, it is seldom identified as striking the *Lexington*. On *Lexington*, a crewmember standing on the port side of the carrier's island raised a still camera and caught the plane three hundred yards away and seemingly heading straight for him. The crewmember was fortunate, but many others were not. Forty-seven perished immediately, another 3 later, and some 132 others were wounded. Had McCain and Thach been aboard, they too might well have been injured. The kamikaze's bomb struck the secondary conning station in the island, and the plane crashed into the starboard side of the island aft within feet of where Mitscher and his staff had been while the carrier was at battle stations. Further evidence that *Wasp* was then the task force flagship is that McCain mentioned nothing about the hit on *Lexington* in his later highly detailed *Saturday Evening Post* articles, and the task force war diary placed him aboard *Wasp*. Also, no comment on the crash appears in any of Thach's extant papers or oral history. Given his detailed account of later damage on *Hancock*, it is unlikely he would have had no comment on such a life-threatening event.

On 6 November, *Lexington* headed for Ulithi for repairs and *Wasp* steamed toward Guam for a replacement air group. While away from the Philippines for the next few weeks, McCain turned tactical control over to Rear Adm. Ted Sherman, who continued hard-hitting raids on enemy airfields and ships. By the end of the month, the task force had thoroughly pounded the enemy and the Army Air Force was finally in position to cover the troops ashore on Leyte. But by the end of the November, *Intrepid* had been struck again by a kamikaze. *Essex* and *Cabot* (CVL-28) were also damaged, and another kamikaze hit close enough to *Hancock* (CV-19), McCain's flagship from 17 November, to damage it. The United States could build aircraft carriers at a rate well beyond the capability of Japan, but it could not build or repair them as fast as they were being damaged.

In the five weeks since the first kamikaze attack on 25 October, one escort carrier had been sunk and five *Essex*-class carriers, two light carriers, and six escort carriers had been damaged. And that total did not count second hits on the same carrier.

DEFENDING THE CARRIERS FROM KAMIKAZES

The kamikaze threat required response on multiple levels. The cutback in pilot training was reversed, marines would soon bring their fast F4U Corsair's aboard carriers to intercept even the fastest kamikazes, additional air groups needed to be formed and trained, and the number of fighters aboard carriers would jump from fifty-four to seventy-three and then ninety-one. Determination was soon made that for reasons of administration that any number of fighters above fifty-four necessitated restructuring. Consequently, fighters were divided into VF (fighters only or predominantly) and VBF (fighter-bombers). Before the end of the war Hellcats were being utilized extensively for bombing and strafing missions. These responses were beyond Thach's domain but nonetheless welcomed. Still, these actions were only some of the answers. Thach and McCain informally charged themselves to address the other defensive needs and before the end of the month Admiral Halsey made the change official.

Soon after the kamikaze threat was known, Admiral Nimitz requested input from Third Fleet, task force, and task group commanders in regard to the optimum composition and size of air groups. Halsey, McCain, Sherman, Davison, Montgomery, and Clark favored as many fighters as could be brought aboard at the expense of both dive-bombers and Avengers, although most favored keeping something aboard that could carry a torpedo. The lone dissenter was Rear Admiral Radford, who wrote that "overall efficiency is being endangered by attempts to enlarge VF squadrons. New pilots are below standard. Until training catches up with demand and pilots are fully qualified, strongly recommend against further change of present group composition.... Maintain training standards. Quality and not quantity is the fundamental."[6]

Although Radford's opinions on any naval aviation subject were respected, the majority opinion prevailed to get as many fighters as possible on carriers as soon as possible. Radford's concerns pertaining to training proved prophetic, particularly given the inordinate number of deck crashes when the marines first transferred from land bases to carriers. But drastic measures had to be quickly implemented to meet the equally drastic kamikaze threat. On 11 November, McCain issued an exercise order to all four task group commanders. Thach termed the exercise "Moose-trap," an expression from a *Little Abner* newspaper comic strip Madalyn faithfully cut out and forwarded to her husband. Again falling back on his days as an athlete for combat solutions, the opening statement in Thach's message to the four task groups read:

> (1.) The most successful football teams obtain the formations and trick plays used by their opponents and duplicate them against the first team in practice until a satisfactory defense is perfected. Exercise Moose-trap is designed to do just that. (2.) Task Group commanders shall whenever the opportunity presents itself, or can be made, conduct one or more parts of training exercise Moose-trap and report results. The task force commander will, when practicable, conduct this exercise as a task force drill to test coordination between task groups, and to determine the required number and disposition of fighters for defense of the task force.[7]

Thach's objective was to have designated pilots from the task force assume the role of individual and multiple kamikazes. Using VSB Hellcats to represent the enemy, all were to turn off their identification signals (IFF) and not dive on any ships of the task force. Although the area around Ulithi was the preferred location for Moose-trap exercises they could be conducted elsewhere, but when within eight hundred miles of an operational enemy airfield planes posing as the enemy were instructed not to approach the task force any closer than six miles. Although other ideas for the training exercises were welcome, eight specific drills were mandated. As the Japanese had already displayed a tendency to send two or more planes in at high altitude while others approached low on the water, that type of approach became a prominent practice exercise. In the end, the Moose-trap practice approaches to the task force helped determine the best disposition and size of combat air patrols.

The ultimate objective of Moose-trap was not just to improve aerial defense. Indeed, the techniques developed from the practice exercise had to be airtight. No kamikaze could be allowed inside the task force. Antiaircraft guns were effective, especially after the introduction of proximity fuses for the 5-inch shells in 1943, but it was now not enough just to damage an enemy plane. The human-guided plane had to be destroyed before it could crash into a ship, and the best way to do that was to have fighters intercept them.

Training exercise Moose-trap rather quickly produced potential solutions. Keeping in mind that the task force's first priority was to strike the enemy, defensive policies had to be adapted to that ultimate objective. Consequently, destroyers were placed like two points of a triangle in the direction of enemy land bases with the task force, fifty miles behind, being the third point. One group of the destroyer pickets was named Tomcat and the other Watch Dog, each having a homing signal plus a fighter director and Hellcats overhead in a combat air patrol. Most Japanese attacks approached via a direct, straight-line route toward the task force. Consequently, as friendly planes returned from a strike they would pass over one of these two destroyer pickets and make a sharp directional change or fly a full circle. Any enemy plane following the friendly aircraft would either not know to make the same turn thereby revealing his presence or it would be seen by the CAP fighters specifically assigned to delouse the formation. The tactic proved successful as some kamikazes did attempt to hide below American planes returning to their carriers, and any enemy plane that survived delousing still had to contend with the CAP over the carriers.

Another tactic that became policy was the fair-game area. Before strikes, pilots were briefed on certain fair-game areas and instructed to carefully mark their navigational charts accordingly. Any plane flying in that zone would be assumed to be enemy aircraft and attacked. Over the respective task groups Thach stacked the combat air patrol so incoming planes could be seen regardless of their altitude. And remembering what had happened at Midway two years earlier, he always made sure that the usual three to four task groups within the task force were close enough for mutual fighter defense.

Tomcat and Watch Dog, with their Dad Cap (dawn to dusk CAP) and Rap Cap (radar picket CAP), proved effective during daylight hours, but another tactic was needed for the seemingly omnipresent enemy torpedo planes, and now kamikazes, that waited until the last minutes of dusk to attack. At that time of day carriers were vulnerable because they could be better seen than the much smaller attacking planes. Too, dusk was the time of day when carriers began to land their daytime CAP. Thach's answer to this problem was a Jack patrol ("Jacks" in the 1944 glossary), a formation of Hellcats vectored out in a fan-shaped sector in the direction of an expected enemy approach. Jack patrols flew at low altitude because the just launched night fighters were too high to see a torpedo plane low on the water, and radar operators experienced extreme difficulty in locating aircraft flying near the ocean's surface.

Thach was both serious and delighted in the matter of naming the measures he devised and presented to McCain. He was serious about the names being phonetic so they would not be misunderstood, and just as important, he did not want the names to identify the function of the measure. For the latter reason he did not favor some terminology used by the Army Air Force—such as Black Widow for their night fighter—because it might reveal its functional purpose to the enemy. Thach picked most of the names for his innovations; Jacks was in honor of his seven-year-old son (his recently born second son, William, was essentially a stranger due to Thach's absence at sea). McCain, however, did get into the spirit of things by christening Tomcat and Watch Dog.

Not all of Thach's ideas translated into immediate action. In one of his pocket books, alongside notations to "get laundry and pay cigar mess bill," were notes to "move pilot ready rooms below the hangar deck and use a battleship as a picket ship." The idea to use a battleship as a picket did not get much further than his notebook, but the idea to move ready rooms below the hangar deck was soon realized to be desirable. Although desirable, it did not occur until after the war when the *Essex*-class carriers were modernized for service in the cold war. One other idea initially committed to a pocket book was approved and implemented when McCain took command of the fast carriers. That idea was for Avengers to not carry torpedoes on the first attack during a new operation. The greater speed and distance Avengers could achieve with bombs enhanced an overall analysis of the target area, and if ships were found they would still be in range for Avengers with torpedoes on subsequent attacks.[8]

A "THATCHED ROOF" OVER JAPANESE AIRFIELDS

Defending the carriers with Hellcats over picket destroyers and around-the-clock combat air patrols plus special searches like Jacks alleviated much of the kamikaze threat when the task force was highly mobile. But another measure was needed for when the task force was relatively stationary while launching heavy strikes on significant targets and in range of enemy land-based planes. Thach's innovation for this problem was a "three strike system" in place of the two deck-load strikes previously employed. Under the two deck-load procedure, carriers launched half of their planes for a strike, and the second half was launched just before the first returned. That system left too much time for enemy planes to take off and retaliate.

Not trusting good ideas solely to memory, Thach initially committed his idea for a three-strike system to one of his pocket books. Again calling upon his football experience, Thach recalled that sometimes the best defense is a good offense. As his idea took form he conceived "a holey [sic] blanket over enemy known operational airfields." He knew he could not keep eighty different airfields covered all day, but by covering the larger fields he would be reducing the number of kamikazes for the destroyer and carrier combat air patrols. His initial terminology of "holey blanket" did not meet either his satisfaction or that of the staff, and "three strike system" did not find favor because it was too definitive and not as creative as other terms the staff adopted. A suggestion from one staff member was "Thatched Roof," a play on his name. But the word "blanket" best fit what the new method would accomplish, and by late 1944 Hellcats sported a Navy-blue color on all surfaces rather than the previous tri-color paint scheme. Consequently, "Big Blue Blanket" became the favored informal terminology for the new three-strike system.

Implementation of the three-strike system was far more complex than the theory, especially for the thirteen to seventeen carrier captains who had to plan their downwind runs just right. Too, the carrier COs had to allow for smaller air group launchings after the initial large launch. At dawn on the day of attack, Thach's plan called for the launch of two-thirds of a carrier's planes in case opposition was strong. Half of these planes carried bombs while the other half did not. After the planes carrying bombs had expended their ordnance, which usually took about fifteen minutes, they returned to their carriers while the other half remained or "lurked" to keep enemy planes grounded. Then as their fuel ran low after another hour over the target, they would return to the carriers just as the third strike was arriving to take their place. This rotation continued throughout the day. And for the few minutes between strikes the enemy could try to pull aircraft out of hiding, fill holes in the landing strips and hope the bombs, rockets, and .50-caliber ordnance of U.S. Navy carrier planes did not find them before they could again duck for cover. Keeping enemy planes on the ground was a must and achieved the objective of keeping them away from the task force. Destroying them when the first strike arrived or finding and burning them later was delicious icing on the cake.

The innovations to defend the task force and to keep enemy planes grounded during major strikes earned Thach the Legion of Merit. The essence of the citation centered on arranging the direction of one thousand planes to cover nearly one hundred airfields while in the process of securing the

Philippines. Broader mention of his critically significant innovations off the Philippines was downplayed at the time to keep the Japanese from fully understanding American tactics. Throughout the war, Thach was obliged not to discuss the names of his innovations, especially Thach Weave, or to describe them in the numerous radio, newspaper, and magazine interviews.

CONQUEST OF LUZON

Although Thach was out of his bunk earlier than most aboard the flagship, many of the pilots and crew were busy before sunrise. Early breakfast for pilots could range anywhere from 0330 to 0530. For the winter of 1944–45, reveille was generally at 0550 and general quarters at 0612, followed by flight quarters at 0615. Air operations normally commenced at 0654, fueling—if planned—was at 0730, and crews ate breakfast at 0745. If not staging combat operations, from 0800 to 0900 was "Field Day," a time for cleaning the ship. Lunch servings began at 1130 and ran through 1330, and supper servings began at 1700 and were usually concluded by 1830. All ships in the task force darkened ship ten minutes before sunset, and torpedo defense was set at sunset.

The flag staff offices were on the gallery deck spaces amidships directly beneath the flight deck. In his office Thach gathered reports and information from other members of the staff and prepared his operations plan for upcoming strikes. Although each plan differed, there were nonetheless similarities. The average typewritten, single-space plans usually began with instructions for each task group to launch a certain number of planes of the several types. Written in clear English, the plans discussed weather in detail, with alternatives for what might be expected, and set the climbing speed required for reaching altitude under routine, fast, and emergency situations. Altitude for the groups was set: high cover was usually twenty-five thousand feet, with orders to not drop below fifteen thousand, with middle-cover fighters usually at twenty thousand feet and low-cover fighters at fifteen thousand feet. When fighters from high cover did descend, six or eight were ordered to remain at altitude in case other enemy units appeared. Ordnance was assigned for the main target(s) and secondary targets designated if early reconnaissance over them resulted in discovery of planes or other worthwhile enemy assets. Nearly always present in the operations plan were underlined statements reminding pilots not to leave the

Operations Officer: Destination Tokyo Bay

main unit of aircraft being escorted to chase individual enemy fighters. Strikes completed, a primary and secondary rally point was designated. Final instructions usually addressed radio call signals.[9]

On 10 December 1944, Task Force 38 left Ulithi to soften enemy aerial opposition before the 15 December invasion of Mindoro. That large island south of Luzon is approximately midway between Leyte and Lingayen Gulf on Luzon, where MacArthur would land on 9 January 1945. Landing on the western side of the Philippine's largest island, MacArthur's troops would then push south to recapture that nation's capital, Manila. But prior to both the Mindoro and Lingayen Gulf landings, the task force had targets on and near Luzon to destroy or neutralize, which it did from December fourteenth into the sixteenth. Hunting was good those three days, and any enemy vessel that could be sunk at sea meant fewer casualties for Allied troops on land. On the seventeenth, the task force steamed away from enemy land-based planes to prepare for refueling. Once fueled, the task force would return to attacks against land, sea, and air targets in support of the Army.

While preparing to fuel west of the Philippines, Task Force 38 encountered a disastrous typhoon beginning on the night of the seventeenth and worsening into the afternoon of the eighteenth. Given the capriciousness of tropical systems and the fact that such a storm could be much worse over some sections of the fifty-mile-wide task force than others, there was little Thach could do but wait for weather information to arrive from Pearl and the respective task groups. In the end it was up to Halsey to make the decisions. Without a full understanding of the storm and a predetermined desire to get his ships back to cover MacArthur as soon as possible, Halsey did not give the order for all ships to turn into the waves as necessary for the safety of their vessels. The result was three destroyers sunk and carrier planes lost overboard or damaged. Thach thought the three lost destroyers, all riding high due to low fuel and maintaining station and course as directed, were unable to take in ballast water fast enough to prevent capsizing. Assessment of storm damage dictated a return to Ulithi for repairs and the task force arrived there on the twenty-fourth.

The only positive outcome of the storm was more time for Moose-trap exercises. But even time for that was short as the task force was back at sea on 30 December for more raids against enemy airfields on Luzon, Formosa, and Okinawa. On 4 January, Thach received word that a single twin-engine suicide plane had mortally wounded *Ommaney Bay* (CVE-79), and kamikazes had slammed into other ships inside Lingayen Gulf. Nonetheless, after task force air strikes on the seventh and eighth, the Army went ashore as planned on the ninth and began their push to Manila.

While the Army Air Force and planes from escort carriers maintained support of troops on Luzon, aboard the task force flagship *Hancock*, McCain and Thach planned for strikes throughout the South China Sea. The presence of Third Fleet in the South China Sea where it could strike at Indochina, Hong Kong, Formosa, and the western Philippines—all within twenty-four to forty-eight hours—had several objectives. First, just being there could disrupt enemy plans for the defense of Luzon thereby aiding MacArthur. Second, Third Fleet was in a position to interdict the movement of troops and supplies to Luzon. Third, any enemy planes attacking Third Fleet would be just that many less that could help in the defense of Luzon. And finally, there would be no shortage of enemy airfields, ports, and shipping throughout the area. For all in the Third Fleet, January 1945 promised a productive month despite known difficulties in navigating the South China Sea with its reefs, bad winter weather, and no friendly surrounding land.

Admiral Halsey was particularly keen on finding and sinking the two old battleships sporting short carrier decks on their sterns that had escaped him off Cape Engano. But *Ise* and *Hyuga* were not found. On 12 January, the task force struck ships off Indochina. On the fifteenth and sixteenth, Formosa and the China coast were attacked with raids on ships at Hong Kong, Hainan Island in the Gulf of Tonkin, and Canton. The geographic setting of Hong Kong made attacks especially dangerous (ten Hellcats, nine Avengers, and three Helldivers lost) as the hillsides channeled planes into a natural flak trap ringing the harbor. Weather continued to be a major problem, contributing to the loss of twenty-seven planes operationally on the sixteenth and hindering refueling on the seventeenth and eighteenth. On the nineteenth, the weather slowly began to improve as the task force moved toward Formosa on the twentieth and by the morning of the twenty-first, the weather system that had plagued operations finally cleared the area. The task force was once again free to strike targets on Formosa, the Pescadores—in the Formosa Strait between Formosa and China—and the Sakishimas, southern Ryukyus immediately to the east of Formosa. The primary targets for the twenty-first were to be ships, particularly those in harbors, plus docks, warehouses, oil storage dumps, railroad yards, and locomotives.[10]

Operations Officer: Destination Tokyo Bay

Thach welcomed the blue skies on the twenty-first so Task Force 38 could once again get on with its deadly business. The priority targets were enemy ships but there was a trade-off in planning. Sinking or damaging of enemy ships was important but for every plane carrying a bomb toward a ship there was one less plane carrying a bomb toward an enemy airfield.

On this same day the Japanese also welcomed clear skies and their planes first appeared over the task force at 0859. Nonetheless, at 1013 Halsey forwarded orders for larger ships to top off destroyers during the day and McCain passed the order to the respective task groups. Although Halsey, McCain, Thach, and everyone else knew that this was not going to be a good day to refuel the tin cans, it was equally apparent that the fast, short-legged ships would be burning fuel at a high rate during combat operations. Shortly after noon, enemy activity put an end to refueling destroyers.

Two kamikazes got through to damage Tomcat destroyer *Maddox* (DD-731) at 1315. Damage was only moderate for *Maddox*, as well as for *Langley* (CVL-27), which earlier took a bomb hit at 1211 well forward on its flight deck. But it was much more severe for the *Ticonderoga* (CV-14). At 1215, CV-15 took a kamikaze, and forty-five minutes later it took another. Casualties amounted to nearly 350 officers and men killed and wounded, and the big carrier had to leave the war for major repairs to its flight and hangar decks as well as its burned compartments on the second and third decks. Both kamikazes were under fire before they hit, but the relatively small confines of the hangar deck, second deck, and third deck left few places to hide.[11] At 2100 that evening, *Ticonderoga* and *Maddox* were ordered back to Ulithi, both too badly damaged to continue in action.

Thach was not a happy operations officer on the early afternoon of the twenty-first. It was one thing for one enemy plane to penetrate the task force and score a bomb hit or kamikaze crash, but too many had come through on that day. He thought he had a pretty good idea what had caused the problem, a line drawn on a map by MacArthur that delineated areas for his planes to handle (south) and for carrier planes (north). Later Thach's concern proved correct when it was discovered that Army planes had been flying too high over one enemy base to know that planes were effectively operating from it. A message from McCain to Halsey resulted in the Third Fleet commander resolving the issue with MacArthur. In short, carrier planes would strike at any target that threatened them. Resolution, however, came too late to help matters on 21 January.

Knowing this would be a busy day and he would be needed on the bridge, Thach ate a few quick bites in the wardroom and then started for his office to change jackets. Aware that his presence might invite conversations he did not have time for, he went directly to flag plot. All was in order there so he stepped out onto the port bridge wing of *Hancock* at 1326 and reached for a cigarette. There he watched a TBM Avenger land and then taxi forward near the aft end of the island. As *Hancock* was heading into the wind for the landings, Thach could not light a cigarette without ducking his head down just below the small section of the steel bridge. Just as the flame touched his cigarette and before he could take in the first puff that would ignite it, there was a blinding flash and blast that threw him against the side of the island. Collapsed on the deck, he could see that one officer who had been standing beside him had taken a piece of shrapnel through his head. The officer on his other side had one shoulder blown off. Looking up he could see the pervasively pockmarked bulkhead behind where he was standing only moments before. Bending over to light a cigarette had saved his life.

The Avenger had landed without difficulty and taxied forward. Apparently one of the four bombs that could be carried internally had hung up and failed to drop during its bomb run. When landing aboard *Hancock*, the jolt most likely dropped the bomb from its rack to the closed bomb-bay door. Then, while folding its wings and opening the bomb bay in preparation for rearming, the bomb fell out and exploded. Seven officers and forty-three enlisted men were killed and seventy-five were wounded.

Peering over the bridge wing, Thach could see a hole in the flight deck. As the smoke and flame cleared, he could see from the bridge wing what was left of his office. He then remembered that he was often in his office at that time of day because its proximity to the rest of the staff served as the best place to gather information and write operations orders. Still, he was never comfortable in that office while planes were landing, in part due to a previous incident. Earlier, bullets from a Hellcat had penetrated into the flag secretary's office and punched holes into orders he was preparing to distribute. That it could happen again was well known as returning planes usually folded their wings amidships thereby placing their guns in a downward position rather than forward when the plane was ready for flight. On 21 January 1945, Thach could not help but wonder how many more miracles there could be.

It did not take long for *Hancock*'s crew to repair the flight deck and continue operations. For Thach and McCain it was apparent that attacks on shipping needed to cease in favor of concentrating all strikes on enemy airfields. Halsey quickly approved McCain and Thach's request and no more enemy planes appeared over the force until after dark.

On 22 January, the task force continued north and launched strikes against northern Formosa and Okinawa before turning toward a fueling rendezvous and Ulithi on the twenty-third. In the early afternoon of the twenty-fifth, *Hancock* was back in the anchorage and at the end of the day McCain relinquished command of the task force. Vice Admiral Mitscher, Commodore Burke, and Commander Flatley had reported aboard *Bunker Hill* to set up for operations they had been planning since leaving at the end of October. Now Mitscher's team, operating as Task Force 58 with Spruance relieving Halsey, would carry the war to the enemy at Iwo Jima, Okinawa, and the home islands while McCain, Thach, and the staff made their way back to Pearl. McCain flew back to Pearl on the twenty-seventh and the rest of the staff rode back aboard the wounded *Ticonderoga* arriving on the thirty-first.

RESPITE

Heading away from the Pacific both McCain and Thach had reason to feel good about the performance of Task Force 38 for the three months it had been their responsibility. A statistical recap for January indicated they had sunk or damaged 403 enemy vessels. Total tonnage claims for sinking and damage amounted to 617,000 tons. For enemy aircraft, the CTF 38 War Diary claimed 149 destroyed in aerial combat, 466 destroyed on the ground, and 671 damaged on the ground. Both officers were aware that postwar analysis would no doubt alter these figures, but they believed the figures would not be far off. Although neither was happy that they had lost 98 planes in combat and 103 operationally, all things considered, it was not an unreasonable price to pay.

From Pearl Harbor McCain and Thach flew to Washington, D.C., for a short visit with Admiral King. Thach was surprised that McCain ushered him into King's office, thinking that the Navy's highest ranking officer and the task force commander would have matters to discuss that should not heard by others. Nonetheless pleased to have been invited in, Thach was further surprised to find that he was indeed hearing sensitive information. Thach and King already had a long history together from Jimmie's days as a test pilot. Still, he was simultaneously uncomfortable and comfortable with some of what he heard. He was uncomfortable for the subject to be broached that other flag officers greatly lusted for the job of task force commander, even though it was a rather open secret. He was comfortable with the reassurance King gave to McCain that he would take his recently gained experience back to the Pacific and resume his responsibilities as task force commander before summer. Being the fighter that he was, McCain very much wanted to resume his task force role, and with that matter confirmed the remainder of the conversation was most enjoyable for all three.

The media took advantage of having the recently returned Task Force 38 commander and his operations officer available. On 1 March, Thach and McCain read from scripts they helped write for a *March of Time* broadcast.[12] And the *Saturday Evening Post* produced an article with McCain that was printed in July after he had returned to the task force. Therein McCain gave credit to Thach for the three-strike system but did not term it as such because he did not want to reveal to the Japanese terminology used in task force communications or dispatches.[13]

On departing Washington, Thach took a short leave to be at home in San Diego before reporting back to Pearl Harbor to begin planning for future operations. Honolulu had been a tourist attraction of long standing and a highly desired residential location for officers before the war. The cost of a home or rental property was high, but the major concern of officers was lack of availability. That aside, Thach considered Pearl a hole in the ground. Home for him was San Diego, particularly since that was where his wife and boys were living.

At Pearl, Thach received daily accounts of the war from official dispatches and occasional personal notes from Jimmy Flatley. While Thach worked at Pearl, Task Force 58 was busy with two major operations, Iwo Jima in February and March and Okinawa from 1 April to the time he and McCain returned in late May.

The distance between the Marianas and Tokyo was fifteen hundred miles. Iwo Jima, approximately halfway between the Marianas and Japan, was contested between 19 February and 26 March. Since 24 November 1944, Army Air Force B-29s had been flying from the Marianas to bomb the Japanese home islands and possession of Iwo Jima had several positives. Once the nearly twenty-five thousand Japanese defenders were neutralized, the B-29s would have a base where planes too badly damaged to return to the Marianas could land. As enemy planes had intercepted the B-29s from Iwo and had jumped off from there to attack the big bombers on Tinian, the capture of Iwo would put an end to that. Additionally, AAF P-51 Mustang fighters could escort the bombers between Iwo and Japan.

Okinawa was only 350 miles from Japan, and both sides knew it was a natural staging location for the invasion of Kyushu, the southernmost of the four main Japanese islands. Consequently, the enemy threw all its power at the invaders. Tied to the beaches at both Iwo and Okinawa, the task force suffered with kamikazes taking a heavy toll not only on carriers and other warships but also on landing craft. Off Iwo, the escort carrier *Bismarck Sea* (CVE-95) was lost to a kamikaze attack, another was damaged, and three more crashed off Okinawa. No fast carrier succumbed to either conventional bombing or kamikazes, but there was a steady stream of them limping back from combat with heavy damage. A conventional bomb attack put *Franklin* out of the war while kamikazes ended the distinguished combat service of Mitscher's flagship *Bunker Hill* and then did the same to his next flagship, battle star champion *Enterprise*. Serious damage was incurred by the *Saratoga, Randolph, Wasp, Hancock*, and *Intrepid*, while *Yorktown, Essex*, and *San Jacinto* (CVL-30) took hits that did not require them to immediately leave for repairs. Thach could only conclude that the assault would be even more intense when he returned and the task force took aim at the home islands. And if the news from the fleet that winter and early spring was not bad enough, word came on 12 April that President Roosevelt had died at Warm Springs in Georgia.

Before leaving Pearl to move back into combat there was some good news. On 28 April, Thach found his name on the list of commanders promoted to captain, his promotion to rank from 30 March. And on 7 May, Germany surrendered, an event many of Thach's fellow officers thought was going to happen the previous November. Soon there would be more ships and men moved to the Pacific for the conquest of Japan. Cut off from resources and surrounded by June 1945, strategically Japan was already defeated, but the language of the 26 July Potsdam Declaration left no alternative but unconditional surrender. Without a more palatable exit and no cooperation from the Soviet Union to assist with negotiations, the Japanese government felt compelled to fight on.

THE LAST THREE MONTHS OF COMBAT

Upon relieving Mitscher on 28 May 1945 at the Leyte Gulf anchorage, McCain and Thach set about to resume operations for the final three of their six months in command of Task Force 38. Spruance had turned the fleet over to Halsey one day earlier so the designation was again Third Fleet. For the next series of operations McCain, Thach, and the staff would move onto the fourth different carrier to serve them as flagship. The newest flagship was *Shangri-La* (CV-38), named for the mythical location the late president had designated as the launching point for the 18 April 1942 Halsey-Doolittle Raid on Tokyo. Although away from the task force for only four months, there were many changes on the task force organization chart. Familiar names among the carriers were absent: *Saratoga, Enterprise, Franklin, Bunker Hill*, and *Wasp*. In their place along with *Shangri-La* were unfamiliar new carriers with familiar names: *Randolph* (CV-15), *Bennington* (CV-20), and *Bon Homme Richard* (CV-31).

The top secret 4 June intelligence appraisal of the enemy revealed mostly unfavorable information for Thach as he prepared his operations plans. Despite heavy losses the enemy was continuing a major air effort against airfields on Okinawa and ships offshore. Inexperienced pilots sometimes using their trainer aircraft were being pooled for suicide attacks while better pilots were being assigned to defend the most important targets within the home islands, especially around Tokyo. Airfields in southern Kyushu were being used primarily for interceptors and reconnaissance aircraft and for staging planes through for attacks on Okinawa targets and the task force. Understandably, relatively safer areas in northern Kyushu, Shikoku, Honshu, and Hokkaido were being utilized for aircraft concentrations. But even in southern Kyushu the intelligence appraisal was particularly disturbing in that the enemy's increased sophistication in dispersing and camouflaging aircraft surpassed "anything previously seen in the Central Pacific or Philippines." Plus, radar, picket boats, and coastwatchers were giving the enemy an excellent early warning system resulting in carrier sweeps being met by "alerted interceptors and a general withdrawal of other planes to rear areas."[14]

The only good news in the intelligence report was that something always seemed to go wrong for the enemy when preparing to initiate a major kamikaze thrust. Weather accounted for some delays, but in any event the enemy to date had not been able "to achieve any considerable degree of tactical surprise."[15] Thach knew that in April and May the enemy had launched over 1000 kamikazes, 355 on 7–8 April alone, toward Okinawa on eight separate occasions in addition to the much smaller attacks on a near daily schedule. Small attacks should not be a problem but for planning purposes he would always have to allow for a major attack of 100 to 350 kamikazes.

Operations Officer: Destination Tokyo Bay

Halsey had hoped to strike the enemy homeland before the end of 1944 but tenacious resistance in the Philippines precluded the hoped for sorties. As success on Okinawa was all but assured (secured 21 June; officially declared 2 July), he would not have to wait long to satisfy his desire to return to the waters off Japan where he last visited three years earlier. But his first operations required him to continue support of the Okinawa campaign, so on 2 and 3 June, Rear Admiral Clark's Task Group 38.1 concentrated on Okinawa while Rear Admiral Radford's Task Group 38.4 raided southern Kyushu. Thach was not overly enthusiastic about the Okinawa assignment writing in his notebook, "are we going to be tied to Okinawa forever?" Enemy resistance for both task groups was minimal but a much more lethal foe was building south of Third Fleet and heading toward it.

Although first observed on 1 June, little information filtered into Halsey's staff for the typhoon that was strengthening and heading north. Although small, this typhoon was well organized, and by the evening of 4 June it was upon Halsey's ships. Thach was involved in discussions throughout the early evening and night as the storm pressed onto the fleet, but for this foe his creative talents were limited. Course change recommendations flowed from the respective task groups both to McCain and to Halsey, but it was Halsey who had the final say. Complicating escape from this particular storm were the usual factors of any tropical storm's capriciousness plus the unfortunate continuation of poor meteorological advice from Nimitz's weather center (then at Guam). Doing what he could to move the entire fleet to the relatively safe northwestern quadrant of the storm, Halsey decided against one course change recommended by Nimitz headquarters and another from McCain and his staff. That unfortunate decision resulted in the fleet turning into the storm.

The effect of the storm on Task Force 38 visible late on 5 June as the storm diminished was major damage to the flight decks and catapults of *Hornet* and *Bennington*, the loss of aircraft equal to a full carrier air group, and damage to varying degrees for other ships. Both *Hornet* and *Bennington* had the forward overhang of their flight decks collapsed in such a manner as to make both look as though they had enclosed bows (which both received when modified in the early and mid-1950s). *Hornet*'s World War II combat days were over, but *Bennington* was repaired at Leyte and was back in action a month later. The conclusions of the 16–22 June 1945 Court of Inquiry and Thach's perspectives pertaining to the storm are treated in the following chapter. For now, however, it is fair to say that Thach's return to combat had a most inauspicious beginning.

Weather was always a critical element in his planning and throughout the final three months of the war bad weather was pervasive and on more than one occasion he had to stop hundreds of planes from launching at the last minute.

Despite storm damage, units of the task force did attack Okinawa and Kyushu from 6 June into the tenth before steaming back to Leyte Gulf to drop anchor on the thirteenth. At Leyte the task force made ready for an extended cruise off the coast of Japan. Three task groups accompanied by eight new battleships, two battle-cruisers, fourteen cruisers, and sixty destroyers would protect the nine *Essex*-class and six *Independence*-class carriers.

Before leaving the Leyte anchorage on 1 July, McCain and Thach emphasized several salient matters in a message to Task Group 38.1, Task Group 38.3 and Task Group 38.4. Target selection was the focus as some old and familiar targets had lost importance or ceased to exist. McCain's letter, drafted by Thach, noted that "the process of separating the valuable targets from the Empire 'junk' requires meticulous care in order to single out the most profitable and appropriate targets for the 'Fast Carrier Treatment.'" The number one priority in target selection was neutralization of enemy aircraft, whether in the air or on the ground, as that was required for the protection of the task force. By the summer of 1945, the enemy could hurt the task force only from the air, as surface units no longer came out to fight and the submarine threat was minimal. If necessary, all planes in the task force would be used for combat air patrol given that destruction of enemy planes was worthwhile in itself.[16]

Assuming the enemy aerial threat was neutralized, emphasis would then shift to "targets of great value and importance to Japanese war strength." B-29s and carrier aircraft had not attacked many industrial, transportation, and infrastructure targets. Now there was an opportunity to strike at vital industrial targets but these targets needed to be hit with enough bombs to preclude effective dispersal.[17]

Thach's major concern was evident in the statement that "the blanket attack is *not* a defensive assignment. It is a *strike* against air strength." It was apparent to him that pilots, especially the strike leaders, needed to be reeducated to understand that targets assigned were more important than the traditional shipping targets that had been emphasized for so long. Stressing that "large and impressive installations do not always house valuable targets at this stage of the fighting," Thach and McCain pointed out that with 90 percent of the enemy surface fleet immobile, bombing idle docks, yards, and repair facilities was a waste of effort. Finally, the message to the task group leaders was that the new emphasis on target selection required education of the pilots beyond the last minute briefings before takeoff.[18]

Although numbers are sometimes misleading, the following numbers indicate which targets found the greatest favor in the final weeks of the war when Task Force 38 had many to choose from throughout the Japanese home islands (while the "destroyed and damaged" totals presented were undoubtedly altered in postwar assessment, target prominence in this period is nonetheless revealed): 268 locomotives, 24 freight and tank cars, 23 railroad stations/yards/roundhouses, 198 hangars and barracks, 87 factory buildings, 28 fuel tanks, 25 radio/radar stations, 24 warehouses, 23 bridges, 20 city blocks (burned down), 10 power houses, 7 lighthouses, 7 docks, 2 gas trucks, and 2 ammunition depots. Not surprisingly, the ammunition depots were known to be such after large secondary explosions were witnessed.[19]

On 10 July, Task Force 38 struck the Tokyo area with carrier pilots finding little opposition except for the expected flak traps of damaged aircraft positioned in such a manner as to draw Hellcats. When enemy fighters did get off the ground, general consensus among carrier pilots was that they were encountering better pilots than they had fought in campaigns to the south. Nonetheless, Thach and all carrier pilots preferred to meet Japanese planes in the air since the prospects for victory were better there than on the ground. Yet Thach had some mixed feelings. He was pleased that the enemy pilots were finally beginning to contest carrier raids in the air, but he had noticed that carrier pilots had picked up some bad habits from so many easy kills. Too often they were running in front of each other to get a kill, only to find a trap sprung by experienced Japanese pilots every bit as good as their 1942 reputation. As a result, Thach volunteered to lead a fighter strike to get a look at new Japanese fighters' performance and observe the tactics currently being used by some of their veteran pilots. But McCain, who trusted Thach's judgment and endorsed nearly all of his operations officers' recommendations and innovations, was not the least bit swayed by this suggestion. Saying no in his most emphatic salty vernacular while laughing, McCain told Thach, "You're restricted to this Task Force." Somewhat surprised that Thach had not already figured it out, McCain sympathetically acknowledged the merit of Thach's desire to directly observe the enemy and get back into aerial combat but told him "You know too much." If Thach had been shot down and captured, the information he might have divulged under torture would have been of inestimable value to the Japanese. Thach did not argue the point because he understood his boss's perspective, but still he was disappointed at not being granted permission to fly a combat mission. Years later Thach would recall that was the only time McCain ever said no to one of his recommendations. The only other occasion the task force commander was as adamant was when Thacher Longstreth volunteered to parachute into China for a firsthand evaluation of whether Chinese farmers were agrarians or truly Communists.

Aerial opposition was again light when carrier planes hit Honshu and Hokkaido on 14 and 15 July. On these dates battleships moved in toward the coast close enough to add the weight of their 16-inch shells. Welcome as they were, Thach and the staff nonetheless had to work some math to see if the battleship role was effective. Making the question more difficult was the fact that there was nothing then existing for a comparison. For Thach the major question was whether 112 VFB Hellcats carrying one 1,000-pound bomb would have caused as much damage to the steel works at Kamaishi as the 14 July battleship bombardment by three Iowa-class battleships (Iowa BB-61, Missouri BB-63, and Wisconsin BB-64). A 1,000-pound general-purpose bomb dropped by a Hellcat contained 500 pounds of high explosives, as opposed to only 150 in a 16-inch HC shell. Four hundred 16-inch shells had been expended on the Kamaishi complex, with a 20 percent accuracy rate, thereby placing 12,000 pounds of explosives on target. This translated to a total of 24 VFB Hellcats of 112 having to place their bombs on target before the resultant smoke obscured the view of the 2,000-by-750-foot target. Thach's conclusion was that low-flying Hellcats could hit the "Achilles heel" of any steel mill by disabling its coke ovens. Kamaishi had four batteries of coke ovens and two bombs on each battery would shut down the operation for at least six months. In conclusion, Thach reasoned that eight well-placed bombs could do the job better than four hundred 16-inch battleship shells. He did not address the psychological effect of battleships moving unmolested along the Japanese coast.[20]

Thach was particularly anxious to launch strikes against Hokkaido, northernmost of the four main Japanese islands, and northern Honshu because both areas had previously been mostly out of reach. Intelligence knew aircraft had been concentrated in the region and Thach was sure the rocket-powered baka bomb kamikaze planes would be found. Range for the baka bombs was extremely limited so they would be carried to within sight of a target by twin-engine Betty bombers or other similar aircraft before being released to dive on a ship at approximately 450 miles per hour. The human-guided rocket was a real threat if it got near a ship, so it was imperative to destroy them to the greatest extent possible on the ground or while being carried. Thach knew that kamikazes at full speed found it hard to turn, and if they had not correctly read the course and evasive action of a ship they could not compensate at the last moment with a radical turn. A destroyer might make a radical turn too sharp for a high-speed baka bomb to follow, but carriers, battleships, and cruisers could not maneuver as fast as a destroyer. Still, many kamikaze pilots were too unskilled to follow the sharp turns of large ships and there is considerable photographic evidence of a multitude of near misses. Be that as it was, Thach wanted enemy planes destroyed before they came within sight of any carrier. After the war he learned that some of the enemy planes his carrier force hit in northern Honshu and Hokkaido were planning for suicide missions to the Marianas in a major attempt to disrupt B-29 bases. Intelligence available to him at the time indicated attacks on the task force and Okinawa.

Thach was particularly interested in targeting railroads and ferries during attacks on northern Honshu and Hokkaido. He knew that a major disruption of transportation stopped everything else. When evaluating strikes he was as pleased with news of a ferry sinking as much as he was with photographic evidence of aircraft destroyed on the ground.

Happy as he was with the attacks on northern Honshu and Hokkaido on 14–15 July, he was disconcerted by news that McCain would be relieved of command one month later. Having been promoted to captain, Thach did not expect to remain with the task force once another vice admiral assumed command of the task force. Too, he expected that the new task force commander would want to choose his own staff.

On 16 July, British carriers joined the task force, and on 17 and 18 July their planes went into action along with American carrier planes over Tokyo and Yokosuka. Following replenishment at sea Task Force 38 returned for strikes that have come to be the most memorable for the last stages of the war. Beginning on 24 July, task force planes struck targets on the western side of Japan and in the Inland Sea (Sea of Japan). The previously secure area between Japan and the China coast took a beating at sea and ashore, particularly Kobe and Kure. At Kure on the twenty-eighth, Halsey finally put battleship-carrier hybrids *Ise* and *Hyuga* on the bottom along with battleship *Haruna* that the American media had claimed sunk early in the war. Other capital ships including the most recent carrier additions to the Japanese fleet were either capsized or sunk in shallow water. While Halsey smiled upon hearing the results of the raids, Thach smiled because an Army pilot flying a PBY had rescued two downed carrier pilots in the

Sea of Japan. Ordinarily this would have been a job for a submarine, but American subs were few in the Inland Sea. Thach requested help for the two pilots, and the Army-piloted PBY was selected, even though it was a long journey for the big, two-engine flying boat. Thach organized a fighter escort to accompany the PBY. The first pilot was picked up without incident but the second pilot in his rubber raft was being shot at from the shore. To further assist in the rescue Thach invited the PBY, by then low on fuel, to fly to the task force. Four destroyers were designated to shine their lights to starboard to assist the night landing by not only showing the PBY where to land but to assist with the pilot's depth perception. Following a safe landing the PBY was intentionally sunk and Thach hurried to McCain to recommend a Navy Cross for the PBY pilot. McCain agreed as he too was aware that the PBY pilot knew he would most likely come under fire and did not have enough fuel to return to his seaplane tender.

The raids on the western side of Japan had everyone in a good mood. An Associated Press correspondent captured the attitude aboard the ships of the task force when he wrote, "If the Japanese Admiral ever expects to inspect his fleet, he'd better buy a diving suit."[21] On 6 August, the attitude was mostly bewilderment upon hearing the news of the atomic bomb on Hiroshima. But the war was still on, and on 9 and 10 August carrier planes were again over northern Honshu. On these dates Thach was again upset over the demarcation line that assigned carrier planes north of a designated line on a map. The thinking was that such a line would keep Navy fighters from attacking B-29s and their escorts as they moved toward their targets. Thach was not impressed with the line because he believed that the possibility of a Navy fighter mistaking a B-29 for a Japanese bomber was pretty remote. Too, the line meant that carrier planes would have to take off from the relatively warm water off Honshu and fly through early morning fog resulting from the cooler water to the north. This worry dissipated somewhat after Thach found that the fog was sometimes a help in keeping coastwatchers from seeing and reporting the direction of the carrier planes. And finally, Thach was upset that enemy planes based on or just south of the line enabled some to escape the "Big Blue Blanket" treatment. With the carrier planes not allowed below the line and the Army unable to reach these bases, a significant hole was created for kamikazes to fly through. More than once planes from the line crashed picket ships, one of the more significant being a 9 August attack on *Borie* (DD-704) that resulted in 114 casualties.

News of a second atom bomb drop on Nagasaki drew little attention from Thach or among the staff due to the necessities of planning and launching strikes. By this time the pilots were so familiar with targets around Tokyo that Thach joked that they wanted to vote in the next election. On 12 August, yet another typhoon moved through the area and Thach thought the enemy would most likely not expect an attack on the thirteenth as the previous tropical storms did bring an interruption in attacks. On this occasion, however, the task force waited for the typhoon to hit and then followed it in on the thirteenth. Better weather forecasting had not only kept the task force out of trouble but also was used as an aid for attacks.

Two atomic-bomb attacks and the Soviet Union's announcement on 8 August that it would declare war on Japan led the government in Tokyo to make known its desire to surrender. On 15 August, intent was interpreted as only that—intent—and strikes from the task force were launched toward Tokyo. A recall went out when word of a cease-fire was received, but some planes were lost over Tokyo and enemy planes bound for the task force were engaged. Celebrations began at sundown within the task force, but all the usual defensive measures were implemented and a combat air patrol was scheduled for the next morning.

SURRENDER

The formal ceremony for the Japanese surrender was scheduled for 2 September. Although McCain's relief, Vice Adm. John H. Towers, would not arrive for several days, McCain prepared to leave for the United States. Admiral Halsey dissuaded McCain and asked him to attend the surrender ceremony aboard the Third Fleet flagship, the *Missouri*. In the same conversation, Halsey requested McCain to invite Thach to accompany him for the ceremony. McCain, though disappointed at his relief and not feeling well physically, acceded to Halsey's request and was no doubt pleased that his good friend Halsey had invited him to bring Thach along.

During the two-week wait for the surrender ceremony, Thach and the staff began to gather statistics from their most recent actions and for all their time in command of Task Force 38. Claims for the period from 1 July to 15 August were impressive. For thirteen strike days there had been 23,556 sorties and 10,678 attacks on enemy targets, with 4,619 tons of bombs plus 22,036 rockets expended. A total of 307 planes were lost, 174 combat and 133 operational, resulting in 126 pilots and 71 crewmen lost. Of 2,423 enemy aircraft claimed,

116 were in the air, 5 were downed by antiaircraft fire, 1,111 were destroyed on the ground, and 1,161 were damaged. For shipping, 247 vessels of five hundred or more tons had been sunk or damaged, including 31 warships, in addition to 618 small craft.[22]

For the period 28 May, when Halsey relieved Spruance, to 15 August, Admiral Halsey's staff claimed 2,965 planes "put out of action." For warships, the claims were 1 battleship, the 2 battleship-carrier hybrids, 1 heavy cruiser, 1 light cruiser, 8 destroyers, and 12 escorts. Damage claims included 1 battleship, 3 large carriers, 2 light carriers, 3 escort carriers, 1 heavy cruiser, 18 destroyers, 44 various types of escorts, and 15 submarines, including 6 midgets.[23]

On 2 September, McCain took his place on *Missouri*'s quarterdeck in the front row next to the white tablecloth–draped mess table upon which the documents of surrender were placed. A few feet away stood Thach, his height being particularly advantageous on this day as he had a clear view of the proceedings. The ceremony was very professionally conducted, Thach thought, and it was soon over.

After the ceremony, Thach went ashore at Yokosuka with intelligence officer Longstreth and others to look into some caves and see some of the planes the Japanese would have used had the war continued. Not surprisingly, the Japanese had hoarded planes, and their use in the hands of kamikazes would have undoubtedly caused great loss of life when used against the Allied landing craft. Interesting to both Thach and Longstreth was a photograph taken over Pearl Harbor on 7 December 1941. Apparently hastily torn up, Longstreth pieced it back together and took a photo of it. Thach kept his copy for the remainder of his life.

LOSS OF A BOSS AND FRIEND

Vice Admiral Towers officially relieved McCain the day before the surrender ceremony. Three days after the cease-fire, Thach received a message relieving him of his duties with the task force and directing him to report to Pensacola, Florida, for his next assignment. McCain had orders to become deputy director of the Veterans Administration, but he wanted Thach to accompany him to Washington to talk with Admiral King and others on some matters important to naval aviation.

Operations Officer: Destination Tokyo Bay

On 3 September, Thach and McCain touched down briefly at Pearl Harbor and soon were back in the air on their way to San Diego, then home for both. On the sixth, McCain and his wife invited Thach and Madalyn to come by for a visit. Not long after the Thaches arrived, McCain excused himself, saying he did not feel well and needed to rest. Just as Thach returned to his father-in-law's house there was a phone call from Mrs. McCain: the admiral had apparently suffered a heart attack and died. Thach immediately returned to the McCain home, but for that day there was nothing more he could do for his task force commander.

Using some of his fifteen-day leave, Thach accompanied the McCain family to Washington for the admiral's burial in Arlington. After the service, Thach met with King to deliver the written report and some oral observations McCain had planned to present. With the war over, Thach was not sure if King would ever read the full report, but he felt duty-bound to deliver it. It was the least he could do for a boss who had also become a close friend.

11

Perspectives on Task Force Personalities

In 1971, Admiral Thach spoke directly to the professional and personal attributes of the more significant luminaries attached to Third Fleet and the task force. As task group and task force operations officer, he worked closely enough with the leaders of the fleet (Halsey and Spruance), task force (McCain and Mitscher), and task groups to offer opinions worthy of consideration.

THACH ON MCCAIN AND MITSCHER

When Vice Adm. John S. McCain offered Thach the job as his operations officer in March 1944, the two men were essentially strangers, knowing each other only by reputation. When McCain died on 6 September 1945, Thach commented that "it was like losing his father for the second time." No other statement could reveal the depth of Thach's devotion to the man he had come to know during the last eighteen months of World War II. Thach considered McCain and Mitscher as different as "night and day." He believed Mitscher to be "very good" but preferred to work for McCain. While not finding Mitscher incompetent in any manner during the time he was attached to his staff, Thach nonetheless had some reservations about him. Among his concerns was that Mitscher "had his own convictions and he didn't see the need to hear from anybody else much." And, he said, "it was hard to change Mitscher's mind if you thought he was doing something that he could have done better. He wouldn't change his mind. He'd say what he was going to do and if somebody came up with maybe wanting a modification or something different, he wouldn't listen."

Thach did acknowledge that during the Iwo Jima and Okinawa campaigns Mitscher had to remain tied to the area due to orders from higher authority (Spruance, Nimitz, and King). Still, he felt Mitscher might have made a more persistent argument to higher authority that a mobile force rendered immobile became nothing more than land-based air. And land-based air or a static task force made location and planning for attacks too easy. Thach's defense against both immobility and kamikazes was the three-strike "blanket" tactic. Being a Thach innovation, Mitscher was not as familiar or comfortable with it as he might have been had the idea come from Flatley and Burke. Thach did know that his counterpart, Flatley, was very much in favor of using the three-strike blanket, but Mitscher never fully adopted it in the manner used while McCain was in command. And Mitscher was not given to arguing with Spruance once a decision had been made whereas McCain had no reservations in speaking directly and forcefully to Halsey in (often successful) attempts to change the attitude and policy of the fleet commander.

Thach was careful to say that he and everyone else worked very hard while under Mitscher's command. "We broke our necks to be sure that we did everything exactly as it was planned," he recalled. "We didn't try to go off on our own in any way." Thach also noted that McCain's task group was commended for action off Formosa, and even though Burke actually wrote the commendatory messages, Mitscher approved having them sent.

Thach did qualify his observations in his oral history noting "I may be too critical of him (Mitscher)." And he pointed out that an incident during the Battle of Midway might have prejudiced his opinion of Mitscher and Mitscher's opinion of him. After *Yorktown* was mortally wounded, Thach first flew aboard the *Enterprise* and the following day assumed command of the fighter squadron on *Hornet* made up of both *Yorktown* and *Hornet* VF survivors. Mitscher, then CO of *Hornet* (CV-8), was justifiably upset when on the previous day Thach's squadron mate, Dan Sheedy, landed his damaged fighter too hard with the result that his guns fired and took the lives of five officers and men and wounded twenty. Thach believed the resulting investigation to show that the master gun switch fused due to combat damage. Mitscher recommended an automatic cutout switch be worked into the system that lowered the hook to keep guns from firing when damaged and landing hard. Thach, however, refused to endorse Mitscher's recommendation while on *Hornet* and later at the Bureau of Aeronautics because he believed fighter pilots did not need yet one more switch that could still malfunction when dirt got into it. Too, Thach worried that the switch might fail in combat, and he considered it highly remote that another enemy bullet would hit and fuse the master switch as happened to Ensign Sheedy. All of this aside, it is highly doubtful that Mitscher would not have adopted the three-strike blanket system because of the 1942 disagreement with Thach. Thach was much more on target when he said that Mitscher "was an old-time aviator and in my opinion he figured that he had the experience within himself and he never took to new ideas or wasn't able to recognize them as well as McCain." Apparently the Thach-McCain three-strike system worked, as the number of aircraft destroyed on the ground was greater than those found in the air, a documentation of its success.

Perhaps one could conclude that both Mitscher (dependent on strong CAP and antiaircraft fire) and McCain's (three-strike blanket) different approaches to defending against enemy aerial attacks were effective but neither was perfect. Fate dictated that Mitscher be a sitting duck for much of his time in command while McCain's blanket was sometimes compromised by the arbitrary demarcation line between Army and Navy planes.

One other experience while still on *Lexington* might have demonstrated a difference in philosophy between Mitscher and McCain. Thach was an early advocate of night fighters in part because of the first successful night interception by carrier aircraft recorded by Butch O'Hare on 26 November 1943. Although that interception cost Butch his life, he had nonetheless proved that night interception using radar was possible. Seven months later, it was apparent to Thach that Mitscher had not fully recognized the potential of night fighters. Consequently, Thach supported Gus Widhelm's plan to not allow the task force combat air patrol to attack one snooper off the Marianas until he was directly over the flagship. When it finally drifted over, Widhelm escorted Mitscher to bridge and then gave the order. Down came the snooper in flames.

Considerations already expressed aside, Thach liked McCain's habit of constantly having pilots with him on the flag bridge. Over coffee, he listened to their experiences and then asked if they thought that current tactics and procedures were adequate to needs. Thach believed this was not only good for McCain but also for the morale of the pilots. And there is no question that the pilots were grateful for an opportunity to share time and thoughts with a leader who had their fate in his hands. To the contrary Mitscher was much more reserved in inviting pilots to the bridge. When he did it was for a special purpose, most often to gain specific information, occasionally for special praise, and very rarely to admonish. Inviting pilots to the bridge to evaluate current policy was not a Mitscher practice. These differences could be assigned more to personality than to professional effectiveness as both men were revered by their respective staffs and greatly respected by pilots for they knew both would steam the whole task force to their rescue if necessary. And there was no distinction between Mitscher and McCain by superiors, contemporaries, or subordinates regarding their will to fight. Pilots who served under both generally admired Mitscher and McCain because they knew both had their best interests at heart, especially in rescue, and because both were fighting admirals.[1]

Mitscher was very clear and reasoned in declining major changes while he was in command of the task force. He was not particularly keen on night operations because they kept both pilots and ship's company from obtaining sufficient rest. He was not in favor of bringing new aircraft into combat during operations. On the other hand, McCain was for most any idea that would be an improvement. He constantly hounded higher authority for the best equipment available, and "he wanted it right now."[2]

Perhaps too much has been entered into the equations of the combat effectiveness of the Spruance-Mitscher-Burke-Flatley team versus the Halsey-McCain-Baker-Thach team. The first is noted for meticulous planning, the second for operating on dispatch. World War II naval literature is replete with examples in which both the meticulous planning and dispatch operations enjoyed success and perceived failure, perceived failure being defined as speculation of how a victory might have been greater. And perhaps it might be agreed that it is not totally fair to overly criticize a meticulous planner (Spruance) for following a well made plan or to criticize a commander depending on dispatches (Halsey) for maximum flexibility. Of course when dispatches lacked clarity or were not sent quickly, criticism is warranted.

The final consideration pertaining to the Mitscher-McCain comparison is that fate decreed that Mitscher be in command during the major invasions of Leyte, Iwo Jima, and Okinawa and that McCain was in command only in support of the relatively easier landings on Mindoro and Luzon. It was Mitscher who absorbed more kamikaze attacks, and it was Mitscher who was in command for a total of fourteen months, whereas McCain was in charge for only six. Both gave everything they had to protect their men and destroy the enemy and despite the personality and philosophical differences between them, both were winners.

THACH IN DEFENSE OF MCCAIN

When Admiral Thach was interviewed for his oral history in 1971, the interviewer, Cdr. Etta Belle Kitchen, USN, based several questions on commentary offered in The Fast Carriers: The Forging of an Air Navy. Written by Clark G. Reynolds and published in 1968, it was the first full treatment devoted to the development of the fast carrier task forces with critical analysis and evaluation of operational techniques and personalities. To date no other study has approached its value to the literature, and given that the author had direct contact with most of the former task group commanders, carrier captains, air group commanders, pilots, Spruance, and Thach, it is doubtful another such study could equal it.

The value of The Fast Carriers stated, Thach nonetheless took issue with some of its conclusions and judgments. As noted earlier, Thach considered McCain a father figure, and his desire to defend McCain's record should always be viewed in that frame of reference. It is apparent that Thach, though a contributor to the Fast Carriers manuscript, had not read the book prior to Kitchen reading pertinent sections to him. For the most part Thach's defense of McCain reveals perspectives that invite reconsideration of several incidents in which McCain may well have performed better than thought. On others, however, Thach's attempt to redeem his boss only further documents conclusions in the book.

Thach was entirely correct in noting that two of McCain's major detractors, Rear Adm. Ted Sherman and Rear Adm. Jocko Clark had reasons to seemingly go out of their way to denigrate McCain. Sherman, although justified on the basis of combat experience, thought he should have been given command of the task force instead of McCain. His book, Combat Command, and wartime diary clearly reveals his thoughts on McCain.[3] Jocko Clark had received several well-deserved acclamations and decorations from Mitscher for his aggressive and effective conduct in battle and would have preferred no one other than Mitscher in command. In his own autobiography, Carrier Admiral, Jocko Clark expressed particular unhappiness with McCain not only for delaying a reply to him during the 4–5 June 1945 typhoon but also for expressing regret over the damage to "your ships." Clark wrote that before the Court of Inquiry on the typhoon that McCain had referred to Clark's ships as "his (McCain's) ships."[4] In sum, Thach considered damaging comments directed toward McCain to have been the result of a "sour grapes attitude," especially if the source was an officer seeking the task force command. Although not mentioning Sherman, Clark, or John Towers—the deserving flag officer that did relieve him—by name, there is little question as to whom he was referring.

Thach's response to the allegation that Halsey demonstrated his displeasure with McCain in bypassing him on the night of 11 January 1945 during the sortie into the South China Sea invites particular reconsideration. With records from the night of 11 January before him, Thach recounted his version of events. Given the bad weather on that date and that fact that the tankers could not get to all the carriers at once, task group commanders were ordered to take charge and refuel their groups rather than hold to the overall speed of advance of the entire task force. With no advance searches off Indochina in front of the task force to advise what exact targets or threats awaited, Halsey wanted to hold to the plan to attack on 12 January. Consequently, Halsey conferred with McCain and the two agreed to send the first task group that completed fueling ahead with the other two groups to catch up as soon as they could. In sum, Thach stated, "I remember it perfectly, the plan worked out just as discussed." Halsey was in no manner dissatisfied or irritated with McCain, as alleged, during the refueling process and he never took a task group away from McCain.

Thach also took issue with the *Fast Carriers*' assertion that McCain was only Halsey's deputy and never had tactical command in any crucial situation. McCain, he insisted, "had tactical command *all* the time." Thach stated that all *tactical* orders emanated from the task force rather than the fleet commander. Tactical orders originated with McCain who always sent a copy of his orders to Halsey to keep him informed on plans, course direction, strike planning, and all other necessary matters. Considering Halsey "wonderful as a fleet commander, the best, as far as I'm concerned," Thach stated the working relationship between Halsey and McCain was excellent. There were occasions when McCain would have preferred Halsey to do something different, but McCain did not debate issues unless they were significant. Often McCain wrote out long messages to Halsey explaining why a plan—especially target selection—should be changed and "often Halsey would change, once he understood what was in our minds." One matter of special concern was Halsey's propensity to change targets at the last minute. Although the flexibility attendant with operating by dispatch rather than meticulous plans was fortuitous, McCain and Thach realized that constantly changing plans late on the night prior to an attack was having a negative effect on pilots. McCain and Thach visited Halsey and asked him not to make such changes unless a target change was directed to meet an unexpected threat because pilots had to stay up later and get up earlier when plans changed. Halsey agreed and ordered his staff to act accordingly.

As to the matter of McCain allegedly not ever having tactical command in a crucial situation, Thach stated that McCain had tactical command of escorting the damaged *Canberra* and *Houston*, both under tow, away from Formosa. Under attack and highly vulnerable for several days after Halsey and most ships retired from the area, McCain was awarded a Navy Cross for his conduct of those operations.

Thach was not the only member of the staff to speak in defense of McCain. Bill Leonard recalled that McCain was quick tempered but was no glory seeker. Leonard found McCain colorful and charming as a person and as a task force commander believed he "was not there long enough to see what he could do."[5]

Through the matter just presented Thach's and Leonard's defense of McCain may alter perceptions to a more favorable remembrance. From this place forward, however, Thach's replies essentially confirm evaluations presented in *Fast Carriers*.

To the allegation that McCain depended heavily on Thach for tactical innovations, the former operations officer attempted to defer credit to his boss but failed. In the two-part article appearing in two July 1945 *Saturday Evening Post* issues, McCain clearly credited Thach for the development of the three-strike blanket system ("constant cap" in the articles). The same article also cited Thach as the staff's "brain truster [sic]."[6] In his oral history Thach described his operations experience in detail sufficient to document that he was heavily relied upon. Thacher Longstreth agreed with others that Thach was the driving force for planning and innovative thinking.[7] Indeed Thach acknowledged that the purpose of the job should have caused a task force commander to rely on him. Thach further stated that he relied heavily on his operations officer later, when he was a task force commander, and that there were far too many details for any one man to handle. McCain gave Thach great latitude, particularly considering that there was no proof Thach's innovations would work and he had the authority to not only originate strikes but also cancel them. Still, Thach stated he "never did anything concerning policy or of great significance without his (McCain's) approval." If weather or other considerations caused Thach to stop a launch, he would "tell the task group commanders to hold everything ... then ... run in and tell Admiral McCain what I'd done."

Thach made an attempt to defend McCain's actions in the typhoon of 4–5 June 1945, but the end result was more a reciting of events rather than a defense. He did not say that experience probably ended McCain's career, but the decision to relieve McCain was made within two weeks of the 16–22 Court of Inquiry and two weeks after that (14 July), McCain received the news that he would be relieved of task force command one month later. His next assignment as the deputy director of the Veterans Administration effectively ended his Navy career.

Thach stated that McCain made proper recommendations during the 17–18 December typhoon. For the June storm, Thach participated in discussions and was involved with the staff aerologist on the track estimate and the recommendation that went to Halsey from McCain. Both Halsey's and McCain's flagships were with Radford's group, which enjoyed relative safety during the typhoon. McCain's recommendation to Halsey was better than what Halsey finally decided, but McCain was cited by the 16–22 Court of Inquiry overseen by Vice Adm. John Hoover, Thach's old nemesis from prewar days, for delaying a reply to Jocko Clark giving him permission to use "his own judgement." This was compounded when Clark delayed giving orders to the ships of his command to chose the best course for their own safety. For anyone who has experienced a typhoon, a minute seems like an hour, and twenty to thirty minutes seems an eternity. Halsey was judged primarily responsible; McCain was secondarily to blame, along with Clark. The court advised McCain's relief (and Halsey's), but Halsey responded with a strong argument that (1) the weather service had failed, (2) the storm made erratic movements, and, most important (3) that orders to serve an entire fleet might not necessarily serve one ship or group due to size of the formation. Halsey got a reprieve; McCain did not.

The closest statement offered by Thach in defense of McCain's actions during the storm was how the task force benefited in the future. The quality of weather reporting immediately improved, and on 13 August, Thach had a plan ready to launch attacks on Tokyo as the storm on 12 August tracked in and curved into Japan. Unlike earlier experiences when storms brought a relief from task force attacks, on 13 August, Task Force 38 attacked right on the heels of the storm to find the enemy not expecting carrier planes so soon. But for McCain, the decision to relieve him was then nearly six weeks past.

THACH ON HALSEY AND THE BATTLE OF LEYTE GULF

Jimmie Thach was unequivocal in his likes and dislikes and very open in discussing controversial issues. Twenty-five years after the Battle of Leyte Gulf, he was well aware that Admiral Halsey had been criticized for pushing north to destroy the four carriers of the Northern Force rather than emphasizing defense of San Bernardino Strait. Despite what had been said and written during that interval, Thach was detailed and emphatic in his support for Halsey's actions.

Whether or not the Northern Force was bait was not a relevant concern to Thach. What was important to him was the potential for any of those carriers to later ferry, shuttle, or launch aircraft at Allied landing sites. Always aware that enemy carriers inherently had the same mobility and striking power of American carriers, Thach believed no opportunity to destroy them should be lost or compromised. Indeed, Thach credited Halsey for looking ahead beyond the immediate day of battle to the problems an existing Japanese carrier force posed both for planning and implementation. The loss of the four carriers to Japan, according to Thach, was far more important than their loss of battleships and cruisers. And as regrettable as it was for the U.S. Navy to lose one escort carrier, two destroyers and one destroyer escort off Samar, the destruction of the four enemy carriers of the Northern Force was much greater compensation.

Thach expressed some regret that Halsey had been not given more credit for ordering McCain's Task Group 38.1 back to Leyte. Halsey knew McCain could arrive there faster than the three task groups with him, and 38.1 in addition to three escort carrier groups should have been enough to handle the Center Force. In sum, had Thach been Halsey at Leyte Gulf in the same circumstances he would have made the same choice as did Halsey.

As for a comparison between Admiral Spruance and Admiral Halsey, Thach did not qualify his opinion on 13 March 1971. "Halsey," he noted, "was wonderful as a fleet commander, the best, as far as I am concerned, and I experienced it under both Spruance and Halsey." Perhaps the most significant of Thach's observations on Halsey's performance at Leyte Gulf was how much worse things would have been in the last year of the war if there had been kamikazes flying from carriers. Certainly, Thach found it difficult enough to blanket enemy land-based aircraft. If kamikazes were also at sea, his problems would have been greatly multiplied and compounded.

THACH ON KING

Admiral King on more than one occasion referred to himself with derogatory words saying that superiors only sent for him and his ilk when there was a dirty job to be done. Most historical treatments of him center on his highly authoritarian but effective leadership of the Navy in World War II. Only miniscule references can be found to indicate he was a man of any warmth. Thach's oral history is a notable exception. More than once he recalls King's fatherly attitude toward him when he made his 1933 experimental seaplane flight to Panama ("Ernie King treated me like a father, like he was my father"). It was King who approved Thach's 1942 Distinguished Service Medal, an extremely high award to so junior an officer, and it was King who spoke openly on sensitive matters to McCain in February 1945 with Thach sitting at his side. It requires no special insight to presume that anyone denigrating King, Halsey, or McCain in Thach's presence after World War II found themselves looking into a very unhappy face that otherwise almost always reflected a smile.

THACH ON MACARTHUR

Thach most likely would never have volunteered for duty with Gen. Douglas MacArthur. Like many others Thach found MacArthur too imperious. Thach was also unhappy on several occasions with MacArthur's tactical decisions, particularly when it came to determining air support, close air support, and the infamous arbitrary demarcation lines separating Army and Navy areas of responsibility. And he certainly did not like MacArthur's manners on one specific occasion.

Not long before the end of the war McCain, with Thach in tow, traveled to Manila to see the damage inside that city's harbor and to talk with MacArthur. Most of the day was to be spent with Commo. W. A. Sullivan who had the job of salvaging sunken ships. However, MacArthur kept McCain waiting nearly an hour beyond the appointed hour. Thach, who before and after made it a point to interrupt a meeting at least long enough to greet someone with another appointment and quickly apologize for running late, was getting more upset with the passing of each minute. After the passage of more time McCain too became upset, especially when the door finally opened and it was obvious that no one else was in MacArthur's office. The lack of consideration within itself was bad enough, but Thach thought it especially egregious that MacArthur would be so discourteous to someone who had been so supportive of him and his troops.

After the meeting, McCain had cooled but Thach had not. After continuous railing, McCain, laughing, finally broke in to say, "You can't afford to let things like that affect your actions."

All the above notwithstanding, Thach offered high praise for MacArthur's "magnificent job" during the 2 September 1945 surrender ceremony aboard *Missouri*. Thach believed MacArthur's speech "professionally written, beautifully delivered... few people... can put across a speech like MacArthur." He saw MacArthur as "a far greater diplomat than he was a tactician or a strategist."

OTHERS ON THACH

There is no question that McCain held Thach in the highest regard. During their period together with the task force, McCain awarded Thach the Silver Star (defense of *Canberra* and *Houston* off Formosa), the Legion of Merit (for development of the blanket system of covering enemy airfields), and the Bronze Star (final raids on Japan). Although Thach was not sure what McCain had in mind, particularly given that each had orders sending him in a different direction, McCain told Thach that he was going to attempt to retain his services in Washington. McCain's sudden death short-circuited whatever he had planned and was working on.

Although McCain would have no doubt asked to have Thach present for the 2 September 1945 surrender ceremony, it was Admiral Halsey who first extended the invitation while persuading McCain to attend. Halsey had approved Thach's request to take Fighting Squadron 3 off their carrier immediately before the war thanks to Thach's cogent oral presentation to Halsey. Certainly, Halsey respected and admired Thach as a fighter pilot, operations planner, and innovator. But it also appears there were many similarities in personality as both enjoyed the company of others, both quickly got to the bottom line of any problem, and both did not delay in making decisions.

Bill Leonard, who flew with Thach at Midway before working for him as his assistant operations officer on McCain's staff, remembers Thach as a man with a keen mind, always sincere and honorable. In short, he was "a natural" for success as a Navy professional and when working with people. Having also worked closely for and with Jimmy Flatley immediately before joining McCain's staff, Leonard stated that he "could not choose between princes" when comparing the two men as pilots, operations officers, or human beings."[8] Thacher Longstreth also greatly admired Thach, and the two corresponded frequently and visited each other after World War II.[9]

12

Hollywood Again, Pensacola, and Unification

In September 1945, Captain Thach was back in California and once again briefly working with Hollywood. Despite a strike that kept studio employees from working at their usual offices, Thach and writer-director Delmar Daves secluded themselves at Daves's home to work out the screenplay for a film entitled *Task Force*. At the end of September, Thach and family moved on to Pensacola, where he completed a lengthy typewritten analysis of the screenplay. Considerable combat footage from the recent war was available, and few veterans other than Thach were in better position to advise on its use.

HOLLYWOOD AND *TASK FORCE*

In his comments to Delmar Daves, Thach asked for a complete rewrite for the Battle of Midway, suggested the disaster of Pearl Harbor was overemphasized, and corrected numerous technical mistakes, especially the use of shipboard equipment. Personal relations did not escape his attention either. Thach suggested a more sympathetic presentation of flag officers who were slow to appreciate the power of naval aviation and even proposed significant changes for the romantic interests in the film.[1]

In 1949, the film was released to the public. Thach was pleased with Gary Cooper's portrayal of a character loosely based on him. Walter Brennen's portrayal of the task force commander was a combination of Mitscher and McCain, and Thach asked that the closing scenes of the movie note that carriers had replaced battleships as the heart of the U.S. Navy. Numerous aircraft carrier movies have appeared over the years and many critics have judged them good or bad with a surprising number noting that little known *Task Force* was the standard for comparison.

TRAINING COMMAND AT PENSACOLA

In late September 1945, Jimmie, Madalyn, and the boys arrived in Pensacola for what would be a four-year stay. Thach's assignment was as director of training, staff, chief of naval air training working for Vice Adm. Frank Wagner. Pensacola was still, and would remain, headquarters of the Naval Air Training Command. Pensacola and Corpus Christi were retained for all flight training, except advanced training took place at Jacksonville. Jimmy Flatley held the same title at Corpus Christi as Thach had at Pensacola for the first year back in training command. Together the two former fighter pilots and operations officers teamed to make necessary changes in postwar naval aviation training.

Flatley was particularly anxious to rid training command of the biplane because that type plane was not utilized in operational squadrons. Thach agreed. And the two officers were not favorably disposed to the three-plane formation being used by the Blue Angels. On 24 April 1946, Fleet Admiral Nimitz, then chief of naval operations, ordered Vice Admiral Wagner to organize a Navy flight exhibition team within the Naval Air Advanced Training Command. The Blue Angels (the name, inspired by a New York nightclub, was adopted soon after) were initially led by Lt. Cdr. Roy "Butch" Voris, one of Flatley's VF-10 Grim Reapers from the Guadalcanal era. In the team's first shows, three Hellcats chased an SNJ painted to look like a Japanese Zero with the SNJ releasing smoke and diving as in a crash. Flatley and Thach reminded Wagner that the basic fighter formation was still four planes in two sections. Consequently, the Blues were instructed to develop a routine using four planes and to add the Thach Weave to their program. Into the twenty-first century, the Blues still use the four-plane diamond formation.

Just before Thach left his training billet in 1950, he was encouraged by Lt. Cdr. John Magda to check out in one of the Blue Angels new jet F9F Panthers.[2] Magda did not have to twist Thach's arm very hard, and after taking the new jet aloft at Whiting Field, an exhilarated Thach wrote Jake Swirbul to inform him that the Panther was even better than expected. Still director of training at the time, Thach was particularly impressed with the Panther's stall warning characteristics, noting that "it not only shakes all over long before stalling, but wags the tail" thereby alerting even the least inexperienced pilots that their speed was too slow.[3] Walking away from a smooth landing, Thach could not help but recall his meeting in Hawaii with Swirbul in the early summer of 1942, when they discussed the then conceptual F8F Bearcat. In 1950, Thach was pleased that Grumman had again produced another winner. And once again they had produced a plane with a drop nose allowing for maximum visibility.[4]

Thach particularly enjoyed being back in training command. His interest in teaching aside, there was another reason he valued the assignment. "Early in the last war it was brought home to me that it is the untrained who are needlessly lost in combat, and who contribute little or nothing to the cause in which they are lost," he noted. "I would rather have twelve properly trained aviators with me to fight twenty planes of an enemy than to have twice or three times that number of half trained or poorly trained pilots."[5] No other statement could more adequately describe why Thach was so pleased to be in the training command.

During his four years at Pensacola, Thach was confronted by one overriding concern, that being the fluctuation of the number of pilots training command was ordered to produce. Throughout the period all the military forces were undergoing a massive reduction in force, but variations in the federal budget impacted the flow of recruits into the naval aviation training system with a consequent imbalance in the number of pilots produced. The ordered alternating increases and decreases left the training pipeline with bubbles. As it took two years to produce a fully trained aviator, the Navy had either too many or too few pilots from 1946 into 1950. Safety became a major problem when too many pilots were unable to obtain sufficient flying time and skills eroded from sitting idle. Thach often found himself writing letters during the period trying to explain that a budget reduction would cause one of the bubbles to have its greatest impact two years later.

With supply and demand for naval aviators varying and auxiliary fields even at Pensacola closing and then reopening, it was apparent that the Navy needed to make the best decisions possible for selecting pilot recruits. Thach was therefore heavily involved in helping develop a battery of mental and physical tests that would provide a correlation with success or failure after being used over a period of years. Of course he would be long gone from training command before the performance of graduates was known and the value of the tests determined. But for the short term he was pleased to see that his old training films were still in use not only in naval aviation but also in the U.S. Army Air Force.

Films seemed never to be very far from Thach's attention while at Pensacola. Again he was instrumental in helping produce a new training motion picture titled "The Naval Aviator." Needing scenes of jet operations from a carrier he wrote to friend Jimmy Flatley, then with the Atlantic Fleet at NAS Norfolk, to seek his assistance. The training carrier at Pensacola did not operate like one of the *Midway*'s based at Norfolk—plus jets did not arrive at Pensacola until the summer of 1949—and he knew Flatley understood "the situation and will help us all you can on this deal."[6] From the end of World War II until Flatley's untimely death in 1958, the two old friends constantly called upon each other for assistance. Often one or the other piloted a plane to pick up the other to head off to meetings and conferences. During the unification period, the two were together in person, on the phone, or via correspondence nearly as much as they had been during their days as alternating task force operations officers. To both, the unification threat to naval aviation was as important as the Pacific war. Interestingly, letters, pamphlets, and associated papers claim more space in the extant files for both than for any other period of their respective careers.

UNIFICATION AND THE REVOLT OF THE ADMIRALS

By 1947, it was apparent to most in the United States that Communists in the Soviet Union intended to make good on their promise to bring the entire world into their socialist utopia. A figurative iron curtain had descended in Central Europe immediately after World War II, and by 1947 a physical barrier was forming as well. But while the Soviet Union was recognized as the threat it was, demobilization continued throughout the U.S. military establishment. In late 1947, most of the carriers Thach had flown from during World War II were in mothballs or gone altogether.

The idea of armed forces unification existed long before 1947, particularly as it related to a single aviation branch. In 1925, the report of the Morrow board—the president's aircraft advisory board—and the House of Representatives' Lampert committee addressed unification of military aviation and rejected it. On 6 December 1928, Rear Adm. J. M. Reeves wrote a long letter to the chief of naval operations declaring that a Department of National Defense with an Army, Navy, and "an additional department, comprising the air forces of the United States . . . [that] are supposed to be 'co-equal'" was "unsound."[7] Other letters similar in essence and tone followed throughout World War II, but on 26 July 1947, the National Security Act created the Department of Defense and established the United States Air Force, effective 18 September 1947.

After passage of the National Security Act, naval aviation found itself in a battle for its existence. The Navy saw unification as domination, given that the Air Force could well take already diminished federal budget funds and naval aviation might even become administratively subordinated to the Air Force. Though now officially blessed, the new U.S. Air Force continued a well-organized public relations campaign with the goal of securing maximum benefit from a shrinking military budget. It appeared only logical to the Air Force that funds should be reduced for the Navy, whose ships were allegedly obsolete due to the promise of nuclear weapons. Whether or not its case was sound, the Air Force was on the offensive, and it had the known support of some congressmen, many media figures, and several renowned aviation authorities. The Navy had a friend in the first secretary of defense, James V. Forrestal, but beyond him they had to look within. Thach did not volunteer to become a significant figure in the Navy's counter-offensive, but neither did he have to be asked twice to join the battle.

From 1947 to the end of his stay at Pensacola, Thach remained director of training but also picked up the additional title and responsibility of special assistant to the chief of naval air training. In the capacity of special assistant, he had the collateral duty to help educate the public why naval aviation was needed. Although he noted to his new boss in 1947, Vice Adm. John W. "Black Jack" Reeves Jr., that his responsibilities as director of training were suffering from the new assignment, Reeves left no doubt that Thach's primary duty was helping save naval aviation. So important was this assignment that Thach's orders to command USS *Salisbury Sound* (AV-13) in September 1948 were cancelled at Reeves's request by Vice Adm. J. D. Price, deputy chief of naval operations (air). Price wrote that "because of the urgent need for the services of Captain Thach on a special assignment which he is especially qualified to carry out," his orders were canceled "at my request." Price continued, "I regard this as evidence of special confidence in the outstanding ability of Captain Thach, and especially in his qualifications for high command."[8]

Thach had seen an example of the debate to come in the summer of 1942 when an article, "The New Queen of the Seas," lauding aircraft carriers appeared in *National Geographic*. The author had sent Thach an autographed copy and wrote that "the only letters of criticism that I have received have been several from aviation fanatics who claim that . . . long range bombers . . . will win the war and that carriers are already obsolete. Strangely, I have not had one letter of complaint from battleship men."[9]

The challenge was clear. The U.S. Air Force argument was that naval aviation was "wasteful duplication." The fear, not paranoia, within naval aviation was that it would either be disestablished or incorporated into the new Air Force. Great Britain and Italy had set precedent by combining their aviation services before World War II. Thach, however, had even more recent experience with the problems posed by having the air arm separate from the Navy. When the British carriers joined Task Force 38 in July 1945, Thach was present for the first meeting aboard Halsey's flagship. After the admirals had spoken, Thach joined his counterpart to discuss tactical procedures and instructions. Thach was pleased that the British operations officer wanted to adopt all policies then in effect with the task force. Thach suggested they start slow and then as they became accustomed to daily practices, they could operate at faster speeds in launching and cover a greater range of targets. But, the British never got up to speed with the result that their four carriers never had more impact than one American carrier. Frustrated at the slow progress, or lack of progress, Thach climbed into a Corsair on *Shangri-La* and flew over to the British flagship to see if their lack of progress was an insistence on having tea in the morning and afternoon or whether there was some other problem. The British operations officer blamed the ship's captain for not understanding his needs while the carrier commander blamed the air group for not understanding the carrier's needs. That experience provided Thach with more than a philosophical argument to offer in his multitude of written and oral education presentations from 1947 into 1950. Having naval aviation incorporated into the Air Force simply would not work, and if the Navy lost funding for aircraft carriers, the entire Navy would be vulnerable to enemy aerial attacks.

The Air Force "psychological warfare against the Navy" continued into 1948 and intensified with misstatements by officials that, seemingly, went unnoticed. A prime example was Maj. Gen. Curtis LeMay's comment that a B-36 bomber could fly too high for radar, this even though the moon registered on radar. A major assist to the Air Force occurred on 24 June 1948, when the Berlin Airlift began in response to the Soviet closure of roads around that city and continued until 12 May 1949. Television sets were just becoming available to the general public in the late 1940s and through that medium, newsreels routinely shown at local movie theaters, and constant coverage in the printed media highlighted Air Force planes making daily landings in the besieged city. Making matters worse for the Navy, Secretary of Defense Forrestal resigned in March 1949 and died two months later.

Prior to late 1947, Thach spent most of his time in Pensacola either working directly with training or hosting groups of influential citizens from across the country. Occasionally he was able to fly a Navy transport to pick up visitors, getting in some flying time in addition to assisting with the public relations effort. While not with the visitors for all their time at NAS Pensacola, he usually escorted them to the training carrier (*Saipan* CVL-48, *Wright* CVL-49, and, later, *Cabot* during his tenure) for an overnight cruise and to watch student pilots launch and land Hellcats, Corsairs, Avengers, and SNJ trainers. Before their departure, he usually served as toastmaster for their dinner at the Mustin Beach Officers Club. The hoped-for result of these visits was that the visitors would return home, talk to others, and write to tell their congressmen that the Navy was understaffed, undertrained, and not ready for any eventuality. And it was hoped that the visitors would also express appreciation for the Navy's hospitality to encourage any young men with two years of college to investigate the opportunities of naval aviation.

Beginning in 1947, Vice Admiral Reeves began sending Thach across the country to make speeches to individuals and groups of influential citizens who could help in the fight to save naval aviation. An early 1948 visit to the Pentagon was at the invitation of old friend Arthur Radford, then a vice admiral and deputy CNO (air). Impressed with Thach's charts and presentation, Radford held him there longer than expected before letting him return to Pensacola. Back in Florida, Thach barely had time to say hello to his family before Reeves had multiple appointments set up for him. As the months wore on, Thach's patience wore thin. He believed heart and soul in the message he was carrying to newspaper editors, politicians, Rotary clubs, and anyone else who could help, but the effort was taking a significant toll on him. He was particularly upset with Reeves calling him on a Saturday night to tell him to be in Chicago or some other distant locale on Monday for one or more meetings. A call on Saturday night not only ruined that evening but also Sunday because he had to prepare for the meetings and pack.

Not all the trips and communications were hardships, especially visits with media leaders who knew as well as Thach that the national security debate would be tried not only in congress but also in the press. He was very pleased with the even-handed treatment rendered by Hanson W. Baldwin of the *New York Times*. Before visiting with him, Thach wrote Baldwin to say that his editorial describing "the dangerously misinformed thinking in the halls of Congress with respect to national security needs, did my heart good."[10] Conversations with William Randolph Hearst Jr. were enjoyed both at the professional and personal level.[11] Thach was particularly pleased to assist Lloyd Wendt of the *Chicago Tribune* with articles on his old Fighting Squadron 3, the Battle of Midway, and Butch O'Hare. Assistance on articles about Butch was needed in 1949, as that was the year discussions centered on naming the large Chicago airport in Butch's memory (Chicago O'Hare International). And it did not hurt to have Wendt to assist in reminding the *Tribune*'s publisher, Col. Robert R. McCormick, that Thach was Butch's squadron leader and that the colonel could help Thach save naval aviation.[12] Old friend and former task force intelligence officer Thacher Longstreth was also in good position to help from his job in the advertising department with *Life* despite being located in Detroit near automobile factories that bought substantial advertising space in the magazine. Indeed Longstreth and Thach traded favors several times as Longstreth needed Thach to speak to his reserve station at NAS, Grosse Ile, Michigan, while each stayed in the other's home when in Michigan or Florida.[13] And even Madalyn got into the fray. Just as talented as Jimmie in placing her thoughts on paper, she especially concentrated on writing long thank-you letters to editors and publishers when their articles favored the Navy's position or offered a balanced perspective on the national security debate. If articles were not favorable or balanced, her use of the English language was even more direct than that of her husband; phrases such as "soap-opera tactics" and "phony bill of goods" found expression in her epistles.[14] Doubtless her husband encouraged her not to drop the pen and lauded her literary endeavors in his own manner. Too, he no doubt wished he could use some of Madalyn's more direct phrases in his own correspondence. But both always made cogent arguments with ample documentation in their respective written documents.

After many months of Reeves keeping Thach on a string at long distance, Thach marched in to tell his boss that he was over three thousand dollars poorer since becoming Black Jack's special assistant. Per diem usually paid only the cost of meals while Thach was away from Pensacola and did not begin to cover hotel bills. Smiling, Reeves replied he had wondered how long it would take before Thach came in to complain. Upon reflection Thach realized Reeves was doing what he had to do and that if Thach had been sitting in that chair, he would have done the same thing. Reeves, Thach, and many others in naval aviation were in pain, but the potential result of the alternative was even more painful.

All came to a head in the spring of 1949. The proposed federal budget contained alarming cuts for the Navy. The new Secretary of Defense, Louis A. Johnson, was a friend of smaller allocations for the federal budget but was no friend of the Navy. He had bought into the concept that a future war would be nuclear and that only the Air Force had planes large enough to carry a 10,000-pound atom bomb. The Navy's attempts late in 1949 (October) to demonstrate that the P2V-3C Neptune and the new AJ Savage could carry such a weapon from Midway-class carriers was interpreted by him to be a duplication of effort rather than a more diversified attack mode. A new aircraft carrier design had jumped from the drawing board and construction began on 18 April 1949. Named *United States* (CV-58), the new carrier would be able to launch large heavy jets that could carry nuclear weapons, but Johnson cancelled construction without consulting either the chief of naval operations, Adm. Louis E. Denfield, or the secretary of the navy, John L. Sullivan. Enraged, Sullivan resigned and a political friend of Johnson's, Francis P. Matthews, was appointed. Without a knowledgeable background of the Navy and with his loyalty assigned to Johnson, the Navy's doom appeared imminent. With their backs to the wall, the Navy called on their best minds to testify before Congress in August 1949 hearings.

The congressional hearings arrived none too soon. On 2 May, Thach received a mild reprimand from an old friend, Rear Adm. E. C. Ewen from the Chief of Public Relations Office, that a speech Thach made at Rensselaer Polytechnic Institute was not cleared by the Secretary of Defense Security and Review Board. Ewen, however, went on to say, "It is extremely difficult at the moment to get clearance on anything but the most innocuous type of presentation. These rules, as you can readily understand, are not of my choosing but I simply have to play ball."[15]

The Armed Services Committee heard testimony from 9 to 25 August and then again beginning 6 October after considerable acrimonious debate in the national press heightened by revelations by Capt. John Crommelin, an older brother of Richard who had flown with Thach at Midway. By October, the turmoil and debate surrounding the Department of Defense took on even more ominous meaning when it became known in the United States the previous

Hollywood Again, Pensacola, and Unification

month that the Soviet Union exploded its first atomic bomb in August. And on 1 October, the Chinese communists proclaimed The Peoples Republic of China as the remaining nationalists prepared to move off the mainland to Taiwan (Formosa). The full significance of these momentous events was not fully appreciated by Americans on either side of the national security debate, but the importance of the debate was better appreciated.

The October hearings centered on whether the U.S. Air Force B-36 bomber could assure national security in view of the cancellation of the super carrier *United States*. Of course the leaders of naval aviation knew that both the question and the answers were far more complicated than just comparing the B-36 and *United States* and they prepared their responses accordingly. Among the active duty speakers for naval aviation were Vice Admiral Radford, Capt. Arleigh Burke, Thach, and Bill Leonard. Retired luminaries participating either in person or by statement included Fleet Admirals King, Nimitz, and Halsey plus Admiral's Spruance and Kinkaid. Among points hammered directly or indirectly by all was that the one weapon answer to national security meant that the B-36 bomber would have to fly for many hours through countless waves of faster enemy fighters before arriving over a target deep in the heart of industrial Russia. Other issues were that Russia had no aircraft carrier force, that the Russians had three hundred submarines that could be effectively countered only by a Navy, and that any land target was easier to hit than a mobile naval task force. Perhaps the strongest argument was that nearly three-quarters of the world was water and to get any military force and supplies to Europe required a Navy with sufficient aircraft to protect surface vessels. Naval officers knew all this. Whether Air Force officers knew it or understood it was not really the question. The question was whether Congress and the taxpaying general public knew it.

Admiral Radford, who before taking command of the Pacific Fleet in May 1949 had set up the task force to defend the Navy and a special office under Arleigh Burke to assist with public relations, insisted that Thach make his educational presentation before the congressional committee. After briefly introducing himself, Thach delivered a lengthy oration on naval air power. The essentials of his presentation are presented below. Though spoken in 1949, what Thach had to say then is almost totally applicable to the first war in the twenty-first century some fifty-four years later:

> It is not necessary to be a technical expert in order to understand the basic principles of aerial warfare.

The basic principles do not change. On the contrary, when dealing with air power, flagrant violation of these principles brings quick and certain military disaster.

Our country's military forces should be designed to exploit to the fullest extent the principles of mobility, flexibility, versatility, economy of forces, concentration of forces, precision, and surprise. A fast carrier task force has each of these principles inherently built into it.

A part of the grand strategy of carrier air power in the Pacific was the principle of forcing aerial combat upon the enemy at the times and places of our own choosing with the idea of depleting his total pilot strength to the extent that we could fly bombers and other types of aircraft over his home territory with little or no fighter escort. This objective was achieved.

Carrier-based air power can destroy enemy atomic bombers over the ocean or on the ground at their advance bases.

The very threat of surprise attack at any time from any direction by mobile air power forces an enemy to spread his defenses thinly around his chosen perimeter.

Command of the air is the heart of air power and the high performance fighter is the key to command of the air.

Carrier air power is the only way we can gain command of the air over beachheads before land air bases are built or captured within fighter range.

Our fast carrier task forces with their tremendous advantages of mobility can take the fighter to the ringside where he can climb in and do the job.

The aircraft carrier multiplies by 25 the bomb load that can be delivered (intercontinental) per pound of aircraft weight in a given length of time.

Because of the great number of atomic bombs that an enemy would have to expend to damage or destroy one carrier, the fast carrier task force is not a profitable target for atomic bombs.

The Navy is not a single purpose force. It has many missions and uses naval air power in accomplishing each one of them.

A strong Navy is an absolute essential to the successful performance of the other military services.

Naval air power is our one clear-cut advantage over any probable enemy.

Because no probable enemy has naval air power it is our greatest assurance that we can fight an aggressor on the other side of the world and not in the United States.[16]

The result of the October testimonials was a feeling within the Navy that they had won the debate but might still lose their war. President Truman fired CNO Denfield on 27 October upon Johnson's recommendation because the

CNO, who was last to speak, added his voice and support to all the sailors who preceded him. The reason officially offered for the firing was that the CNO was obligated to give his loyalty to Secretary of the Navy Matthews. In a letter to Truman, Matthews had written, "There can be no twilight zone in the measure of loyalty to superiors and respect for authority existing between various official ranks."[17] Newspapers announcing Denfield's firing also carried stories in the same issues that the Navy and Marine Corps manpower would be cut by 54,891 by 1 July 1950.[18] It can be deduced from the above that the question of who or what deserved loyalty was on every naval officer's mind.

Another officer singled out for retribution for helping spearhead the Navy's public relations effort against the Air Force was then Capt. Arleigh Burke. Working directly for Vice Admiral Radford, Burke's OP-23 office was specifically charged to counter the Air Force effort. His reward from Matthews was to have his name on the selection list to rear admiral be crossed through in red ink, an illegal action law school trained Matthews failed to appreciate. Writing to a friend, Thach stated, "I have known Arleigh Burke for a long time and ... he is ... the best material for Flag Rank that could possibly be found." Continuing Thach wrote, "there is no legitimate reason for passing him over. It could be nothing but an underhanded political purge. It is evidence that the Armed Forces are being bludgeoned into a path leading toward complete thought control and resulting in a militarily weak political organization instead of a capable military force."[19] Indeed, the action was so egregious that President Truman refused to return the selection list to Johnson and Matthews until they saw the light and returned Burke's name to the list. They quickly saw the light.

Despite all the concern, fear, and frustration during the unification debate and the "revolt of the admirals," naval aviation not only survived but thrived. In June 1950, North Korea invaded South Korea and carriers quickly came out of mothballs for another war. Three months later, in September, Johnson was replaced, and on 30 October a new supercarrier (CV-59) was laid down and named for James V. Forrestal, a former secretary of the navy and the first secretary of defense. In mid-June 1950, Thach had reason to be as pleased within for his role in defending naval aviation during the unification process as he had for any of his World War II combat service. In mid-June 1950, however, he did not know his combat days were not over.

13

Korean War Combat on USS *Sicily*

After World War II, the great reduction in force applied not only to personnel and aircraft carriers but also to ships of all types. By the summer of 1950, manpower was a major problem for the U.S. Navy, as there were insufficient numbers to keep ships fully manned and adequately maintained. Deployments were missed because ships were not ready for sea and safety problems were pervasive throughout the fleet. Naval aviation had forty-one thousand planes in 1945 but only fourteen thousand in 1950. Early in that year, fifteen carriers were in commission, but only one, *Valley Forge* (CV-45), was in the Far East. Three other *Essex*-class carriers were on the West Coast, along with one escort carrier and another en route from the Atlantic.

PREPARATION FOR A NEW COMMAND

The Navy-wide problems with inadequate personnel and training would have affected Jimmie Thach even had Vice Admirals Reeves and Price not retained him at Pensacola in late 1948. While Thach was anxious to have a ship command, the decision to keep him at Pensacola turned out to be a blessing in disguise. The officer who then received orders to command the seaplane tender *Salisbury Sound* declined because it was not in shape to turn over to a new command. It was standard procedure for a prospective commanding officer to inspect a ship and

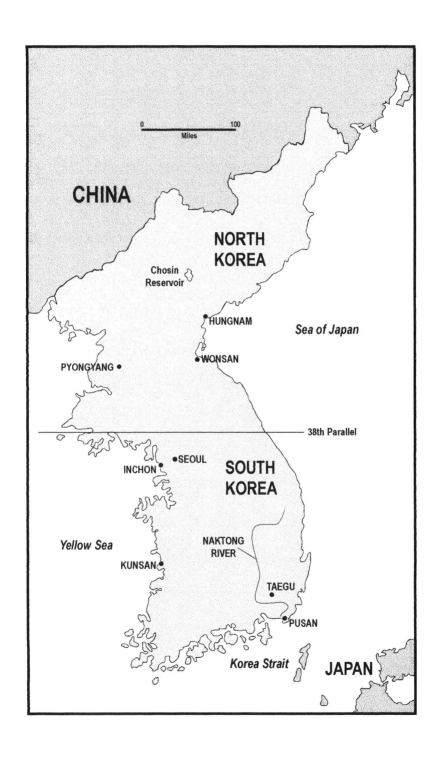

observe the crew at various drills. If he deemed the ship unacceptable, it was both his right and duty to refuse command and request an investigation. But the problems with AV-13 became moot when Reeves and Price succeeded in keeping Thach from going to sea in 1948.

Although the congressional hearings that pitted the Navy and Air Force against each other in the fall of 1949 were over, Thach and other naval officers were still busy into early 1950 attempting to influence the public and their congressmen not to make deep cuts in the Navy and Marine Corps budget. In late February 1950, Thach learned he was slated to get command of one of the antisubmarine carriers in the Pacific. To qualify for that command, however, he had to attend the Sonar School at Key West, Florida. The next course began in late March, so he had to start canceling speeches and let others know they would have to continue the public relations battle without him.[1]

Though scheduled to report to Naval Operating Base Key West on 27 March, Thach nonetheless still had his thoughts on the federal budget and the U.S. Air Force public relations that continued "sowing of the seeds of contempt for a sister service."[2] Top priority was to forward a mailing list of publishers to about a hundred Navy reserve officers throughout the country that in civilian life were in the public relations business.[3] Mental time not devoted to the continuing worry about interservice problems was devoted to studying for his impending antisubmarine warfare assignment. Chief among his concerns was a report indicating that the German-inspired Russian Mark 21 submarine had a top underwater speed just over seventeen knots.[4] That was nearly as fast as the escort carrier he would be commanding. Still, however, his major concern was that the Soviets had three hundred submarines.

Taking command of a warship was no small challenge for anyone. Not given to leaving anything to chance, Thach sought the opinions of other officers in regard to courses other than the required Sonar School that he should attend. Consensus was that the Combat Information Center course and then Fire Fighting and Damage Control should follow the four-week Sonar School.[5] For the time being he did not know whether he would take command of the *Sicily* (CVE-118) or *Mindoro* (CVE-120), and part of that equation was how long it would take him to complete the Sonar School. The sooner he completed that school, the sooner Washington could cut his orders. Thach then wrote the Fleet Sonar School's commanding officer to find out if the course could be successfully completed in three weeks.[6] The reply was that given the absence of previous antisubmarine experience, he should allow four weeks, including the usually skipped air-surface coordination segment.[7]

On 22 April, Thach completed his course work and the seagoing phase at Key West, believing he had learned the essentials of his new command. However, he was not totally satisfied with everything he experienced, his "assumption that the eleven odd different Commands involved in ASW are suffering from lack of a strong organizational tie-up." Deducing that the cooperation among antisubmarine warfare (ASW) commands was primarily due to personal relations, Thach outlined several needed changes to the Navy Department. Advocating a fleet command billet under a vice admiral reporting directly to the CNO, he pointed out, using several examples, how the current organizational structure was impeding implementation of technical advances. Among the examples was the UOL (underwater object locator), a short-range instrument that provided high resolution of a submerged object that was then subject to a tug-of-war between the Bureau of Ships and Bureau of Ordnance. Another problem was the failure of submarine skippers and pilots to appear for the sea phase critique at the conclusion of the Sonar School course. The equipment stalemate, training problems, and other similar problems could be quickly resolved if one commander had the authority and charge to increase the efficiency of the ASW program.[8] Thach's letter to the Navy Department pertaining to the ASW program was similar to those forwarded during World War II and for the remainder of his career. Never did he offer a complaint without also offering a detailed explanation of what was wrong and exactly what was required for correction or improvement.

On 30 April 1950, Thach was detached from Pensacola and on 22 May, he reported to the CIC School at Point Loma. That course completed he then traveled to Treasure Island at San Francisco for Damage Control School on 5 June, and a week later he was in San Diego to assume command of his first ship.[9] The assignment was welcome in more ways than one. San Diego was Madalyn's home, and for the first few days all four members of the Thach family stayed with her parents as they had done during and between earlier assignments. As the San Diego area would be the place of their retirement, the Thaches looked for a house in Coronado and soon found one at 721 Eighth Street. Reading about himself in the newspaper, Thach wondered if he would ever be remembered for anything other than inventing the Thach Weave. Seemingly every press release he saw concerning him referred to his 1942 innovation even more than the usual acknowledgment of him being a "famed wartime fighter pilot" and McCain's operations officer.[10]

COMMANDING OFFICER, USS *SICILY*

When Thach saluted the flag and stepped aboard his new ship at North Island in early June, Capt. Clifford H. Duerfeldt and other ship officers assured him that *Sicily* was the happiest ship in the fleet. To prove their point, they suggested Thach ask any member of the crew. That proved unnecessary because it was quickly apparent that the four-year-old escort carrier was indeed a happy and effective ship. Fresh from overhaul at the Boston Naval Shipyard, *Sicily* had been with the Atlantic Fleet since 1946 before arriving at San Diego on 28 April. At 10,900 standard tons, the 557-foot *Commencement Bay*–class escort carrier could reach a top speed of only nineteen knots, but that was then fast enough for ASW duty. Tankers inspired the *Commencement Bay* escort carriers' design and they therefore had exceptional range due to their ample fuel storage capacity and could easily refuel the destroyers that operated with them. At slow speed in a harbor, Thach thought the carrier had a tendency to be capricious and whimsical and sometimes balky. And once the anchor broke loose from the bottom, the ship seemed to want to turn to port. But in rough weather *Sicily* had little roll or pitch, and it sailed smoothly. Still, there were rare occasions when Thach had to use the pinwheel maneuver to dock the carrier, especially in the narrow confines of the harbor at Yokosuka, Japan. With planes on the starboard and port bows, the starboard and port quarters (with their tails facing outward), and, of course, tied down to the flight deck, the wind driven by the planes' propellers effectively moved the ship alongside the pier.

As with all aircraft carriers, its main battery was its aircraft. The two primary antisubmarine aircraft Thach expected to come aboard in late 1950 or early 1951 were the new Grumman Guardians. These single-engine planes would supplant the older TBM Avengers but operate essentially the same. One plane carried the equipment to find an enemy sub; the other carried the ordnance to destroy it.

Having an opportunity to go to sea aboard the escort carrier a couple of days before assuming command, Thach had an opportunity to gain an early acquaintance with both the ship and crew. The unofficial orientation and official inspection left Thach feeling very good about both the ship and crew thereby providing him a positive frame of mind when at 1000 hours on 15 June he assumed command. For the following ten days, he prepared for the scheduled antisubmarine exercises and was totally absorbed in that until late on 25 June, when the very disturbing news shot around the world that North Korea had crossed the thirty-eighth parallel and was pushing into South Korea.

Sicily was not the only escort carrier based in the Pacific. *Badoeng Strait* (CVE-116), another *Commencement Bay*–class escort carrier nicknamed "Bing-Ding," combined with *Sicily* to form Carrier Division 15 under the command of Rear Adm. Richard W. Ruble. Neither Ruble nor his boss, Vice Adm. Thomas L. Sprague, commander Naval Air Pacific, were strangers to Thach. But there was little time for the three to discuss World War II days with the new crisis on their hands.

WAR IN KOREA

Victory by Japan over the Chinese in the 1894–95 Sino-Japanese War ended centuries of Chinese influence over Korea. After the Russo-Japanese War of 1904–5, the victorious Japanese announced a protectorate of Korea followed by annexation in 1910. The harsh protectorate continued until Japan's defeat in 1945, when Korea was divided into a northern zone occupied by the Soviets and a southern zone occupied by the United States. The U.S. government and general public focus on Europe and the Soviet Union in the late 1940s left both largely surprised when the Communists took over China in late 1949. With two strong Communist allies at its back, North Korea desired to unify the country by force, but both China and the Soviet Union declined permission or support. That changed, however, when Secretary of State Dean Acheson publicly announced on 12 January 1950 that the United States would defend Japan, the Philippines, and Okinawa from Communist expansion but made no mention of South Korea (or Taiwan). In retrospect Thach and many others viewed that omission as an invitation for the Communists to move into South Korea. Since the end of World War II, expectation of all parties had been that the country would eventually be reunified. But the Communist dictatorship in the north and democratic south led the respective divisions in two opposite and increasingly divergent directions, especially after the Republic of South Korea was formally established on 15 August 1948 and the Democratic People's Republic of Korea (North) in September 1948.

Upon hearing the news of the invasion on 25 June, Thach remembered his recent conversation with Col. McCormick in his Chicago office regarding statements indicating that the United States' sphere of interest did not extend into Korea; they were, he believed, an invitation for trouble. And he wondered why General MacArthur, who had been caught off guard in the Philippines in 1941, would be caught off guard again in Korea. Intelligence indications were that the North Koreans were assembling troops just north of the thirty-eighth parallel and infiltrations had recently increased. In sum, Thach figured that the Communists had taken heart not only because of Acheson's diplomatic faux pas but also because of the severe reduction in force throughout the U.S. military.

TO PEARL HARBOR, GUAM, AND JAPAN

From the late afternoon of 25 June, thoughts of diplomatic and political oversights were relegated to distant recesses of the brain because there was too much to do to prepare *Sicily* for sea. On the twenty-seventh, Sprague and Ruble asked Thach how soon he could get under way with both the flight deck and hangar full of aircraft and supplies. The conversation ended with agreement that *Sicily* would be ready to steam from San Diego on 4 July and *Badoeng Strait* would follow when ready. Happily, *Sicily*'s supply officer, who was getting his information from local newspapers, got a running start on loading supplies the carrier and crew would need. But in the rush to respond to the crisis and to orders from Washington, the *Sicily* would have to sail without two necessities. First, because of the extra aircraft and supplies, the carrier would not be able to defend against any submarine threats. Second, there was no time to line up the usual destroyer escort. This not only hampered its defense against submarines but also disallowed air operations for lack of a plane guard. Given the emergency, however, these normal safeguards were set aside.

Although this was Thach's first time in command of a ship, the usual challenges of taking a ship to sea for the first time were the least of his new worries. Not helping his concerns about possible submarine contacts were constant pips on radar that did not seem to make sense. Closer investigation revealed that the radar operator had tuned the equipment so thoroughly that the carrier's radar was picking up birds that had swallowed pieces of metal. The decision on the bridge was to adjust the radar for less sensitive readings.

Although he had experienced combat before, there were many unknowns in this new war. Newspapers referred to the fighting as a crisis and, later, a police action, but to Thach it was a war. Attempting to learn all he could about what was happening on the Korean peninsula, he ordered his communications officer to decipher all the traffic his codes allowed. News therefore became available, and the more that came in, the worse the situation seemed to get. It was apparent that the South Koreans were in retreat and the only question seemed to be how far south they would be pushed. But for the time being, all Thach could do for his part in the war was to think about encountering a hostile submarine as no one was sure what actions the Soviet Union and China might take. Without planes and escorting destroyers, his only option was to outrun subs and evade their torpedoes, and he thought he could do that thanks to the exceptional range of the sonar on *Sicily*. Destroyers had the same equipment, but they operated with a more shallow depth than the *Commencement Bay*–class carrier with its sonar dome nearly five feet beneath the keel of the thirty-foot-draft ship. The result was that the more stable carrier could receive long-range readings under the surface thermal layer.

Arriving at Pearl Harbor, Thach received a thorough briefing while some supplies and aircraft were unloaded and other planes lifted aboard. Although he was aware the news from Korea was bad, at Pearl he learned the logistics problem was monumental. The central questions were (1) where was the equipment required to stem the North Korean tide? (2) how soon could it be delivered? and (3) could sufficient men and materiel arrive in time to stabilize the rapidly receding front? North Korea had a tremendous advantage of having relatively short supply lines in addition to the advantages of the first initiative plus tanks that the South Koreans lacked in quantity. Calling on old friend Arthur Radford, then a four-star admiral and commander in chief, U.S. Pacific Fleet, Thach volunteered the idea that if there appeared to be no serious submarine threat that the *Sicily* might take on a Marine fighter squadron for close air support. Smiling, Radford asked if Thach had been reading his mind. Radford had already considered that option and told Thach that by the time he arrived at Guam a decision would be made regarding a change in *Sicily*'s mission.

Under way from Pearl Harbor in less than a day and steaming toward Guam, Thach was pleased to know there was a possibility of taking his ship into combat. Unquestionably, U.S. Navy submarines were already heading to designated positions off Taiwan, China, Japan, and Korea and none would be available for the earlier intended ASW training for his ship. Given that fact and the possible lack of hostile submarines to track, Sicily might be used only as a transport, which was for the moment exactly how it was being utilized. Although critically necessary, the role of transport was not what Thach wanted.

Soon after leaving Pearl, Thach received the expected orders that Sicily would indeed have to hold in abeyance its function as an ASW carrier and prepare to become an attack carrier. As it had been at Pearl, the scheduled stop at Guam on 20 July was a matter of hours rather than days. ASW equipment and aircraft had to be offloaded in Apra Harbor so the decks would be ready to pick up a Marine squadron in Japan and head directly into battle. Immediately upon receipt of the orders Thach and his staff began thinking about how to shorten the time at Guam. If planes could be flown off, that would save time, but with some seventy-five crowded onto a carrier that normally operated thirty to forty-five the idea required a major effort. All aircraft on both the flight deck and hangar deck were moved as close as possible, but still there was insufficient clearance to use either catapult. Just as the thought was germinating as to which one plane would be shoved over the side to clear the necessary space, Thach decided the one last plane impeding his plans could be lifted by the crane and swung outboard. Once that was done, Thach asked Guam to send its fastest vessel to meet him to serve as a plane guard. That fastest vessel turned out to be a submarine, probably the only one ever to serve in such a role. Although it could not keep up with the 18.5-knot speed of the carrier, it stayed close enough for rescue if needed. Fortunately, it was not.

Instead of two or more days at Guam, Sicily was there for only hours and was then back at sea making its best time toward Yokosuka. On 25 July, the carrier had to slow down as it encountered very heavy seas, but six days later the ship was just south of Tokyo at Yokosuka. Orders were placed in Thach's hand as soon as he arrived. In retrospect, he considered the orders given him on 31 July the best he ever received: "Proceed to Korea via Kobe. Render all possible support to ground forces. Direct air support or interdiction at your discretion. Keep ComNavFE informed of actions, intentions, and position." Given complete freedom to go where he could help anybody who needed it, Thach planned to do just that.

Although all his ASW equipment and aircraft had been offloaded at Guam, it was necessary to steam south to Kobe to pick up the Marine squadron's equipment and support personnel. The pilots would bring their planes aboard immediately after the carrier returned to sea. But Sicily almost did not make it to the dock at Kobe. Early on the morning of 1 August, Sicily had just entered the narrow channel when a merchant ship turned toward her.[11] Assuming an inexperienced helmsman had turned the wrong way, Thach immediately ordered five blasts, emergency. Although there was water to his right, it had not been cleared of mines sown during World War II. Happily, the merchant ship reversed course, but the big S-turn right in front of the carrier did little to soothe nerves already raw from dealing with the world's new crisis. Regrettably, things did not get much better for Thach as he later discovered some of the Japanese harbor pilots could not speak English. On those occasions, Thach informed all on the bridge, "I'm taking over."[12]

Expecting to spend one or two days loading spares and necessary equipment from two supply ships already docked at Kobe, Thach had barely finished tying up at 0850 on 1 August before he got a call from the office of Vice Adm. C. Turner Joy, commander, U.S. Naval Forces, Far East. The question was, "How soon can you get under way?" Thach replied he needed to load spares but could get under way early the next morning. The voice on the other end said Thach had misunderstood and told him he had to sail "now," and that if he did not, there would be no reason to sail at all. Understanding the seriousness of the moment, Thach took the 1MC (ship's loud speaker) in hand and announced to everyone on the pier who could hear him that *Sicily* was leaving for sea in thirty minutes. "I don't care what you are, unless you're on watch, go over to those two vessels, pick up something that's good for a Corsair and bring it back over to the *Sicily*," he said. "Make one trip. I'm leaving right after that." Thirty minutes later, at 1601, the carrier cleared the lines and began to move away from the pier with Thach and nearly everyone else aboard amazed at how much materiel had been brought aboard in those thirty minutes. Officers and men from all the ships near *Sicily* had picked up everything from carburetors to rockets and carried them aboard.

The following day, 2 August, the flight and hangar decks of the carrier had been cleared of all the equipment brought aboard the previous day and the carrier steamed south of Kyushu and then northwest toward Korea. There were a few anxious minutes early that morning when a sonar contact was made, but nothing developed and the base course was resumed. On this day, immediately before and for months thereafter, Thach was concerned that the Soviets or Chinese might open submarine warfare.[13] With its two escorts, the carrier was off the southwestern tip of Korea near Pusan on the third. At 1336 that day, twenty-four F4U-4 Corsairs of Marine Fighter Squadron 214 (VMF-214) flew aboard from Japan. Thach warmly welcomed the pilots, but there was little time for introductions, orientation, or acclimation. Quickly fueled and armed, eight planes were launched west toward Korea at 1638 and flew directly into combat.

With VMF-214 on board, Thach began to feel that some of the pieces were beginning to fall into place that would allow the Navy and Marine Corps to make a meaningful contribution to the Korean crisis. Organizational structure prior to the crisis remained in place for the most part, but changes were frequently made in August as naval units got close enough to the action to fight. Until 12 August, the *Sicily*, *Badoeng Strait*, and escorting destroyers were units of Task Element 96.23. This task element was part of Task Group 96.2 commanded by Rear Admiral Ruble, who was commander air, Japan while Vice Admiral Joy commanded Task Group 96. On the twelfth, Joy ordered a designation change converting Task Element 96.23 to Task Group 96.8, Ruble commanding in flagship *Badoeng Strait*. *Sicily* then became Task Element 96.82. Thach knew that designations and redesignations were necessary, but he could not help but think that there were more force, group, and element designations than there were ships in August 1950. Seventh Fleet had only one carrier in the area with a small number of escorts. With orders to keep the two Chinas from invading each other, plus calls for help from Korea, Seventh Fleet was then not equal to either requirement. At least Thach had only the responsibility off Korea, and for the first few days in August he would have to look no further than to his own ship to address needs there.

VMF-214 ENTERS COMBAT

It was appropriate that VMF-214, the famous Gregory "Pappy" Boyington Black Sheep Squadron, was still flying Corsairs as probably no squadron other than VF-17's Skull and Crossbones squadron had done more to popularize the Corsair during World War II. The newer Corsairs did have improvements over the F4U-1s Boyington's flyers fought in over the Solomons, but the old and newer inverted gull-wing fighter-bombers were fearsome sights in the eyes of foes. Aerial combat would not be the major function of the Corsair in this war, but a bright new chapter would be written for close air support.

Lt. Col. Walter E. Lischeid was the commanding officer of VMF-214 along with Maj. Robert P. Keller, who had joined the squadron in August 1949 and was already acquainted with Thach, serving as executive officer. By 1 August the relentless North Korean push south had forced the remaining effective troops fighting a delaying action into a small corner of southeastern South Korea around the provisional capitol of Taegu and nearby Pusan, South Korea's best port. When Thach received the request to "get under way right now," elements of the Eighth Army under Lt. Gen. Walton Walker were in danger of being destroyed before they could secure their defenses along the Naktong River. The river provided a natural barrier for defense of the Pusan perimeter, and if that perimeter near the river could not be held, United Nations forces, authorized 27 June, would then have to fight the war from Japan.

The U.S. Army was now paying for its support of the U.S. Air Force during the unification debate. With the Air Force emphasis on heavy bombers, the Army found itself with a great need for air support to counter the Communists in July and August. The Fifth Air Force did provide F-51 Mustangs (P-51 during World War II) and these fast propeller-driven fighters proved indispensable to the delaying actions being fought on the ground. North Korean human wave tactics were new to the inexperienced U.S. Army troops thrown into the unexpected conflict and Mustangs frequently prevented disaster by attacks on massed troops and by interdiction of troops and armor streaming down from the north.

Although the Mustangs of the Fifth Air Force were deserving of the laurels bestowed on them, close air support—when strictly defined—was generally inadequate. Communication from Army troops on the ground to the Air Force, with many of its planes flying from Japan, and Navy was very poor, and messages that did get through usually were too delayed for air support to be helpful. *Valley Forge*, the first carrier to launch planes over Korea, could only interdict enemy units on their own and were dependent upon what they could see from the air given the absence of ground intelligence and communication. Making things worse was Air Force insistence on controlling all aircraft over the battlefield, plus the absence of experience and education for the ground-support role. Even given the emergency, the Air Force did not want to allow anyone on the ground to control their aircraft. But as the North Koreans pushed farther south, the Air Force was being pushed off its bases in Korea. As the general combat situation deteriorated throughout July and into early August, it was agreed that Navy planes would provide close support and fly interdiction sorties south of the thirty-eighth parallel while the Air Force concentrated on striking the type of targets usually associated with bomber commands. B-29 heavy bombers and B-26 medium bombers were never intended for close air support, and that role was never the strength of the Mustang. Later in the war, whether admitted or not, the Air Force F-80 jet fighters were ill suited for close air support even if pilots had the education and experience to conduct it. With an ordnance load only a fraction of what a Corsair or AD Skyraider could carry, the F-80 was little more than a morale booster. The F-80 was an excellent fighter for its time, but when it began to encounter Russian-built MiG-15s after 1 November 1950, its future was past.

Eight Corsairs were launched from *Sicily* on 3 August, expending sixty-four rockets and eight 500-pound bombs against targets just west of Taegu. It was a modest beginning, but it was a beginning and no pilots or planes were lost. On 4 August, *Sicily* was in the Yellow Sea on the western side of Korea from which it launched sorties. Results were disappointing with only two of thirteen bombs dropped on two bridges finding their mark. On 5 August, results were much better as the Corsairs struck the Inchon port area just southwest of the capital of Seoul that had fallen to the Communists on 28 June. Among targets hit were hangars at Kimpo airfield, bridges, railroad yards, factories, oil storage facilities, gun positions, and the omnipresent trucks-on-the-road. Corsairs also flew escort for two P2V Neptunes on a photo mission over the traditional capitol. The most interesting mission, however, was one in which a low-flying marine, Maj. Ken Reusser, spotted a building with a number of tanks and vehicles with U.S. markings on them. Correctly deducing that the building was an assembly and repair shop, he returned to *Sicily* for rearmament and friends. Although he was not required to fly the next mission, he asked Thach for permission to lead because he knew exactly where the target was located. Thach quickly agreed, and before the end of the day the building and its contents were burning.

Not all days were as productive as 5 August, but there were some wherein even more damage was inflicted upon the enemy. Flying nearly every day during the month of August, the Marines usually flew about twenty-four sorties a day with a high of forty-seven. Ordnance loads were mostly 500-pound bombs, five-inch rockets (four on each wing), napalm tanks, and, of course, the heavy, hard-hitting 20-mm cannons. Occasionally, 1,000-pound bombs were delivered, but these were saved for special missions. Although the 5-inch rockets could destroy a tank, napalm proved to be the weapon of choice against tanks. A rocket had to make a direct hit, but napalm did not. Napalm proved very effective not only due to the larger area it covered but also because it would stick to a tank upon which there were many parts that easily ignited in addition to the ordnance within. Too, there was no shrapnel from napalm to damage a low flying Corsair as sometimes happened when dropping bombs or firing rockets. While briefly in the area later on 29 September, Thach had an opportunity to see some of the tanks his Corsairs had destroyed. Some did not even have a dent on them, but they were completely scorched inside and out. Few had remains of enemy troops as they abandoned "like popcorn" when they saw the wall of fire heading toward them.

On 6 August, *Sicily* moved south of Inchon and launched planes against Kunsan and Mokpo. The next day, it joined with *Badoeng Strait* to share duties of strikes and combat air patrol. On 8 August, the 1st Marine Brigade moved into action, and for the first time the Marines on the ground and in the air began the coordination that quickly became the envy of the Army. With a tactical air controller in communication with the Marines on the ground, strikes were directed to eliminate enemy tanks, artillery, and troops resisting the move to expand the Pusan perimeter. From the beginning, the coordination was excellent and the Marines steadily advanced. When targets could not be easily located, the low flying Corsairs often generated targets by drawing ground fire while flying just above the enemy.

Greatly impressed by the coordination between the Marine aviators and their counterparts on the ground, Thach complimented all on their teamwork and declared that they and *Sicily* were "now doing it right." The professional and personal relationship between Thach and the pilots of VMF-214 developed quickly and strongly. Unlike most commanding officers on carriers, Thach left the bridge and descended to the ready rooms after pilots returned from strike sorties. Rather than have the pilots drag their charts and maps to the small bridge, Thach left the running of the ship to his competent staff and was usually in the ready room just as intelligence officers began debriefing pilots. Nearly all the pilots of VMF-214 were decorated World War II veterans, experienced and older than squadrons that would follow during the Korean War. All had infantry training and fully understood the role of close air support and how best to conduct it. They also knew Thach's reputation as a pilot and further understood that his presence in their ready room and his dialogue with them was primarily intended to understand how he could best assist them. Serious talks in the ready room were soon followed by good-natured jokes back and forth between Thach and the VMF-214 pilots.[14]

On 10 August, *Sicily* steamed to Sasebo, Japan, for replenishment, but being only 150 miles from Pusan, the respite from battle was temporary. On

12 August, the carrier was back in action, and next day the most significant target once again was captured U.S. military equipment that had fallen into enemy hands. A tropical storm on the fourteenth delayed strikes, but not for long. Having found a book on typhoons that indicated only one storm had made its way into the Yellow Sea over the past fifty years, Thach, then off the southern tip of Korea, conned the carrier into the Yellow Sea instead of the Sea of Japan. Then, as he had done while operations officer for Task Force 38 off Japan in 1945, he began launching strikes just behind the receding storm. *Sicily*'s planes found many targets on the road as the enemy thought no one would be flying that soon after the storm and with the skies still overcast. Once again Thach demonstrated for his Air Force friends the strength of mobile air power, as even they were surprised to find he had planes in the air. Although he understood that when supporting Army troops he had to work through the Joint Operations Center (JOC) at Taegu, he much preferred working directly with the Marines. Years later, he recalled that elongated communications resulted in more failure than success when naval air tried to work through the Air Force on ground support for the Army.

The weather cleared on 17 and 18 August, and the Corsairs from *Sicily* (call sign Goose Bumps) had two of their most productive days. But these were days they had to be productive, because the North Koreans appeared to have better than an even chance to take key areas of the Pusan perimeter. A major bulge in the perimeter required the Marines to counterattack and secure the front. The Corsairs concentrated on tanks and artillery before turning their attention almost exclusively to troop emplacements. In this period the North Koreans' major objective was to cross the Naktong River—and they did in the bulge. The maneuver could remove the last significant natural barrier in their drive to push United Nations forces, mostly the U.S. Eighth Army and U.S. Marines, off the Korean peninsula. On 18 August, the Marines on the ground and Corsairs drove the enemy back into the Naktong River while several thousand were attempting to cross. According to the *Sicily*'s war diary that day, the planes "strafed the retreating Reds with 20mm explosives and incendiaries killing them in such numbers that the river was definitely discolored by the blood."[15] The diary recorded that "300 enemy dead and many more wounded were left in the river bed."[16] Strikes continued the next day with the primary target being enemy troops near the river. But on the night of the twentieth, Thach again had to turn his attention to yet another tropical storm.

On 26 August, *Sicily* replenished again at Sasebo and was back at sea the following day. Strikes were still concentrated around the Pusan perimeter, and it was apparent that the enemy's objective to collapse the last pocket of resistance had failed. Each day brought more allied troops and critically needed tanks to the region plus several of the *Essex*-class carriers from the West Coast had arrived in Korean waters. Although pleased that *Leyte* (CV-32), *Philippine Sea* (CV-47), and *Boxer* (CV-21) had either joined *Valley Forge* or were en route, Thach did experience some anxiety soon after their arrival. Operating from farther out than the two escort carriers, their captains did not think about the problems their planes caused when launched in the Yellow Sea. At first Thach did not know whether the planes appearing to the west on *Sicily*'s radar were friendly or Chinese. Consequently, he requested the larger carriers route planes to targets without flying over *Sicily* or *Badoeng Strait* which were closer to the coast. Once advised the big carriers informed Thach that they would do what they could for their "little sisters."

Optimism that the Pusan perimeter could be held began to grow after 18 August and increased toward the end of the month. That optimism was tempered, however, on 31 August when VMF-214 suffered it first casualty. Capt. James English crashed following a rocket attack on a troop concentration, the probable cause being small arms fire at the low-flying Corsair. Until this date, the only meaningful damage to VMF-214's planes had been shrapnel from their exploding targets.

Although the loss of Captain English dampened spirits on the last day of August, Thach, the Marine Corps pilots, and squadron personnel plus the crew of *Sicily* could look back on their intensive efforts with justifiable pride for the month of August. Those ashore in southeastern Korea were equally impressed. During the Marine Corps' first days in the fighting, several wounded marines needed to be evacuated from the battlefield but could not due to heavy enemy fire. A call to the *Sicily* solved the problem, as four Corsairs laid down an intense weaving fire to keep enemy heads down while an ambulance roared toward the wounded. Quickly loaded, the ambulance headed back down the road, Corsairs just above spraying both sides of the road with covering fire. A newspaper account on 14 August read, "Ground troops have named the Marine fliers the 'Thatched roof' air cover in honor of commander of the Marine carrier group—Capt. James Thach." Thach had long since become accustomed to being called "James" instead of "John" and seeing word play with his last name. But that did not take away from his pride, as he particularly appreciated the fact that the media account came from marines on the ground.[17]

The effectiveness of close air support during August was a matter of special professional pride for all aboard *Sicily*. During the critical engagements on 17–18 August, the coordination between ground and air was so good that there was some levity tossed into the mayhem. While driving the North Koreans into and away from the Naktong River, some marines on the ground were watching the action by peering over a ridge with enemy troops on the other side. Diving on enemy personnel, one Marine Corps pilot asked the ground controller, "Would you please have the people in the front row be seated. I can see the back of their heads in my gun-sight and it makes me nervous." Another incident reflecting how close the Corsairs were to the ground was when a ground controller guided one of four Corsairs down to hit a well-camouflaged enemy gun emplacement only a few feet from him. After blasting the target into oblivion, the pilot called back to the ground controller to tell him that as he was diving he spotted a tank just a few feet away and was preparing to dive again to get it. The ground controller quickly replied, "I told you we were close; that tank is me."

The only problem with the success of the tactical air controller in the air and those on the ground was that there was some concern, from Thach on down, that ground controllers were becoming a little too confident. When written reports with phrases such as "Shoot at the top of my head; I'll duck and let it go by" began showing up, and accounts of tanks being destroyed under "carports" and because they "failed to stop for a red light" began to appear, Thach felt compelled to warn against overconfidence. However, he did not issue his warnings until November, after he was away from Korea. In early and mid-September, all the confidence that could be had was welcome because United Nations forces were about to take the biggest combat gamble of the war.

14

Inchon, Chosin Reservoir, and Final Korean Operations

Some twenty miles from the occupied traditional capital of Seoul, Inchon was (and is) the largest port city on the western coast of South Korea.[1] The North Koreans had captured Seoul within three days of their invasion and Inchon soon after. Too important to completely ignore, the harbor and port city were guarded by a small North Korean force. That United Nations forces—still just South Korean and United States troops plus a small British contingent by early September—might attack there in strength was highly unlikely given the extreme tides (twenty-nine to thirty-six feet) and fast-running currents through the harbor. In addition, the North Koreans believed they could overwhelm South Korean and United Nations forces before any meaningful counterattack could be mounted.

INCHON PRELIMINARIES

The Joint Chiefs of Staff in Washington also recognized the problems of an amphibious attack at Inchon, and General MacArthur's first recommendation in early July for the plan was denied. Not one to be denied, MacArthur persisted on 23 August, when the Joint Chiefs flew to Tokyo to dissuade him. Final approval came on 29 August—after preparations had already begun. With most of North Korea's army deep in South Korea around the Pusan perimeter, they were extremely vulnerable to having their flank turned at Inchon. Victory there would not only take pressure off the Pusan perimeter but also seal the North Koreans in the south as a result of having their supply lines cut. And if the trapped army could be destroyed before making its way back north, the war could be won in short order.

From 1 September to 6 September, *Sicily* was at Sasebo to refuel, rearm, and make repairs to the boilers, and in this time Thach learned of the impending counterstroke at Inchon and his role in it. Coming aboard along with all the usual materiel were new rockets with a shaped charge that caused the explosion to move forward and push the nose right through thereby providing a higher magnitude of penetrating power. The new rockets would be welcome for their greater killing power, but unfortunately there was nothing new to help accuracy.

By the time he took his carrier back to sea on 6 September, another organizational change was initiated. With United Nations forces growing and more U.S. Navy ships arriving in the region, commander, Naval Forces, Far East placed TG 96.8 (*Sicily* and *Badoeng Strait* plus escorts) under operational control of commander, Joint Task Force 7. On 12 September, TG 96.8 became TG 90.5 within Task Force 90 and was designated the air support group of the Attack Force for the amphibious landing at Inchon.

While Thach and *Sicily* were at Sasebo, the pilots of VMF-214 were supposed to have a rest period. The squadron flew to Itami Air Force Base near Kobe, Japan, but got only one night's sleep ashore before being called back into action due to heavy fighting around Pusan. On 2 September, the squadron moved to nearby Ashia Air Force Base and continued making strikes from there until the *Sicily* and its three escorting destroyers moved back to sea on the sixth.

On 8 and 9 September, the ship's war diary noted, "*Sicily*'s planes stepped out of their usual role of close support for our troops and conducted attacks on enemy logistics, supply lines, and communications by interdiction bombing and strafing."[2] Attacking from off the west coast of Korea, these strikes against railroad bridges, locomotives, and rail cars between Pyongyang, the North Korean capital, and Inchon on the west coast were in preparation for the approaching invasion in addition to the ongoing effort to disrupt enemy supply lines. Napalm and rockets were still the ordnance loads of choice, but 500- and 1,000-pound bombs were preferred for use against bridges. Despite some success against railroad bridges, the North Koreans were repairing them nearly as fast as they were hit. After primary targets had been attacked, the Corsairs usually flew back to the carriers by making a last pass over enemy airfields to ensure no activity by enemy planes.

With the invasion of Inchon fast approaching, a new worry was introduced to planners. A typhoon threatened the invasion fleet and because the invasion could take place only one day in a month when the tides were sufficiently high, a month rather than a day measured a delay. Consequently, the invasion force went to sea earlier than planned. For Thach, however, the storm was only a minor inconvenience as *Sicily* rode steadily in heavy weather. The carrier did lose a lifeboat but did not have to slow down.

On 10 September, *Sicily* and *Badoeng Strait* moved in close to Inchon with orders for their planes to burn with napalm the west half of the island of Wolmi-do, an island from which a causeway ran into the harbor at Inchon. Planes from the two escort carriers put ninety-five thousand pounds of napalm onto the target area and burnt it naked. Surprise for the invasion was not lost because the area had been constantly pounded before plus there were similar attacks on the east coast of Korea to confuse the enemy as to the exact location of the coming invasion.

On the eleventh and twelfth, *Sicily* took time for another replenishment and on the thirteenth was sixty miles southwest of Inchon and ready for the landing on the fifteenth. On the fourteenth *Sicily*'s Corsairs hit gun emplacements around the landing area but only with 20-mm while other planes flew combat air patrol to protect the advancing invasion force. At 0530 on 15 September, *Sicily* launched its first aircraft in support of the First Marine Battalion Landing Team. Landing craft hit Wolmi-do beach at 0630, and within an hour and a half the Leathernecks secured the beach. Caught by surprise just as MacArthur had prophesied, enemy troops fought briefly before fleeing across the causeway or swimming toward the mainland. With enemy troops heading for cover, Corsairs concentrated on machine gun positions and antiaircraft weapons. Throughout the day *Sicily* kept two tactical air observers on constant patrol over the city and beaches for call fire from cruisers and destroyers to eliminate artillery. That accomplished, the Corsairs then turned attention to strikes on the main highway between Inchon and Seoul. Despite flying forty strikes for the day at the usual low altitudes, only one plane was hit. The early afternoon casualty resulted in a lost plane but the uninjured pilot ditched in the channel and was rescued by *Pledge* (AM-277).

The element of surprise gave the invading Marines a relatively easy time on the first day ashore, but the consequences of the landing were fully appreciated by the North Koreans and they began moving all available forces toward Inchon. On 16 September, the "lack of immediate opposition by enemy ground forces permitted the advance of our own forces with little close air support required. Therefore, this ship's aircraft were diverted from close support to attack various targets between Seoul and the front lines."[3] At 0550 on the sixteenth, eight VMF-214 Corsairs discovered six T-34 Russian-built tanks advancing from Seoul and engaged them five miles from Inchon. By the time the squadron's second strike left them, all six were burning.[4] Buildings near the tanks were also hit as the tank crews were seen running to them. Thach was especially pleased with the results of forty strike sorties on that day, as the enemy tanks could have caused considerable havoc given that allied armor was not yet available to assist the Marines as they pushed inland from the harbor. But the success of the day was dampened when news arrived that VMF-214's second pilot had been lost in combat. Capt. William F. Simpson had already dropped his bomb on one of the enemy tanks and was strafing when he was hit by ground fire and crashed.

On 17 September, four more tanks were found and attacked by *Sicily*'s planes. Interesting news arrived on the *Sicily* during the day: the large Kimpo airfield near Seoul was partially secured and several propeller-driven Russian-built Yak fighters had attacked ships at Inchon. Although they had arrived undetected, no meaningful damage was incurred and ship's guns shot down one. But the attack confirmed for Thach that his usual combat air patrols were not a waste of energy. Other news was more disturbing, as pilots reported enemy troops were dressing in white civilian clothes over their drab olive uniforms. Spotted moving toward the front, the civilian-dressed troops began turning in the opposite direction when the planes arrived over them, a dead giveaway of their identity. When the Corsairs began attacking, the enemy quickly discarded their white attire, scattered, abandoned their pushcarts filled with rifles and ammunition, and returned fire.[5]

As it had been for the past three days, the pilots of VMF-214 were simultaneously engaged in close air support for the advancing Marines and interdiction of enemy armor, troops, and supplies moving toward Inchon. By 18 September, however, pilots noticed that the movement of enemy troops was in the opposite direction. Their retreat was not disorganized but it was a retreat. In Seoul the enemy began moving out of the center of the city as pressure on them mounted. Without meaningful air cover and the loss of much of their armor, the North Koreans had to conduct holding actions like those the South Koreans and Americans had been using since the war began. Enemy resistance stiffened for the next three days, and VMF-214 maintained constant pressure.

Unlike the previous month, when *Sicily* averaged about twenty to twenty-four strikes a day, the carrier averaged nearly forty sorties a day during the first week after the Inchon landings keeping "support aircraft over and ahead of the troops."[6]

On the twenty-second, there were no strikes while the carrier fueled her escorts. The following day brought a mixture of good and bad news. The good word was that the enemy was being pushed out of Seoul. The bad news was that Maj. Robert Floeck had become the third member of VMF-214 to lose his life in combat.

On the twenty-fourth, *Sicily*'s designation as part of Joint Task Force 7 was dissolved and the carrier was again designated part of Carrier Division 15. But the daily operations remained the same, except that on that day a feeble two-knot wind made it difficult for recoveries. On the ground, the Reds were using smudge pots to lay a smoke screen over themselves. In combination with ground fog, these pots reduced visibility and hampered early morning strikes. Although progress of the Marine Corps and Army troops on the ground was steady, the need for close air support required another heavy day (thirty-eight) of sorties, especially against tanks.

September twenty-fifth proved to be another bad day for VMF-214 with the loss of the squadron commander, Lt. Col. Walt Lischeid. Older and very gung-ho, the popular officer, who sported a fiery red mustache, was strafing enemy positions north of Kimpo Airfield when his plane was damaged by enemy ground fire. Initially unaware of the damage, the commander had to be told by his wingman that the after portion of his plane was on fire. Lischeid jettisoned his 500-pound bomb and tried to save the plane rather than bail out. Two miles short of the airfield, "he just didn't quite make it." Although sad to have to do so, Maj. Bob Keller again assumed command of the squadron. Lischeid had reported in time to relieve Keller, but Keller was held over as executive officer due to the combat emergency.[7]

On 26 September, *Sicily* proceeded toward Inchon for a week of upkeep and maintenance. Anchored in the harbor about twelve miles out from the city, the crew experienced considerable difficulty and delay in loading from small craft. This was due to the swift three to five knot current (Thach estimated seven and a half knots on the day the ship left the harbor) and tides plus the wind rose by midmorning each day, making the harbor unsafe for small craft loading activity. Still, it had been nearly a month since the ship had not been either engaged in combat or quickly replenishing for more combat. Ashore, friendly troops retook Seoul on the twenty-eighth and two days later they were back to the thirty-eighth parallel.

On 29 September, Thach and all ship captains in the harbor were invited to Seoul by General MacArthur to witness the ceremony turning over the capital of the Republic of Korea back to President Syngman Rhee. All the special guests rode in jeeps escorted by some fast combat vehicles. Along the hot and dusty roads, Thach bounced around in his Jeep, as did the British officers, dressed in white uniforms soon turned tan, and other United Nations representatives. South Korean troops lined the roads with their backs to the dignitaries to ensure no snipers disrupted the event and as a sign of respect for their visitors.

Thach thought he was probably the most junior person in that caravan, and with his immediate superior, Ruble, back in Japan, he was for the time being tactically the task group commander. Before the day was over, he was happy to have been junior to most of the special dignitaries because his relatively junior rank placed him far enough back in the crowd to be in less danger than others. As Thach expected, MacArthur broke into one of his usual flowery speeches, although it was appropriate for the historic moment. But just as the general hit his stride, the wind picked up and glass in the damaged dome began falling onto the rostrum. Marines present, almost as though "a signal had been given, picked up their steel helmets and put them on and folded their arms." MacArthur, however, "didn't bat an eye, he kept on making his speech." The bare-headed MacArthur, almost always at the top of his game when at the podium, continued on, his only change being to raise his voice a little so he could be heard above the noise of glass flying down and crashing on the floor.

THE FRUIT OF VICTORY GOES SOUR

After a week in Inchon Harbor, *Sicily* raised anchor on 3 October and steamed to Sasebo for a much-needed rest. Arriving on the fourth, the carrier remained until the sixteenth, the crew alternating rest and recreation with general maintenance and catapult repair. On 5 October, Thach's Task Group 96.9 was temporarily placed under the operational control of Vice Adm. A. D. Struble's Seventh Fleet in preparation for an amphibious assault that would place the Army's 10th Corps and 1st Marine Division ashore at Wonsan. Directly across the Korean peninsula from North Korea's capital, Pyongyang—captured on

19 October—the Wonsan assault was intended to add pressure on the northeast coast to complement the Eighth Army's northerly push on the west coast. On 16 October, *Sicily* was heading back to the war zone to provide strikes, cover, and gunfire spotting for Task Force 90, the attack force commanded by Rear Adm. James H. Doyle. However, some of the preparation for the amphibious assault proved unnecessary, as United Nations troops were advancing north so rapidly that troops put ashore on the twenty-sixth enjoyed an orderly unloading at Wonsan Harbor rather than an amphibious assault, and *Sicily*'s designation within Seventh Fleet was dissolved on the thirty-first. Still, the enemy was present in the region and now they were again attempting holding actions like those used by American and ROK troops from the opening day of the war until the landing at Inchon.

Fixed and floating mines in Wonsan Harbor proved to be more worrisome than combat to the arriving ships and troops. From 16 October into November, Thach conned *Sicily* in circles to remain approximately fifty miles offshore from Wonsan. As UN troops advanced west and north from Wonsan, there were enemy pockets of resistance, but to the end of October, VMF-214's Corsairs expended most of their 20-mm ammunition while testing guns on combat air patrol and reconnaissance flights. Marine Air Control Parties were ashore, and on 2 November they began to find more work for themselves and VMF-214. The controllers reported on 3 November "increased activity by enemy ground units, reportedly containing Chinese Communist Forces, resulted in extensive close support work this date."[8] Close air support missions became more numerous for the first half of November, but sorties averaged sixteen to twenty a day and as often as not ordnance was dropped in the water rather than on enemy positions.[9] With snow and ice now becoming a common occurrence, the possibility of deck accidents increased and the release of ordnance in the water or designated "dump targets" before landing became daily practice when no enemy targets were attacked.[10]

On 13 November, Thach received orders to disembark VMF-214 no later than the following day. Winding his way through the partially mine-swept harbor, Thach dropped anchor just before dark on the fourteenth and the unloading began. The cold, overcast harbor was an appropriate setting for the "sad parting of two members of a team that had worked so beautifully together, the marine squadron and the crew of the *Sicily*." As the marine plane crews and technicians went over the side into the small boats sent to carry them to shore,

Sicily crewmembers began "taking off their own gloves and pitching them into the boat to the Marines. So I [Thach] took mine off and threw them in also. This sort of illustrated the close feeling between the Marine squadron and the Sicily crew."

The feeling was mutual. Major Keller did not think an oral statement was sufficient. He wrote that as an "unusually and extremely close, friendly relationship has existed between squadron and ship ever since our first joint effort against the enemy on 3 August 1950, only something as informal as this note can begin to express the warmth we feel toward the Sicily and all her good crew." He added, "We consider that our joint 'job' has been well accomplished."[11]

While VMF-214's pilots and technicians set up at their new base ashore near Wonsan, Thach took Sicily back down the narrow channel and headed to Sasebo. The stay there was brief but necessary as the carrier was nearly out of food from its long month off Wonsan. On the twenty-first, Sicily arrived at Guam, where it picked up twenty TBM Avengers of Air Antisubmarine Squadron 21 (VS-21). Having been detached on the sixteenth from control of the Seventh Fleet, Sicily's new orders directed it to "enter into a period of intensive training for the purpose of establishing proficiency in hunter-killer operations for itself and destroyers assigned."[12] On the twenty-second, the carrier departed Guam, anchoring at Yokosuka on the twenty-seventh.

When Major Keller wrote his farewell note to Thach on 13 November, he neither knew nor suspected that his comments were premature. While Sicily was steaming to and from Guam in preparation to return to its antisubmarine warfare duty, VMF-214 was experiencing major challenges from both the weather and a new enemy. Few in the military or in Washington or other United Nations capitals paid much attention when Red China began making pronouncements on 30 September that the country would not sit by while one of its neighbors was invaded. Warnings of intervention became frequent throughout October, particularly after United Nations forces crossed the thirty-eighth parallel with the stated objective of destroying the North Korean Army or forcing its surrender. On 16 October, Chinese troops secretly began entering North Korea and on the twenty-sixth engaged ROK troops near the Yalu River separating China and North Korea. There was general knowledge among United Nations military units on 2 November that the Chinese were involved. Things were relatively quiet until 26 November, when the Communists began a major push down both the east and west coasts. While Sicily rode quietly at anchor in

Yokosuka the last three days of November and Thach studied his new ASW orders, VMF-214 was in a desperate fight for survival along with all United Nations forces then in North Korea.[13]

Thach was aware of what was happening back in Korea, and he knew he might have to shelve his plans for some new innovations in the ASW training that he was about to commence. Before he could place his plans and thoughts on the shelf, orders came to immediately offload VS-21 and to proceed to Hungnam (just north of Wonsan on the east coast of North Korea) at best possible speed and reembark the Black Sheep squadron (VMF-214). Best speed proved to be one knot faster than the carrier was supposedly capable of thanks to Thach using currents he had experienced while in the area. Arriving late on 6 December, Thach was concerned that enemy artillery might be encountered from the hills overlooking the long, narrow, and crooked channel. That did not occur, but another ship lost its bow to a mine. All ships and personnel were moving as fast as possible because the Communist steamroller was approaching.

Thach conned *Sicily* as close to shore as possible to shorten the distance the Marines would have to travel in the small boats. As soon as the carrier was in place, the Marines began loading and continued through the night. The pilots stayed until morning so they could get in one last badly needed mission before flying aboard after the carrier was back at sea. Although the carrier's crew was happy to see their marine friends again, the marines were much happier to be back aboard *Sicily*. One even gave the hangar deck a big kiss as soon as he stepped aboard. Conditions ashore were nothing less than horrible, the presence of extreme cold and absence of showers adding to the misery. Dust, dirt, and mud mixed with snow and ice to clog engines. Near frozen hands had difficulty accomplishing elementary functions whether directed to personal needs or to the planes. Back aboard the carrier, Thach learned that the Corsairs had not been using their 20-mm cannons because they had gotten so dirty they did not work. Immediately he offered additional manpower from the carrier's crew to assist in cleaning and reworking the guns and within three days all the cannons were firing again.

Ashore, the 1st Marine Division was surrounded at the Chosin Reservoir. The Air Force was flying in replacements and bringing out casualties, but to get back to the coast for evacuation, the marines would have to fight their way out. The only positive for the surrounded marines was the presence of their air artillery, naval aviation. From *Sicily*, VMF-214's Corsairs flew mission after mission to blast the enemy from their roosts on the mountain ridges above the only road the marines could use to escape back to the east coast. The number of combat sorties soared from the November lows back to the summer highs when the Pusan perimeter was in danger of collapse. Rockets and bombs were liberally expended, but for the missions of December the weapon most used, as it had been around Pusan, was napalm. The pilots of VMF-214 and ground controllers did have one advantage over the summer days in that the presence of the enemy was almost always known given that they would be immediately behind the location of the retreating marines and other United Nations forces.

By 4 December, United Nations forces were in full retreat on both coasts and all fronts. On 10 December, the 1st Marine Division broke out of the Chosin Reservoir encirclement and began marching to join the rest of 10th Corps at Hungnam, which was evacuated on Christmas Day. Throughout the month of December, *Sicily* and VMF-214 supported the withdrawal, and other than the usual CAP, all missions were close air support. Pilots from *Sicily* and other carriers stayed over the retreating ground troops "like a tent," with those in the air and on the ground aware that if the Communists massed enough troops to stop them, they would become "a jackpot target."[14] For the days of this retreat *Sicily*'s Corsairs—carrying more ordnance than usual due to the short distance offshore—seldom dropped or fired ordnance more than five hundred yards from the ground troops, and they were again flying at "bush-top," not "tree-top."[15] To Thach, anything beyond five hundred yards was strategic bombing, and the major consideration for close air support was "customer" satisfaction.[16]

The mobility of the carriers was proving the UN's greatest strength against the enemy's greatest weakness resulting from not being able to effectively counter the aerial threat to mass forces.[17] Surprisingly, the Chinese did not commit any significant air power to support the powerful ground offensive. They had "a definite air capability," but into early December it had not been committed.[18]

With the Corps' departure from Hungnam on Christmas Day, *Sicily* was ordered to round the tip of the peninsula and steam up into the Yellow Sea and operate off Inchon—the irony not lost on anyone—to fly support for the Eighth Army. Arriving on 27 December, the Corsairs went into action and again began flying nearly thirty strike sorties a day ("hot days"). Trips below deck for pilot debriefings and information received on the bridge confirmed Thach's fears that air force coordination of air support for the Army was again not equal to the need. From arrival on 27 December until relieved less than two weeks later, the carrier's planes worked with Army controllers. That a Marine general had commented that his troops never would have made it out of the Chosin Reservoir "if the close support hadn't worked as well as it did" assuaged Thach's feelings somewhat in view of his disappointment while trying to help the Eighth Army. Again he pointed to the elongated line of communications within the air force. Too often the official record for the day noted, "But due to excessive traffic on the central net, [VMF-214 Corsairs] were unable to get an identification of the troops or be cleared to attack" or "No activity sighted."[19] Too, it proved unfortunate that the joint operations center was located in Seoul that had to be abandoned again on 4 January.

On 8 January 1951, *Sicily* was relieved by a British carrier and ordered to Sasebo and arrived there on the following day. After six days of rest, orders to both *Sicily* and *Badoeng Strait* on 15 January directed them to Yokosuka where VMF-214 and VMF-223 were to be offloaded. By sunset on the seventeenth VMF-214 had disembarked from *Sicily*, again and for the last time, claiming over three thousand enemy troops killed, ninety-one artillery pieces, and thirty-seven tanks destroyed during its time with Thach. For some of the Marine fliers and the two escort carriers, there would be a break from combat. Other carriers had arrived to take their places, and on 24 January, *Sicily* was on its way to San Diego, where it arrived on 5 February for a period in drydock at Long Beach. Unexpectedly, Admiral Radford was passing through San Diego on his way from Washington back to Pearl Harbor and he came aboard *Sicily* to congratulate Thach and offer a "well done" to captain and crew.[20]

BACK TO KOREA

Captain Thach spent the late winter and early spring of 1951 overseeing work on his ship and making numerous speeches in the San Diego–Los Angeles area, where there was considerable interest in his "Blacksheep Jeep."[21] With a clean bottom, new zinc to protect its rudder bearings, and some new communications equipment, the *Sicily* headed back to the war zone in May for blockade duty and air support. On 2 June, VMF-323 reported aboard but Thach was not at all enthused with the worn-out Corsairs they flew aboard. When one of the long-serving Corsairs burst into flame on landing, most likely due to a gas leak, he wasted no time in calling up the line to get the squadron some newer aircraft.

By the middle of June, new orders were on the way to Thach, but before he was detached six weeks later he was again present for some interesting moments in Korean War history. In June and July, the combatants finally began the first steps that eventually led to a cease-fire. Carrier planes still had a combat role in the stalemate that developed in the summer of 1951 and continued for two more years. Naval aviation continued its paramount role of close air support and interdiction while faster air force jets, particularly the excellent F-86 Sabre, battled the high-performance Soviet-built MiG-15. In early July 1951, Thach led a task element of nine ships representing five nations in the Yellow Sea to attempt recovery of a MiG that had crash-landed in shallow water near China. While the recovery mission was under way *Sicily*'s group was also in the process of retrieving a downed aviator by helicopter and wondering if enemy planes might attack. A large number of enemy planes were spotted on radar but no attack ensued.

Late on 28 July, *Sicily* and the ships with it developed a submarine contact in the Yellow Sea. Already authorized to defend against any submarine threat, Thach pulled his carrier out of the way to allow attending destroyers to attack. Oil rose and was tested with the conclusion that it was diesel oil, a strong indication that the target below was a submarine. By the time the protracted attack broke off on 29 July, word had arrived from Vice Admiral Joy's headquarters that the ships had most likely been attacking a World War II wreck. Thach accepted that verdict, but the issue is still open primarily because records indicate the wreck in question was not powered by diesel oil and during the early stages of the attack it was believed to be moving. And if it was a submarine, the question remains whether it was Soviet or Chinese.

The *Sicily* remained off Korea until mid-October and returned for a third tour of duty in 1952, earning a total of five battle stars for its Korean War service. But in the first week of August 1951, Thach was detached as commanding officer of the carrier (relieved by Capt. William A. Schoech) and immediately reported for duty as chief of staff and aide to commander, Carrier Division 17. Carrier Division 17, like Carrier Division 15, in which he had served while CO of *Sicily*, was also an antisubmarine warfare unit. Soon after Thach reported for his new assignment, Rear Adm. Herbert E. Regan assumed command of Carrier Division 17. Although Thach was chief of staff for just over three months, he and Regan initiated some improvements of past practices. The biggest problem was the timely reporting of antisubmarine warfare exercises. Deciding too much detail was delaying reports and hindering a full understanding of exactly what was being accomplished in the exercises, Thach and Regan decided to curtail the detail and length of reports and publish only significant information. The two officers also agreed that the bigger, faster *Essex*-class carriers were much better suited to antisubmarine warfare than escort carriers. The *Essex*-class carriers had the speed to outpace improvements in newer submarines and could launch and recover larger aircraft. Although not immediately implemented, Thach and Regan had reason to feel justified only a few years later when retiring *Essex*-class attack carriers were converted to antisubmarine roles.

END OF THE KOREAN WAR

Captain Thach moved on to two additional assignments before the conflict in Korea came to a halt in the summer of 1953. Fighting concluded with a cease-fire agreement on 27 July 1953, but fifty years later there is still no peace treaty and the threat of more violence on the Korean peninsula persists.

The Korean War placed the unification conflict in abeyance for reasons beyond mere economics. Coming so soon on the heels of the August and September 1949 congressional hearings, Korea presented hard, cold facts to counter the theoretical arguments presented by the Air Force. First, despite geographic obstacles, a hostile population, and lengthening lines of communications (including lack of control of the air and no Navy worthy of the name), North Korean forces had marched some two hundred miles south in the first two months of war. U.S. Air Force B-29s rained tons of bombs on the enemy's troop columns and transportation nodes vulnerable to air attack, but strategic bombing did not stop the advance. Second, neither the Air Force nor Army could get to Korea without the Navy, and the hastily assembled naval armada was the force that protected the flow of logistics. From early in the war, no ground troops wished for a diminished carrier force and the consequent absence of tactical support. The Navy's seagoing artillery, so recently forgotten by the Army, again demonstrated its value, often turning the enemy's flank near the coast in addition to disrupting rear area communications and destroying tanks and artillery. Indeed, close air support had saved both marines and soldiers on many occasions both while advancing and retreating. The economics of casualty rates also drew notice, given that surface ships were all but immune from enemy reprisal.[22] The lessons of the war made an impact on politicians, officers in all branches of the military, the general public, and Thach. Having served on active duty for twenty-four years when he left the Far East in December 1951, he would soon be in position to incorporate those lessons into significant policy changes over the remaining sixteen years of his service to the Navy and United States. And he would carry a second Legion of Merit for his exemplary service off Korea to subsequent assignments.

15

Commanding Officer to Flag Rank

In early December 1951, Captain Thach was preparing to leave the West Pacific for Washington, D.C. The Korean War raged on until 27 July 1953, when a truce ended major combat operations. From 16 January 1951 to February 1953, Thach would serve as the senior naval aide to the assistant secretary of the navy for air, John F. "Jack" Floberg. Although he confided to a friend that he had never been an aide and would require some coaching, his time with Floberg proved advantageous—not only for the assignment itself but also for how it prepared him for greater responsibility in Washington.[1]

NAVAL AIDE TO ASSISTANT SECRETARY OF THE NAVY FOR AIR

Frantic to find a place "preferably in Virginia to park my family," Thach had little time to find a new home for his family and new schools for his sons.[2] Although he did not know it as 1951 rolled over into 1952 and he prepared for his new job, he would have little more time with his family in Washington than he had while at sea aboard *Sicily*.

Having been at sea, Thach expected shore duty, but not as a senior naval aide and not necessarily in Washington. Given his recent experience with civilian appointees during the unification debates, he initially held some natural trepidation about his new boss. Floberg, however, was himself a Navy veteran who had combat experience and wide knowledge of the Navy. During World

War II, he commanded a sub chaser in the Atlantic and Mediterranean and participated in operations at Tunisia, Sicily, and the Salerno beachhead landings. He then moved to the Pacific with USS *Goss* (DE-444) for the invasion of the Philippines, Iwo Jima, and Okinawa and Third Fleet operations against Japan. Rising to the rank of lieutenant commander, he commanded USS *Bivin* (DE-536) before leaving active duty.

Appointed to his position 5 December 1949, Floberg was responsible for everything aeronautical in the Navy at the secretary's level. Appointed by President Truman although a registered Republican and declared political independent, Floberg also accepted responsibilities, especially personnel problems that required legislation, that either fell through the cracks or were not handled by other assistant secretaries.[3] Though Thach found his new boss a pleasant and well-educated person, he also quickly learned that Floberg was a workaholic. During their first meeting, Floberg told Thach that he wore out one Navy captain a year, and as a consequence of that he would not keep him longer than twelve months. Thach later admitted that the highly energetic Floberg almost killed him. Worried about his weight, Floberg announced to Thach, as he had to all others around him, that he planned to give everything to his public service position before returning to the private sector. With Floberg constantly running up staircases and to meetings as part of his self-imposed exercise regimen, Thach—omnipresent cigarette in hand—felt compelled to ask his boss to walk rather than run to meetings both had to attend.

As senior naval aide, Thach had help from a staff of three officers, a civil servant, a lawyer, and two secretaries. Usual daily procedure was for all correspondence to first cross Thach's desk at which time he would write advisory notes of recommendations before passing them on to Floberg. Arriving at the Pentagon about 0715 every morning, Thach nearly always found his boss already at his desk. Quitting time was 2200 hours, though on numerous occasions the hands on the clock moved past midnight before Floberg announced that he and Thach should go home. Madalyn cooked for only three during this assignment, and then prepared a meal for one between 2200 and midnight. Thach did not like the length of the work day, primarily because he felt Floberg was assuming too many responsibilities that could and should have been handled by others. And Floberg had a propensity to accept additional responsibilities before telling Thach what he had done. Consequently, their large stack of paperwork, involving everything from nuclear weaponry to boards, courts, and even decorations, grew even higher. Still, it did not take long for Thach and

Floberg to become close friends. "The better I get to know this man Floberg the more I realize what a fabulous capacity he has," Thach wrote. "You know he is holding down three major jobs and any one of them would take the full ability of an ordinary person. They are not just titles—he really runs these jobs and knows exactly what is going on. In addition to being the boss of Naval Aviation he is the Comptroller of the Navy and the Navy member of the Defense Department Research and Development Board which administers everything in research from atomic energy to a better ball bearing."[4]

Weekends were little different as that was when the two men jumped into a small transport and flew to Naval Air Reserve stations all over the country, often arriving unannounced. But time away from home was the only down side for this assignment. On the professional side of the ledger, Thach always knew that his advice and recommendations were acted on. Consequently, there was never any doubt in Thach's mind that he was in a position to have real and meaningful influence.

Floberg always knew he could depend on Thach to give him a straight answer and was not surprised when Thach did not think his idea to make a parachute jump was inspired. Despite Thach's protests that he needed specialized training Floberg exercised his prerogative and nearly killed himself by waiting too long to pull the cord. The event was no publicity stunt, and Floberg was particularly anxious that his wife not learn of it, especially the part about his helmet being jerked off. Thach promised silence and expressed thanks that his boss did not invite him to join in the exercise.

Not long after the parachute jump, Floberg asked Thach if he thought the person holding his position should be a naval aviator. Thach replied that he thought it would be helpful but not necessary. Sensing what was next, he told Floberg that he was doing a good job. That qualified response was good enough for Floberg as he immediately assigned himself for a short course at Pensacola and was off to learn how to fly. So enthusiastic when he flew out for carrier qualification, the commanding officer thought he might have to shoot him down when he kept flying aboard after the necessary four traps. A good student, Floberg confided to Thach that he did not have time to learn everything such as navigation. Later on, one of their weekend visits to a Naval Reserve station, Thach sat quietly in the back of their borrowed U.S. Air Force T-28 while Floberg searched in vain for Washington. Not until the neophyte flyer appreciated the value of navigation training did Thach offer tips on where the nation's capital and their landing strip might be. Upon landing, Thach suggested to a receptive Floberg that he not attempt a cross-country flight in a single-seat plane.

While Thach fully appreciated the enthusiasm Floberg brought to his office and his talented hard work, he especially appreciated being totally involved with all the political and bureaucratic issues then before the secretary. Despite his considerable knowledge, Floberg often needed Thach's advice during meetings with the secretaries of the other armed forces and various boards. On other occasions Floberg invited Thach strictly for Thach's benefit.

In Korea, many lessons were being taught and learned. The appearance of Russian-built MiG-15s in December 1950 soon brought an end to regular daylight B-29 sorties over North Korea. The only American fighter capable of countering the MiGs was the U.S. Air Force F-86 Sabre jets, but even these could not climb high enough to dictate battle. No extant Navy fighter had the speed, climb, or ceiling of the heavily gunned MiGs and therefore engaged them only when the MiGs descended to challenge. The Navy needed newer and better planes and newer and larger aircraft carriers to handle the heavier and faster planes, and it was clear that research and development had to accelerate. At the center of all these needs were Secretary Floberg and Captain Thach.

Interviewed for *Aviation Week,* Thach noted the losses of the big U.S. Air Force bombers over North Korea and advocated the need to develop all-weather and night capability in Navy fighters as any atomic attack would most likely occur under these conditions.[5] Although the Department of Defense was a reality and the value of aircraft carriers apparent, some newspaper writers were reluctant to let go of the Air Force–versus–Navy debate while the federal budget was being produced. The supercarrier *Forrestal* (CV-59) had been authorized in 1951 and was to be laid down 14 July 1952, while the second member of the class, *Saratoga* (CV-60), was scheduled to be laid down 16 December 1952. The *Saratoga,* however, became a volleyball in the budget debate during the spring of 1952 with funding an open question. Thach summed up his feelings on the issue to the commanding officer of Fleet Air at Jacksonville: "I see on every hand opportunities missed to inform the public in the proper manner." He added, "This stems from a failure of most of us to realize that we have public relations every day whether we like it or not."[6]

Politics was the name of the game in Washington, and both Floberg and Thach worked to address the continuing public relations challenge pertaining to the federal budget. Both men knew that the Navy would be in deep trouble if it could not sell its needs in time of war. Knowing that friend Jimmy Flatley, then

Commanding Officer to Flag Rank

commanding officer at NAS Olathe in Kansas, had the ear of both newspaper editors and congressmen Thach wrote, "I am sure you will know how to handle this."7 To nationally known writer Davis Merwin, Thach was specific in his argument:

> I can hardly name offhand the forthcoming planes that would be too heavy to operate off the *Midway* class. The weight problem is only one of the many reasons why our country needs to continue progress in mobile air bases [aircraft carriers]. We have not a carrier in commission that is younger than a ten-year-old design. [We need] the ability to launch fighters from four catapults simultaneously, [the] height and weight of [a] hangar deck to take the increasingly high tail surface of modern high-speed aircraft. [And we need] water tight integrity [plus an] engineering plant space design to accept the atomic power plant.8

Other new carrier needs were for increased fuel capacity due to jet propulsion (at least five times greater than for piston-driven aircraft), more aviation ordnance space, and better protection against enemy ordnance.9

Thach best summarized the significance of Floberg's contributions in a letter to an acquaintance at *Collier's* magazine as the congressional debate on the budget drew to a close in the early summer of 1952:

> Floberg . . . more than anyone else is responsible for pushing through at the higher levels (Atomic Energy Commission, research boards, etc.,) development of the tactical A-bomb and the atomic-powered submarine. There was a tremendous amount of opposition from the Pentagon Air Force on these two projects. He really cut corners on that one, and as President Truman said at the keel laying of the *Nautilus* (14 June 1952) "in a few short years we have opened the door to nuclear power plants—a step comparable to the revolutionary change from sail to steam." You know what the next step has got to be. It hasn't been stated officially yet but you may have seen a slight leak in the press recently concerning an atomic-powered carrier. Putting atomic propulsion in a carrier will be much easier and pay greater dividends. Supersonic fighters carrying A-bombs are the only things that anyone can depend on to shake off the defending Migs and get to the targets. Everything else depends on this. If we can't do it the Army and the Air Force cannot survive overseas. Even so, there is going to be one h―― of a fight over the carrier building program and in spite of the fact that the second *Forrestal* class ship can be atomic-powered. The strong opposition to this at times seems almost sinister. There are some people here in Washington whose decisions are setting our country so far back that we may have to give up the Far East and Europe without much of a

struggle. They are preparing for the wrong kind of war and that on the basis of last priority first. Their heads are surrounded with great slabs of armor-plated stupidity and it's hard to penetrate. Well, I didn't mean to make a speech, but the lid is going to have to blow off this business soon.[10]

Only two weeks after forwarding the letter, Thach jubilantly wrote a brief memo to Floberg stating that the "Navy is authorized to build a second *Forrestal*-class carrier by utilizing funds from items of a lower priority."[11] Letters of appreciation immediately flowed to supporters who had used their influence to get the carrier put back into the appropriations bill. Still, a few months later Thach was writing to many of the same people, noting that "there are all kinds of evidence of a concerted attack on the building program of the *Forrestal* class carriers." He added, "We are going to need all the help we can get in circulating the factual evidence that supports the necessity for our country to build modern aircraft carriers."[12]

COMMANDING OFFICER, USS *FRANKLIN D. ROOSEVELT*

When Secretary Floberg selected Thach as his senior naval aide, Thach was then first on the list slated to go to the National War College. Before his tour with Floberg ended, others at the Pentagon advised him that he was too old for assignment to the college. Thach was happy working with Floberg, especially since the secretary was so loyal to the Navy and it was apparent that he liked Thach too much to push him off to another assignment. As the one-year mark came and went, neither Thach nor Floberg mentioned the one-year time frame. But Vice Adm. Matt Gardner, then deputy CNO (air) did notice, and he told Thach that the only way Floberg could be convinced to relieve Thach was for Gardner to get Thach a job as a carrier commanding officer. Handing Thach a list of the upcoming big carrier commanding officer slots, Gardner invited him to take his pick. Obviously the best choice for any Navy captain in such a position was one of the three *Midway*-class battle-carriers, and the USS *Franklin D. Roosevelt* (CVA-42) would soon need a new CO. On 27 February 1953, orders were cut for Thach to relieve classmate Capt. George Anderson Jr. To take command, all he had to do was pass a physical, a process complicated by a recurring ulcer. For nearly a month he had to endure a special diet, but it got him through

Commanding Officer to Flag Rank

his latest bout and allowed him to pass the physical. Floberg made no attempt to derail the assignment, as he was in the waning days of his tenure, which came to an end on 23 July 1953. Thach left Washington with the secretary's blessing and best wishes.

Thach believed that the two best jobs in the Navy any officer could have was commanding officer either of a fighter squadron or aircraft carrier. To have one of the three newest and largest carriers in the world was a special delight. Commissioned on 27 October 1945 during Navy Day celebrations tied to the end of World War II, the carrier's name was changed from *Coral Sea* to *Franklin D. Roosevelt* (FDR) in honor of the late president, who had died 12 April 1945. Initially the carrier was classified as a battle-carrier (CVB), but that designation changed 1 October 1952, when it was designated an attack carrier (CVA rather than CVS for antisubmarine carriers). At forty-five thousand tons standard, a length of 968 feet, a speed of thirty-three knots, and a complement over four thousand, the big carrier could handle the biggest and fastest aircraft then in the Navy's inventory. Too, the Navy had pinned their strategic role hopes on the three *Midway*-class carriers during the unification debate while they argued for newer and even larger carriers.

Thach assumed command of the FDR on 22 May 1953 at Norfolk. Aboard during his time on the bridge were mostly F-2H Banshee fighter-bombers, F9F-6 Cougar fighters, and AD Skyraider attack planes. The Banshees and Cougars were jets, and like all other jets they were notorious fuel burners. As large as the *Roosevelt* was, a combat situation would have greatly taxed the carrier's ability to remain on station for more than a few days. Happily for the time Thach was commanding officer of the big carrier there was no threat of combat.

In the late spring of 1953, Thach and *Roosevelt* were on their way to the Mediterranean. Forward deployments to the Mediterranean were normal for the three *Midway*-class carriers. None were dispatched to the waters off Korea during the war. The first carriers capable of launching large, multiengine planes for nuclear strike, they were more valuable near Eastern Europe as a deterrent in the early days of the cold war.

Reporting to the commanding officer of the Sixth Fleet, the *Roosevelt* immediately began training exercises. Operating with jets for the first time, Thach quickly acclimated to their peculiarities. For instance, the Banshees had wing-tip tanks that had to be fueled after the wings were unfolded. Always competing with the *Midway*- (CV-41) and *Essex*-class carriers also in the Mediterranean, Thach worked with his staff to improve launch and landing times. Conservation of fuel and time had always been practiced in carrier operations, but with jets being such fuel hogs, the practice could mean the difference between completing or scraping a mission. Ideally, the *Roosevelt*, still a straight-deck carrier in 1953–54, catapulted twenty-four planes in four minutes, or one every ten seconds. To accomplish this, Thach applied the tricks of the trade that came with long experience and he began launching while the carrier was still coming into the wind. With as many as a hundred planes aboard, getting the planes back down was critically important. Again timing was the answer and Thach promoted the idea of having returning aircraft in the landing circle while the carrier turned into the wind, and once a plane landed it was to taxi as far up the deck as possible. Being forward up the deck allowed for a delay in folding wings or some other problem. The bottom line was not to foul the deck in any manner that interrupted getting everybody out of the sky.

With over one hundred pilots aboard, Thach did not get to know everyone the way he had when he commanded the *Sicily* off Korea. He did make it a point to get to know the squadron commanders and usually restricted his visits with pilots to evaluations after exercises and simulated war problems. Believing it was important to know people socially as well as professionally in order to get a full appreciation of their capabilities and limitations, Thach, at no expense to his personal inclinations, made rounds throughout the ship.

Informal social visits throughout the carrier were not required or expected of a commanding officer, but social life off the ship was very much a part of the CO's role. Constantly on the move throughout the Mediterranean region, with stops in many ports, Thach maintained a heavy social schedule while in port. While the carrier division commander (usually a Rear Admiral) was most often aboard the *Midway*, Thach was automatically the senior officer aboard when the *Roosevelt* docked. For these occasions he especially desired to have his wife join him, but she was still in Arlington and the boys were in school. Nonetheless, Thach asked Madalyn to join him, and several letters on the subject passed between the two, the main issue being that Madalyn did not think they could afford the cost of transportation to Europe, the cost to travel between ports, and the associated living expenses. Her concerns later proved correct, but she made the journey and treasured the experience of assisting her husband. With Madalyn at his side during port calls, Thach particularly enjoyed the social demands, protocol visits, and calls on officials. Local officials, both political and military, royally entertained the Thaches, and Madalyn was particularly helpful in making friends with the wives and families of various officials. Together the couple was able to go places and see things no ordinary tourist could.

At sea, Madalyn could not help her husband, and the demands there were in no manner as pleasurable as in port. The cold war permeated all aspects of thinking and planning, and during Thach's tour aboard *Roosevelt* there were border raids between Greece and Bulgaria that drew special attention. On 13 August 1953, a major earthquake in the Ionian Islands west of Greece drew the carrier to the area to provide supplies and treat many of the injured. And when not involved with international problems or nature, there was always concern for the operation of the ship. Fortunately, the most challenging of these moments came when Thach's boss, Rear Adm. Hugh H. Goodwin, was aboard and had opportunity to see Thach handle what could have been a disaster. The worst possible scenario for a carrier while fueling was a steering breakdown, because tankers often carried highly volatile aviation fuel in the bow and that section of the tanker would certainly be damaged in a collision. But on Goodwin's first visit with Thach on the bridge, the tanker alongside *Roosevelt* had a steering breakdown. Seeing the tanker beginning to turn toward the carrier and knowing that such a maneuver would never happen unless the tanker's rudder control had failed, Thach immediately took the con. Complicating the situation was a destroyer delivering mail on the port quarter, but with all boilers on line the big carrier responded instantly as Thach ordered "emergency back full." Hoses between the carrier and tanker broke but the two ships just "kissed" instead of colliding while the alert destroyer captain quickly backed away from both. No injuries from broken cables or hoses resulted from the emergency and Thach earned a "well done" from Goodwin. Later he reminisced that probably no other carrier captain ever received a "well done" for nearly having a collision.

In their original configuration the *Midway*, *Franklin D. Roosevelt*, and *Coral Sea* (CV-43) handled beautifully and could turn so tightly that even destroyer captains envied them. But the joy of ship handling was considerably tempered by equipment and personnel problems. In the transition from piston aircraft to jets, new equipment was needed, and often the first issue did not meet the needs, flight deck sound helmets for after-burner tests being notable. With personnel, the usual problems of venereal disease were never ending and reenlistments seldom met quotas. Thach could offer only the same advice as other commanding officers on VD but thought the basic problem with retention was too much time at sea and that not enough thought was being devoted to personnel needs.[13]

Retention took a big hit aboard the *Roosevelt* when rumors began circulating that the carrier was scheduled to travel in early 1954 to Bremerton, Washington, for a major overhaul. Many of the crew did not want to go to the Pacific because their homes were on the East Coast. That matter was resolved by allowing fifteen hundred to be detached with five hundred new sailors coming aboard in the trade.

Even in its near-original configuration, *Roosevelt* was too wide to pass through the Panama Canal, so the voyage would take the carrier down the eastern side of South America to the Cape of Good Hope and then up the western side. The State Department advised the Navy that Thach could make any port call except Chile. On 7 January 1954, *Roosevelt* pulled away from a pier at Norfolk and began its 15,600-mile odyssey.

One of the highlights of the voyage for Thach came early in Brazil. The stopover in Rio de Janeiro was enjoyable but very busy. The schedule included many luncheons, the most important of which was in the presence of Brazil's naval luminaries. Thach had little time to prepare the remarks he was to make at the luncheon after a Brazilian admiral made his. Fortunately, his Brazilian liaison was able to obtain a copy of the admiral's speech. Thach wrote his remarks, had it translated into Portuguese, and then memorized it. Knowing Thach was not fluent in Portuguese, the admiral and luncheon guests rose with thunderous applause after Thach unexpectedly spoke directly to the points just stated by the admiral. Thach considered that event the epitome of his diplomatic endeavors.

Moving on to Uruguay, Thach and the crew were concerned about confrontations with Communists, so plans were made with the ambassador to bring selected guests aboard the carrier. The captain's cabin could seat sixteen comfortably and had been used on several occasions for similar gatherings while the ship was in the Mediterranean. As with the best of plans, they had to be changed at the last minute when the water around the anchored carrier proved too rough for transportation of the guests by boat. Consequently, a helicopter was dispatched to collect the distinguished visitors. As helicopters were previously unseen by many of those waiting ashore, some decided to skip the formalities. Too, Thach was previously unaware that the country had more than one president and the one that did come aboard had an appetite only for ice cream.

The diplomatic puzzle of Uruguay was not repeated during the visit to Argentina, although that visit too was a challenge. There the Argentine government would not permit the visiting ship to land any of its own shore patrols because they said they had plenty of policemen. Thach managed to get around that by issuing blue cards to trusted petty officers that went ashore with all liberty parties. Sailors were instructed to walk away from all provocation, especially from Communist sympathizers attempting to start riots, and any sailor approached by a petty officer with a blue card would know that he had the authority of shore patrol personnel.

Unlike Uruguay, Argentina had only one president, Juan Peron. Never one to pass up an opportunity to host representatives of other nations, Peron invited Thach to an air show and lunch, after which the president presented Thach with a pair of Argentine Navy wings and an autographed picture of himself. In 1971, Thach recalled the event and pointed to the picture of Peron on a wall in his home next to a signed photo from the late John F. Kennedy. Both photos had been given to him without him being asked whether or not he wanted them, and Thach acknowledged he was not an admirer of either president. Nevertheless, he displayed the two pictures in his home to the end of his life.

Roosevelt made another stop off Peru, offered that country's president a twenty-one-gun salute while he hovered over the flight deck in a helicopter, and then steamed north to Bremerton, where the carrier arrived on 5 March 1954. For the next six weeks, Thach and crew engaged in a well-organized, well-executed program of material removal, machinery inactivation, and cleaning and preservation of the ship. The Puget Sound Naval Shipyard had requested the carrier's crew to prepare the ship for conversion and applauded the valuable technical assistance to Design personnel. In a letter of appreciation, officials at Puget Sound noted the high degree of cooperation and resultant saving of a considerable sum of money, "which most assuredly will be reflected in added improvements to the ship during conversion."[14]

On 23 April 1954, the carrier was decommissioned, and for the next two years it remained in dry-dock while growing in size and capability. Emerging for recommissioning on 6 April 1956, the rebuilt carrier had an angled flight deck, steam catapults, and hurricane bow, among many other improvements. In 1956, *Roosevelt* was a more modern and capable warship, despite giving up some earlier attributes, such as a tight turning radius. Somewhat sadly walking away from his last ship command, Thach was cheered by the thought that his family would be reunited. In the past year, the financial cost of assisting with social functions in the Mediterranean had been a stress equal to Madalyn's loneliness.[15] Things would be better for the family during the next assignment.

COMMANDING OFFICER, NAS JACKSONVILLE

In the first week of April 1954, Thach and his wife knew that new orders were being prepared for them to report to Jacksonville, Florida, and that they had until 30 June to take leave, travel and get set up in their new quarters. On 1 July, Thach relieved Capt. Burnham C. McCaffree and settled in for what became one of his most enjoyable tours of duty. For the next year and a half, he was the commanding officer of Naval Air Station Jacksonville in addition to being commander, Naval Air Bases, Sixth Naval District. Jacksonville was also the location of the headquarters of Commander Fleet Air, Jacksonville, a position held by a rear admiral.

Thach's most significant naval contribution was a recommendation that Mayport, located at the mouth of the St. John's River, be upgraded. When Thach arrived, NAS Jacksonville, then in service only fourteen years, was still more a training facility than full-scale naval air station. The potential for full development, however, was great. Mayport Harbor was perfect for large ships and the only deep-water turning basin between Norfolk and Guantànamo, Cuba, that could accommodate the forthcoming *Forrestal*-class carriers. With Cecil Field already in place several miles inland and Mayport developed to moor supercarriers, the area was perfect for basing carriers and conducting training. Thach's proposal was not new, but a combination of realities caused the Pentagon to once again look at the area's potential and this time a decision was made to act. On 1 July 1955, Mayport was commissioned, and Thach believed even more at that date that the Jacksonville Naval Air Complex should be fully developed as soon as possible.[16]

The only memorable problem with the Jacksonville tour before it ended on 31 August 1955 was one woman who constantly called Thach at work and home to complain about naval aviators flying too low. Thach always responded by asking her to get the numbers off the plane and telling her that if she could not, they were not too low. On one occasion, she called to announce that a flying saucer had just landed on one of his planes. In that she was partially correct, as the new E1B Tracer carried a radar dome that looked exactly like most artists' conception of a flying saucer. That one continuing challenge aside, along with surgery for knee cartilage in May and June 1955, the Thaches' time in Jacksonville was especially happy. Thach enjoyed an outstanding relationship with the city fathers and business and civic leaders. All bent over backward to support the Navy, and Thach was only too happy to support community projects. It was during this tour that he and Madalyn became grandparents, and in July 1955, he learned that he had been selected to flag rank. Congratulations flowed in from scores of friends and colleagues. One well wisher noted, "The 'Thach Weave' seems to have woven a good pattern," to which Thach replied, "I... will endeavor to continue 'weaving' in a progressive manner."[17]

WEAPONS SYSTEM EVALUATION GROUP

Although selected in July 1955 for flag rank, Thach would not officially attain the rank until 1 November. Arriving in Washington, D.C., again in search of living quarters and a school for his youngest son, he underwent two weeks of intensive briefings for all newly selected flag officers. Knowing his first assignment as a flag officer beginning 12 September would be the senior naval member of the Weapons System Evaluation Group (WSEG), he was concerned about joining the interservice group wearing "colonel" wings rather than two stars. Fortunately there was authorization for wearing of higher rank in advance of promotion in cases where prestige was involved.[18]

Lt. Gen. Sam E. Anderson, USAF, director of the Weapons Systems Evaluation Group, welcomed Thach to his new duty. Anderson informed Thach that "our mission is to provide the Joint Chiefs of Staff and the Secretary of Defense with rigorous, unprejudiced and independent analyses and evaluations of present and future weapons systems under probable future combat conditions. We have a small, well-blended team of civilian scientific analysts and specially qualified military personnel from all the Services for this purpose, and you were carefully selected to fit in with them."[19] Additionally, the group assessed known enemy weapons systems and assisted in the selection and evaluation of counter or deterrent systems. Before his first meeting with the WSEG Thach understood he was somewhat like a delegate to an international congress who had no power to make a decision or any agreements. And he soon learned that the group's effectiveness would go only as far as the willingness of the people in authority to pay attention.[20]

In his office at the Pentagon, Thach had every reason to believe this assignment would be exceedingly worthwhile in serving his country. Dwight D. Eisenhower was president, former squadron mate and friend Arthur Radford was chairman of the Joint Chiefs of Staff, and his old friend from the last two years of World War II, Arleigh Burke, became CNO on 17 August 1955. Calling on Burke just as both assumed their new duties, the two officers discussed Thach's response to Burke's invitation to suggest ways of improving the Navy. Repeating his earlier written comments, Thach stated that he thought the Navy was undernourished rather than overcommitted and that justification for funds had to be made more convincingly.[21] Then he asked if the new CNO had any special instructions pertaining to his new assignment with the WSEG. Burke replied, "You're being ordered to that job because of your experience. The other services send people there for the same reason, presumably they're not representatives of their service, but they bring together a combined experience that will be able to be sure that all of the opinions and factors involved and problems they study are heard. I want you to forget about the navy when you go down there. You're there to do a job for the WSEG. Shed any parochial ideas and just do the best job you can, and we don't need to hear from you at all." Thach left Burke thinking, *Wonderful*.

After meeting with Burke, Thach reported to Admiral Radford. Radford told him pretty much what he had heard from Burke and added, "We in the Joint Chiefs of Staff have to make decisions about certain things. Sometimes we have to make them sooner than we would like because they may involve advice to the president who has to make a decision about something and circumstances may make it so that he can't wait and we can't wait, either. So we want answers from WSEG as soon as we can get them . . . somewhere between hysterical and historical!"

Thach soon found in his assignment that WSEG had a good balance between the services. The director was a "three star" from one of the respective services, Lieutenant General Anderson, relieved later by Vice Adm. John H. "Savvy" Sides. Each service had a senior member assisted by eight other officers down to the rank of lieutenant commander or major plus civilians, usually Ph.D. level. Although working administratively under the Office of the Assistant Secretary for Research and Development, the Joint Chiefs nonetheless assigned most projects. Still, WSEG had to be responsive to the secretary of defense and its members could themselves initiate up to 25 percent of their work. Internal initiative was helpful because on numerous occasions the group could see some questions coming before they got there and it was beneficial to be a little ahead.

Reports went directly to the chairman of the Joint Chiefs with a copy to the secretary of defense. The Joint Chiefs could write a dissenting view, approve, or disagree, and Thach recalled that during his tenure the JCS agreed with nearly all of WSEG's studies. Most often the reports were large and consisted of text, summary, conclusions, and sometimes recommendations; in addition, documentation being extremely important.

One of the studies completed during Thach's duty that later drew special attention was on limited war, with Southeast Asia having been chosen by coincidence. Later Thach recalled that people talk about limited war and fall into the illusion or trap that it is limited in effort and commitment of forces. But, he continued, if you have a limited war, you have a limited objective, but in any war if you do not attain the objective you have not done the job. In sum, he thought any war important enough to fight was important enough to win. As expected, considerable attention was placed on studies of air defense in Europe. As a result of these studies, certain actions were taken. Communication and detection in Western Europe were vastly improved, the location of early warning stations changed, plus certain gaps were filled and procedures recommended for objectives to be met with less expense and greater effectiveness.

Usually about five studies were ongoing at any one time. Each study had a project director and the project reviews were very important. The project review board was composed of a director, senior service members, the civilian scientific director, and project director. All information was highly classified and all knew that it would never be kicked around publicly so there was no point in trying to make one's service look good. Without public relations battles and end runs, bias was washed out and objectivity prevailed. Senior service members like Thach learned early to take hard looks at the inputs and the assumptions for a study. If the assumptions were sound, the facts fell into place and documentation sealed the conclusions. As much of what was being studied was still only on a drawing board, the desired quantitative analysis was not always available, but qualitative analysis could serve as reasonable logic—in regard to concepts of mobility for example.

The WSEG and higher authority always hoped to have contingency plans in place, but five years after Thach left WSEG, there was no exact study available for the missile crisis in Cuba. Yet other studies pertaining to Cuba made it clear that control of the sea, which was implemented in 1962, proved its worth. When orders were cut on 19 July 1957 terminating Thach's WSEG duty in October and sending him back to sea, he knew that the group had improved the military's capability and saved a great deal of money by studying proposed weaponry before production began. Unquestionably, his experience at the Pentagon in 1955–57 strengthened his foundation for a job of greater responsibility when he returned in 1963.

16

Task Group Alpha

Those who had hoped the 5 March 1953 death of Soviet dictator Joseph Stalin might bring a higher degree of rationality to the Soviet state and a lesser degree of danger to the world were sorely disappointed toward the end of the 1950s and early 1960s. On 4 October 1957, the Soviets launched Sputnik 1 into orbit and followed up a month later with the launch of an even larger satellite. The race for space was on, and at the end of 1957, the United States was clearly running second. The 1960 presidential election debates between Senator John F. Kennedy and Vice President Richard M. Nixon focused significantly on the possibility of a missile gap between the two superpowers. In the spring of 1961, a failed U.S.–backed effort by Cuban exiles to overthrow the incipient Communist regime on the Caribbean island was followed by a Soviet perception of American weakness. The consequence of that perception was a more vigorous challenge to Western interests in East Germany. The increased tensions, especially in Berlin, resulted in the activation of reserves and increase in active-duty troops. Throughout much of the five-year period, Rear Admiral/Vice Admiral Thach was to serve as an antisubmarine warrior; the fear of an accidental start to World War III was surmounted by an apparent inevitability that war would begin intentionally.

The seriousness of the possibility of open war was barely appreciated by the American public, although several opinion polls indicated a positive response to using military power in many of the world's flash points. Most civilians' primary interests were family, work, and leisure. When not discussing these interests, conversation often centered on movies and television shows, and there was far greater debate on whether Ford had made a mistake by building a two-seat Thunderbird than there was on world affairs. Conversational comments were much more numerous in arguing foreign aid than the implications concerning construction of the Berlin Wall, and few Americans were gravely concerned about nationalist movements in colonial Africa. Only the few days in late October 1962 during the Cuban Missile Crisis significantly diverted American attention from domestic matters to the international stage and the reality of possible open conflict.

In November 1957, Rear Admiral Thach was detached from his duties with the Weapon Systems Evaluation Group in Washington and ordered to a command at sea. The move was easy, only ninety miles east to Norfolk, Virginia. The commander in chief, Atlantic Fleet at that time was Adm. Jerauld Wright but Thach would be reporting directly to the commander, Anti-submarine Defense Force, Atlantic. His new title was commander, Carrier Division 16, with the additional responsibility of commander, Hunter/Killer Force, U.S. Atlantic Fleet, titles he still held after the initially unexpected collateral duty as commander, Task Group Alpha was added. Although the Navy consistently used "Alfa" for the spelling of the new experimental group, Thach preferred "Alpha." As this was the first group of its kind, he preferred that the first letter of the Greek alphabet be used.

Soon after receiving orders to his new assignment Thach completed Norfolk's Antisubmarine Warfare (ASW) Tactical School on 22 November 1957. Other formal training was completed in early April 1958, when he attended the U.S. Navy Underwater Sound Laboratory at New London, Connecticut, and other antisubmarine warfare studies at the Naval Aviation Development Center, Johnsville, Pennsylvania, plus a visit to Key West's Weapons Testing Center.

To a meaningful degree, Thach was familiar with the task that lay ahead of him. While at Jacksonville from 1942 into 1944, he followed the Battle of the Atlantic from its darkest days to the turn of fortunes in 1943. During his tour of duty as Vice Admiral McCain's operations officer, the threat of Japanese submarines received his direct attention every day. Off Korea he patrolled against submarines and worked with higher ranking officers to improve the effectiveness of ASW training exercises. Indeed, his tour as commanding officer of the *Sicily* would have been exclusively antisubmarine warfare had not the unexpected beginning of that war required his escort carrier to have been pressed into service as an attack and ground support platform. And he well remembered that the Germans had opened World War II in 1939 with only fifty-seven submarines. Forty-nine of that number were operational, and only twenty-six were capable of sailing well out into the Atlantic in September 1939, with ninety operational after the United States entered the war in December 1941.[1] By contrast, the Soviet Union had more than four hundred submarines in 1957.

The other degree of familiarity with his new task was not as well known but nonetheless appreciated as being daunting. Nearly three years earlier, on 17 January 1955, *Nautilus* (SSN-571) got under way on nuclear power, and from that day it was apparent that both the U.S. Navy and the Soviet Navy were in transition to an era in which the nuclear submarine would be the capital ship. The types of boats, ordnance, tactics, and strategy that had prevailed since World War I were quickly moving into a time when the newly proved technological advances in submarine propulsion plus progress in ordnance and delivery systems could result in a strategic stalemate. That could mean a return to use of conventional tactics. Consequently, Thach's new challenge was twofold. First, he had to anticipate and resolve tactical considerations for the present and near-term transition period for antisubmarine warfare, and second, he had to anticipate an antisubmarine response for the approaching day when then understood technology would be totally inadequate for the antisubmarine warrior.

SUBMARINE WARFARE AND TECHNOLOGICAL ADVANCES

The potential of the submarine against a large surface ship was confirmed during the American Civil War, when CSS *Hunley* attacked and sank USS *Housatonic* on the night of 17 February 1864 in Charleston Harbor, South Carolina. Fifty years later, a small number of German U-boats with top speeds of sixteen knots on the surface and eight knots submerged—only for about an hour to conserve battery power—wreaked havoc on the Allied cause. Although the initial targets of the U-boats were warships, they later enjoyed success against merchant ships. Usual practice was to attack alone in daylight, often while submerged.

Between the two world wars, the German Navy studied tactics used from 1914 to 1918 and concluded that (1) the primary target for U-boats should be merchant ships and (2) group (wolf-pack) tactics should be employed. Less than a dozen U-boats would spread out over a section of the Atlantic searching for convoys. Once spotted, the other U-boats, traveling at speeds similar to their World War I cousins, would sail to a designated position in the path of the six-to-nine-knot convoy. Attacking on the surface at night, the U-boats would then run ahead of the convoy on the surface to prepare for the next night's attack and submerge only when threatened.

From 1939 into 1943, having too few escorts for convoys hampered the Allies, and their aerial surveillance was effective only in areas near the English coasts. By May 1943, however, the United States and Great Britain not only had longer-range aerial surveillance—aircraft accounting for nearly half of all U-boats kills—but also a competitive number of escorts. Additionally, advances in radar allowed for triangulation location and for finding surfaced U-boats and their supply vessels at night and under cover of fog. These advances plus the introduction of hunter-killer teams of an escort carrier with accompanying destroyers and destroyer escorts placed German submarines on the defensive.

The German Navy foresaw the defeat of mid-1943 and moved to develop submarines that would enable them to counter the advances in radar, ship construction, and greater aerial capability. The paramount need was a submarine that could stay submerged. Extendible breathing tubes (schnorkel) that allowed diesel engines to be run while under water were first installed in 1943. These had limitations, however, especially in rough water, and visible exposure while near the surface dictated that it be used mostly at night. The answer was a new design fueled by liquid hydrogen peroxide that promised underwater speeds near that of many convoy escort ships. This design became available too late to change the course of the war, but other innovations did see service. Magnetic and acoustical torpedoes in combination with hydrophones that allowed a U-boat to track and attack from depths of 150 feet proved their worthiness and even greater potential.

The use of German V-1 and V-2 rockets and the United States' introduction of atomic ordnance in the last year of World War II inspired thinking that nuclear weaponry could someday be attached to a missile. Although the Army and Air Force led the U.S. military in developing ordnance-bearing missiles, the Navy was not slow to experiment. Late in the war, small, short-range rockets had been fired from a surfaced submarine on targets near the Japanese coast. On 12 February 1947, the first launch of a guided missile from a submarine took place. Looking very much like a German V-1, a Loon rocket was launched from USS *Cusk* (SS-348). In the same year, Chance Vought, builder of the famed F4U Corsair fighter, began work on the Regulus surface-to-surface missile, and by 1951 test launches from submarines began.

While the idea of a nuclear-tipped missile had obvious potential, the technical problems for the Navy often seemed insurmountable. The World War II rockets fired from submarines had an effective range of less than five miles, and the Loon was basically just a test weapon. The Regulus I and II missiles carried research, and a warhead of nearly two tons, much further from 1947 into the early 1960s and greatly improved guidance and flight technology. But even though the Regulus missiles were placed aboard both surface ships (1955) and submarines (1959), they were not ideal for submarine operations. The missiles were bulky and were carried in a deck-mounted hangar that impaired a sub's underwater maneuverability. Further, alert fighters could potentially shoot down the early subsonic Regulus missiles, and the missile's maximum range of five hundred miles limited targeting. But their most worrisome feature was their gasoline or kerosene fuel. Given the natural instability of a small vessel in any turbulent ocean, the potential for accidental ignition was unacceptable. This same factor, plus its large dimensions, precluded Navy acceptance of the otherwise successful Army–Air Force Jupiter missile. The solution was solid fuel propellant, which led to rapid development of the Polaris missile.

THE SOVIET SUBMARINE THREAT

While many in the late 1950s and early 1960s Free World worried about the Soviet submarine potential to sever sea lanes and disrupt maritime commerce, the Soviets themselves looked at their nearly five hundred submarines' first priority as a defense against U.S. aircraft carriers bearing nuclear armed aircraft. Their reasoning was justified, as nuclear strike was the primary role of U.S. aircraft carriers from 1951 until about 1962.[2] A second priority was the development of antiship, cruise, and ballistic missiles that could be submarine launched. Indeed, when Thach assumed command of Carrier Division 16 in November 1957, the Soviet Navy was closely paralleling the U.S. Navy in the development of both nuclear-propelled and ballistic-capable submarines, except for the fact that the Soviets were producing a greater number but with somewhat less reliability. The disparity was not as overwhelming as the numbers might have indicated, given that many of the Soviet submarines were basically intended as short-range coastal defense craft. Still, the number of Soviet long-range boats capable of threatening maritime interests caused justifiable concern.

Thach assumed in 1958 that Soviet submarines with short- to medium-range nuclear missile capability were already in place. Certainly he thought they were building.³ And the U.S. Navy fully expected the Soviet Navy to progress technologically along the same track it was moving. Although the Soviets were slightly behind in submarine ballistic-missile development, that expectation proved correct. It was foreseen in the Free World that war might not be necessary for the Soviets to achieve their aims. Should they move dramatically ahead in submarine and ordnance technology they could succeed with blackmail and intimidation. As many of the United States' most significant population, industrial, financial, military and government centers were either coastal or short distances inland, and that no point in the United States was more than 1,500 miles from a coast, the new threat would negate the country's former insular security.

TASK GROUP ALPHA

As the winter of 1957 rolled over into 1958, Rear Admiral Thach went to sea several times with his new command and returned to port disappointed each time. Although not a historian, he knew enough history to think that the 1958 Navy was perhaps prepared to fight another Korean War or World War II Battle of the Atlantic as the tactics and technology were essentially the same in early 1958 as in 1945. But the early 1958 antisubmarine warfare Navy was not prepared for the revolutionary technology then so rapidly evolving in newer submarines. Soon into his new assignment he had an opportunity to convey his disappointment and concerns to an old friend who also shared the same concerns.

Well settled in as chief of naval operations since 1955, Adm. Arleigh Burke held a conference in the winter of 1958 to discuss the Navy's readiness in antisubmarine warfare. He wanted to know from his commanders whether or not they believed the current tactics, ships, ordnance, personnel and equipment, and all the other elements required for ASW were satisfactory or not, and if not, why. It had been thirteen years since the U.S. Navy had faced a major undersea threat wherein a measure of success or failure was tested in combat: the experience off Korea proved more a threat than test. Thach had shared combat with then Commodore Burke in 1944 and 1945, and Burke was well aware of Thach's capabilities as a warrior and tactician. And soon into the winter conference Burke was not surprised to see Thach rise and speak at length in responding to the CNO's questions.

Thach began his discourse by pointing out the instability within the antisubmarine forces organization, noting that rapid turnover in personnel and ships meant that lessons learned had to be constantly relearned with the result that well intended effort repaid insufficient dividends. Another major issue was that all the ships comprising a hunter-killer team were also attending to other tasks, drills, and assignments, a practice that diverted antisubmarine warfare focus and lessened readiness. Knowing better than to offer a complaint without a remedial course of action, Thach concluded his impromptu speech by suggesting the CNO create one permanent, or at least semipermanent, hunter-killer group to concentrate only on developing new equipment and tactics for meeting the current and coming Soviet submarine menace. And before retaking his seat, he suggested that the commander of this group have two divisions of destroyers (eight), two submarines, and one squadron of land-based patrol aircraft in addition to the aircraft available on the aircraft carrier flagship.

Admiral Burke, a workaholic more renowned for his good-natured humor than his volcanic temper, listened to other comments and suggestions after Thach sat down. When all had spoken, he rose and said, "Jimmie Thach just made an unfortunate speech. He talked himself into a job. I'm going to give him the job that he just outlined." Thach was stunned, but happily so.

The idea of creating a special task group to concentrate solely on antisubmarine warfare was not greeted with enthusiasm by everyone. Especially unhappy were destroyer commanders, who believed their crews required training in a number of proficiencies. And few commanders were happy to have any type ship, equipment, or authority taken from them. But Burke reminded detractors that those same arguments had been offered before, that some drastic action was required, and emphasized his decision by stating that the new group would get the permanency it wanted. Promising beefsteak for breakfast if necessary to ensure success, Burke left no doubt as to his support for the new experimental group.

Following the winter conference, there were other discussions between Burke and Thach. Purportedly, Thach insisted that he had to have both ships and men for two years or there was no reason for him to accept the appointment.[4] Burke agreed, and orders flowed from his office to the Bureau of Personnel, type commanders, and others directing that Thach had complete operational control of the new experimental group. Naturally, there was some personnel turnover, but it was controlled to a satisfactory degree primarily by staggering detachments of critical officers and ratings. And when ships or planes had to leave for maintenance or refit, they were replaced.

Quickly organizing his new eleven-ship, fifty-aircraft, five-thousand-man command, Thach's two-star flag was raised on 1 April 1958 aboard the *Valley Forge* (CV-45), the only large carrier to enter hostilities off Korea before he arrived in *Sicily* in the summer of 1950. After overhaul and refitting in late 1953, CV-45 (nicknamed the "Happy Valley") joined the Atlantic Fleet in January 1954 as an antisubmarine warfare support carrier having been redesignated CVS-45. Now as Thach's flagship, it would operate with the eight destroyers of Destroyer Squadron 28 and two submarines he had requested. The submarines, *Sea Leopard* (SS-483) and *Cubera* (SS-347), would operate with their long-range "ears" as a target for other ships and aircraft of Task Group Alpha. And they also assumed the role of attack submarines against other subs. Additional forces were made available for limited exercises such as target ships, nuclear submarines, and even some blimps.

On 11 April 1958, Task Group Alpha's first operations order was issued to accelerate the development of antisubmarine tactics, doctrine, and equipment in order to improve the ASW readiness of the Atlantic Fleet. The plan was to be accomplished in three phases: team development, tactical development, and equipment improvement. Thach's thinking was that once the team was developed it would be properly focused on tactical problems, and the result of the tactical exercises would demonstrate what equipment worked, what equipment needed improvement, and what the Navy and private industry needed to develop that was currently unavailable.

As a former athlete and fighter squadron commander, Thach was optimistic that a team spirit could be developed because all personnel designated for the new experimental group would have the same goal and their training would be directed toward that goal. He moved to create a new concept of teamwork by stating to all in the task group that they were first and foremost sub hunters. Then they were aviators, submariners, and destroyermen only incidentally. Knowing there was a need to balance the strengths and weaknesses of the air-surface-subsurface members of the team, he ordered a policy of cross-pollination to form an integrated antisubmarine weapons system. This practice sent submarine personnel aboard destroyers and then to aircraft. Destroyer personnel spent time aboard submarines and aircraft, and so forth. Even Thach went aboard the rolling and pitching destroyers, bruised his shins and bumped his head inside the submarines, and occasionally took the controls of an S-2F Tracker during an exercise. These exchanges were not merely for orientation. The experience gained by all personnel was expected to assist understanding how to detect and counter an enemy submarine, and the combined experience was to find heretofore-unknown problems and possibilities. Submariners did not like to share secrets, but Thach convinced them that they must become the airmen and destroyer crews' professors with their tactics and equipment to deceive. Additionally, they were the only ones who could effectively measure a destroyer's effectiveness. Naturally the submariners did not like being the bad guys, but they opened up on how they could break a contact, and they shared other tactics with destroyer crewmen and aircrew. From the beginning Thach believed the sum of this mutual knowledge grew not in a straight line but geometrically, as each officer and rating saw what could and could not be done, especially with equipment and in tactics. And after his team began to think along the same lines as their other teammates, all could begin to try to know the mind of their adversary and thus how to confront him.

To elevate and maintain morale for his team, Thach decided his ships and planes would go to sea for two weeks and then spend two weeks in port for ship and aircraft maintenance and minor refits. This would make both Navy and domestic planning easier for everyone. While ashore, the lessons learned at sea would be discussed, incorporated into training, or modified for further experimentation. And as always, the two weeks in port were welcome.

When at sea, the team's schoolhouse was a selected 10,000 square mile area of ocean off Norfolk. To fully develop Task Group Alpha's overall team, Thach created a three-team concept in each ship. Explaining to his men that an enemy submarine could show up at any time of the day, night, or year, it was necessary for everyone to move toward round-the-clock operation. Informing all that he had observed respective ship captains on the bridge during refueling operations whenever they occurred, he directed that they delegate this and other major responsibilities to members of the second and third teams. The eventual goal was to have three "first teams," any one of the three as effective as the other two. He ensured compliance with his directive by watching the bridges during refueling. Not only did this three-team policy negate the constant need for general quarters, but it also ensured that whoever was on the bridge or manning sensitive equipment was alert. To Thach's thinking, a relatively inexperienced sonar operator who was alert was a better servant of his ship than a more experienced sailor who did not have full command of his senses due to lack of sleep. Ship captains, initially opposed to the idea for fear irreparable damage to their careers would be done if some ensign messed up, were assured by Thach that he would accept blame and cover for them in writing. Practicing what he preached aboard his flagship, Thach shared responsibilities with his chief of staff and operations officer.

For Task Group Alpha, a typical at-sea training period might start with destroyers working at basic sonar exercises in one quadrant, aircraft working with magnetic anomaly detectors (MAD) in another, helicopters dipping sensors in a third, and the killer subs tracking each other for practice. Then helicopters might be sent to work in coordination with destroyers and fixed wing aircraft to work with friendly subs. On the next day they would switch off, helicopters working with subs and aircraft with destroyers. Then destroyers might begin controlling both helicopters and fixed wing aircraft simultaneously in smaller scale coordinated exercises. Finally, the ten-thousand-square-mile schoolhouse was cleared and the group's ships and planes turned to a two or three day practice to track down and hold a target submarine. Evaluation revealed lessons from the coordinated exercises, and the individual members of the group then fell back on simpler specific exercises to iron out problems.

The goal of antisubmarine warfare was, and is, to deny an enemy effective use of his submarines and Thach never failed to remind any teammate, visitor or listener of that ultimate goal. From 1957 to 1959, the challenge was relatively manageable because even nuclear propelled Soviet submarines had to surface to fire a missile, to communicate, or to use surface radar. Any submarine was vulnerable when running shallow (visible), fast (easy sound detection), or snorkeling (visible to radar). Probing for the enemy was not easy, but when found in peacetime, the friendly or unfriendly sub must know he has been contacted. The crew within the discovered sub must also know that the discovering vessel or aircraft knows what he is doing and is prepared to counter any hostility. Discovery itself in the late 1950s would have kept an enemy submarine from surfacing to launch missiles unless it first destroyed the vessel(s) or aircraft that discovered it.

Thach learned from his exercises in the at-sea schoolhouse that it was easier to catch, identify, and kill a submarine when it had to move. Before his command of Alpha ended, he was able to borrow a nuclear propelled submarine and test his theories and practices against what was believed to be the maximum threat. To counter nuclear subs, his objective was to cut down their speed on any transit or on any maneuver as even a fast (thirty-knot-plus) submerged sub could not outrun aircraft. In keeping with the stated goal of antisubmarine warfare, he was always looking to give an enemy sub some disadvantage. Frequently it appeared there were strangers in the schoolhouse (termed "goblins" in Norfolk's ASW plotting room) which provided some real excitement beyond the usual tiring drills. As a submerged sub was blind and partially deaf, Thach would supply them with some interesting noises, the objective being to get a curious sub skipper to raise his periscope and thus increase the opportunity to detect him. Having observed submarine commanding officers' curiosity during his early 1958 educational visits to them, Thach particularly liked "the other shoe" approach of pinging only once and then waiting for the curious sub commander to ascend to find out why there was no second ping. Then he would "drop the other shoe."

During his exercises at sea, Thach was a little surprised to find that his own two submarines were not very successful in detecting a borrowed sub, but there was always hope that when the borrowed sub avoided his own submarines, his aircraft might get it. This was a matter of special concern as many who were studying the problem of tracking the new nuclear submarines suspected that the day might come when the only effective deterrent to a fast moving, deep diving nuclear submarine would be another fast, deep diving nuclear submarine.

Orders for a typical exercise at sea emanated from the flagship, a mobile command and communications center and logistics depot in addition to being a fast moving airbase. Aboard *Valley Forge* was Antisubmarine Squadron 36, built around twenty Grumman-built S-2F Trackers. These twin-engine antisubmarine planes crewed by two pilots and two radar operators combined search and strike functions. Capable of a maximum speed of 250 mph, its slower cruising speed of 150 mph allowed it to stay airborne for about seven hours when necessary although normal searches lasted less than five hours. Within the large assortment of electronic gear the S2Fs carried were search radar and electronic countermeasures (ECM) devices to detect submarine radar signals. A large wing-mounted searchlight added night capability. For attack, the planes were armed with an equally wide assortment of ordnance, including conventional and atomic depth bombs, two homing torpedoes, or rockets. Even with a full 4,800-pound allowance for ordnance, up to sixteen sonobuoys—miniature listening devices parachuted to the water that lower a hydrophone, raise an antenna, and broadcast to the planes whatever they hear below the water—could be accommodated (thirty-two in newer Trackers). In the tails of the Trackers and the six larger Norfolk shore-based Lockheed-built P2V Neptunes (described earlier) of Patrol Squadrons 8 and 21 were retractable magnetic anomaly detectors that registered disturbances caused in the earth's magnetic field by large metallic bodies such as submarines. Also on board CVS-45 (and later USS *Randolph* CVS-15) during the Alpha experiments were five AD-5W Skyraiders (Guppys) from Detachment 52 of VAW-12 to provide wide area surveillance around the group with its radar. These planes were essentially elevated antennas for radar and communications control stations.

Helicopter Antisubmarine Squadron 7 operated fifteen Sikorsky HSS helicopters. Despite the impediment of the helicopters' own deafening engine and rotor noise, for hunting the two pilots used ASDIC dipping or dunking short-range sonar providing an underwater search capability not possible with a fixed-wing aircraft. For attack, two Mark 43 homing torpedoes served as their primary weapon. Quick to arrive over a target, the HSS helicopters had a top speed of approximately 125 miles per hour but were limited by a range of less than 200 miles. The time they could remain airborne was further shortened if they were armed, otherwise their dunking exercise usually lasted a little over two hours.

The eight destroyers of Destroyer Squadron 28 were the highly mobile workhorses of Task Group Alpha. Relieved by Destroyer Squadron 36 on 1 October 1958 for overhauls, Squadron 28 returned on 1 May 1959. Given the improving submerged speed (twenty-five to thirty knots) of newer submarines, the thirty-five-knot destroyers were required to keep pace, their more maneuverable World War II twenty-four-knot destroyer escorts not best suited for the new challenge. The noisy destroyers could send out sonar waves and get back echoes from objects, but they had problems in rough seas and when running at high speed. While stopped or moving slowly, they enjoyed some success with passive listening for sub-generated sounds without putting any sound energy into the water. Good coverage could be obtained since sound travels five times faster in water than air. When attacking, the destroyer-borne ordnance was varied and lethal with conventional and atomic depth bombs and homing torpedoes.

The greatest advantage the submarines of Task Group Alpha had was their ability to listen for another submarine at depths beyond thermal layers and at depths beyond the capability of some other devices. Once located and identified, a homing torpedo was the weapon of choice. As long as the hunter submarine was stopped or moving slowly, it could rely on very long range passive sonar to detect an enemy submarine. Still, a submarine could back through its own wake and lose itself in the echo.

The concept of teamwork was well developed and appreciated during tactical exercises. To use and maximize their detection devices, both helicopters and Trackers operated within fifteen to thirty feet of the ocean surface, elevations below the masts of the destroyers. Consequently, everyone involved needed to be on the same page and understand exactly what other units were doing. But as good as the team became, very bad weather grounded planes and rough seas upset personnel efficiency and distorted sonar returns.

Although not an operational development force to evaluate equipment or perform controlled tests, the limitations of existing equipment became evident early in Alpha's exercises. The all weather Tracker (and Neptunes) still required too much of its pilot's attention to remain airborne at low altitude in bad weather; therefore, the need for an automatic navigation and control system to enable concentration on ASW problems. The cables aboard the group's helicopters were ninety feet long, but to dunk the sonar below temperature layers required additional cable. This, however, would have added considerable weight to the helicopter limiting its endurance and making it more difficult to move across the water to follow a submarine. Reliability of much equipment was inconsistent and experience was often lacking in the operators of the equipment. It was quickly apparent to Thach that equipment development was his most significant challenge and he became very proactive in announcing this need, especially to the scientific and industrial communities.

Where equipment could be improved, it was. During exercises the destroyers encountered difficulties switching radio frequencies while operating with aircraft because destroyer radios were not designed for such operation. Therefore Thach had aircraft push-button frequency changers installed on his destroyers.[5] Other similar changes were made when found necessary.

Justifiably confident that the challenges in developing a team concept for his antisubmarine people could be met without undue effort and that at sea exercises would reveal tactical modifications, Thach turned his attention to the heart of Task Group Alpha's reason for being: equipment development. He was particularly disappointed in research and development in Washington regarding deception devices for antisubmarine warfare. Told there was research for ASW he visited the Washington office only to find emphasis limited to deceptive devices for submarines to use against antisubmarine devices. From that point early in his command Thach attempted to enlist support for what Task Group Alpha was trying to accomplish by seeking the help of newspaper reporters, all with the Navy's blessing.[6]

The message to reporters was to alert the general public that success in antisubmarine warfare was a matter of national survival and that it had to be approached on a national scale. While acknowledging the need for early warning radar, then developing antimissile missiles, the U.S. Air Force's Strategic Air Command, aircraft carriers and other weapons systems, he warned that the country could not relax behind these vital deterrents. As valuable as these deterrents were, the "cellar door," as he termed it, was ajar. He cautioned that Hitler had gone around the Maginot Line and that the Soviet Union's new submarines could go under our missile-lined "Maginot Line." To ensure maximum exposure of his message, he not only invited reporters to *Valley Forge* or Norfolk but also used some of his limited time ashore traveling to present speeches. Although he made some stops at local chambers of commerce and universities, most visits were to industrial entities urging them to send representatives for two weeks at sea with the objective of interesting companies in research for ASW purposes. On land or at sea Thach advised industrial representatives of the long list of new or improved equipment required. Included was everything from improved sonobuoys and precise automatic tactical navigation to behavioral scientists studying how long aircrew could maintain optimum performance while serving in the antisubmarine warfare environment. And industry would need to invest some of its own money as Thach knew the Navy could not in 1958 fund all the research needed for ASW. Further, in producing new ASW equipment, industry needed to create devices that were fully automated, lightweight, inexpensive, and easy to maintain—goals rapidly becoming reachable with the transition from vacuum tubes to transistorized circuits.

In 1959, Thach announced that analog computers would give way to digital computers, and that this electronic evolution would enable a detecting vehicle to almost instantly send necessary information to another vehicle's fire-control system for attack. There is no documentation (unless classified), however, that he foresaw the utilization of satellites for either communications or navigation assistance during his Task Group Alpha duty.

While equipment development was at the center of his interest he knew that new detection and tracking devices were first dependent upon knowing what problems needed to be solved. And the first problem that demanded attention was the sea itself. New submarines and those building would have submerged speeds in excess of thirty knots and would be able to dive to one thousand feet or more. Soon neither U.S. Navy nor Soviet submarines would have to surface to fire their missiles, and certainly both sides hoped for a breakthrough that would allow search and communication without surfacing or having to travel so near the surface that they could be spotted from the air.

Thach advertised his perception to all listening that the Soviet Union in 1958 was well ahead of the United States in oceanography. The mysteries of the sea, he said, must be revealed before equipment could be invented or modified to view into it or adequately decipher its sounds. Ocean depths were not silent or empty. Frequently Thach referred to the sea as a "jungle" because, like Tarzan, one had to know the friendly sounds from the unfriendly. Detection devices had many problems, among them the fact that the ocean floor is rugged with mountains and ridges. Seawater itself attenuated sound, while temperature gradients—or layers—bounced back sonar transmissions or bent them. In addition, many areas of the sea bottom are littered with sunken ships, and pings from them could deceive even the best 1958 equipped sonar operator. Off Cape Hatteras, North Carolina, there was a historic graveyard of ships, making it a great place to hide. This area was not only a section of Thach's schoolhouse but also a near perfect place for a Soviet submarine to find a high degree of security. And Thach often reminded scientists and industrial visitors that MAD equipment started its ASW career in an industrial attempt to locate oil off the Gulf Coast and that radar was first thought of as interference until its utility was realized.

With many specific equipment requirements known, one of Thach's recommendations did not survive the censor's pen. In a speech titled "Electronics in ASW: Yesterday an Aid—Today a Necessity," the admiral was prepared to say that the answer for part of the detection and classification problem was "fingerprinting the underwater intruder . . . to develop a distinctive signature." The Department of Defense on 10 August 1959 cleared the undated speech but deleted the sensitive phrase above. The technology was not then available, but "fingerprinting" was an objective that science, industry, and the Navy wanted to achieve. Later it did.[7]

REPORT TO BURKE

Before the middle of 1959, Thach had a good grasp of the current state of readiness for antisubmarine forces, what had been accomplished with the Alpha experiment and what needed to be done. On the one hand, he was pleased that the other three antisubmarine groups of the Atlantic Fleet had shared in the knowledge gained in the past year and a half. The word was getting out how important teamwork was and that the weakness of the individual ships and planes of an ASW group could be largely overcome by each unit balancing and complementing the others with respective strengths. The objectives to detect, classify, identify, hold contact, and, if necessary, attack were known and appreciated. Frequent visitor Admiral Wright was pleased, and infrequent guest Secretary of the Navy Thomas S. Gates Jr. was interested and excited.

Near the top of Thach's priority list was the need to encourage a massive scientific effort on behalf of antisubmarine warfare, a goal he believed could be obtained only if the problems were made known. Newspaper reporters and editors from across the country who visited wrote extensively and well about the exploits of Alpha. *Time* magazine was so impressed with their 1 September 1958 article that Thach himself was on the cover. That the media was impressed both with the ASW problems and approaches to solutions was no accident. Thach gave all he visited and all visiting him written summaries of what was going on. Many of his written presentations and most all his orations combined the technical with semiserious levity. When speaking of new electronic requirements for planes or ships he often ended the list with "a coffee pot" or "better way to cook food." Permission to visit any of the ships or planes was encouraged, and plenty of his time was devoted to guests on the bridge of the carrier plus detailed orations during meals in his cabin (where Jack Daniels was an omnipresent guest).[8] All the favorable press and progress made in optimizing the current antisubmarine state of the art with Alpha to the contrary notwithstanding, it was time to present Admiral Burke with specific recommendations for the Navy to address. Those "Ten Most Urgent Needs for Antisubmarine Defense" are listed below, Thach's exact language recorded in italics.[9]

First: Burglar alarms—*Warning systems to indicate the presence and locations of submarines in deep ocean areas, in the moving areas around convoys or task forces, and in special areas not covered by other systems.* (Planting devices throughout the ocean accomplished this, a top-secret project not fully revealed until after the cold war and apparently one of the few major secrets kept from the Soviet Union. No mention of it was offered in Thach's oral history, but general statements alluding to deep-fixed, semifixed, and portable and mobile detection systems appear in his papers from the period.)

Second: "*Identification systems that will ensure that when our warning system flashes, it has been triggered by an enemy submarine, that there is in fact a burglar and not a stray* cat (fish). (Thach was especially interested in classification systems as the false contact rate during Alpha experiments was too high and he realized this could waste ordnance. He was particularly concerned that a nuclear depth charge not be used unless necessary.)

Third: *A prediction system, similar to existing weather prediction systems for forecasting sound propagation conditions at and below the surface over wide areas.* (Thach asked for a weather prediction type system to study, record, and predict changes in all the discontinuities and phenomena in the ocean to better understand the conditions under which submarines could operate. A favorite comment in his speeches was, "In a war of submersibles, cold or hot, the one who best knows the sea controls it." Again, Thach did not elaborate in his oral history on the considerable progress made in this area of research from 1958 to 1971 as much of it was still classified.)

Fourth: *Central stations and prowl squads, mobile ASW forces, to ensure apprehension of submarines located by the warning systems.* (Thach knew the U.S. Navy needed better fixed systems along with mobile ASW forces to ensure apprehension of subs located by the warning systems. He foresaw that detection would still be possible in the near future because even nuclear submarines would need to use radar and send messages from near or on the surface.)

Fifth: *Certain detection of any submarine which exposes any part of itself above the surface.*

Sixth: *Provision in each ASW vehicle for the capability of finding, recognizing, and destroying submarines.* (Thach championed the idea of an electronic submarine simulator to help train operators in distinguishing underwater sounds. Also needed were research in sound frequencies, improved communication systems between ships and aircraft, and an automatic data display that not only replaced hand drawn charts but also increased the speed of disseminating command intelligence.)

Seventh: *Discovery and utilization of phenomena, other than sound, for location of submerged submarines.* (Although Thach was not sure what phenomena might replace sound, he nonetheless reminded industrial representatives that they must consider size and weight and keep their inventions and innovations as small and light as possible. He also promised that those who could produce on this need ensured not only their own fortune but also a significant portion of the nation's security.)

Eighth: *Weapons with range and accuracy equal to the detection ability of our equipment's, so that a submarine found can be a submarine destroyed.* (In 1958, Thach quickly learned that detection occurred much farther away than the ability to attack, therefore new weapons were required with range and accuracy equal to the detection equipment. Faster and more accurate underwater guided weapons were a priority.)

Ninth: *Forces sufficient to accomplish the mission. Increase in the capability of our vehicles and equipment's will result in proportional decreases in forces required.* (Thach knew that the U.S. Navy would never have sufficient forces to cover the large number of Soviet submarines, but he believed that an optimum number with adequate equipment could achieve more than the numbers required using the 1945–58 technology. Although not included in this list to Burke is what he said in so many speeches, Thach constantly said that ships of Task Group Alpha were of 1930s designs, were 1940s built, and were wearing out fast from war service and patrols off Korea, and that development of the submarine had outstripped the equipment for defeating it. Of course Burke, and nearly every other officer on active duty in 1959, was well aware of the fleet's composite age as well as the difficulty in obtaining sufficient personnel for the Navy.)

Tenth: *Program to meet these requirements along two parallel lines: a. Interim— Improvement of present equipment's to counter the immediate threat. B. Long Range— Research and Development to counter the future threat.* (For the immediate future Thach pointed to a need to stimulate the development of new techniques and equipment for continuous surveillance in finding, classifying, and keeping track of Soviet submarines. For both the short and long term Thach strongly advocated an even closer relationship with the scientific fraternity knowing it could contribute to ASW progress if aware of its needs. Observing that 1958 submarine detection was registered in yards he stated the need for detection in miles. Continuing, he noted there must be preparation for the day when a submarine would have nothing above the surface to detect, and that electronic countermeasures equipment was easily foiled even in 1958 by the simple expedient of not radiating.)

Toward the end of his tour with Alpha, Thach was satisfied with the knowledge that several years' worth of progress in advancing antisubmarine warfare had been made in a period of months. Still he knew by mid-1959 that some of his recommendations to the CNO were beyond the technological reach of current science. But he also knew that scientific barriers had to be surmounted or there would be no effective answer to the antisubmarine challenge. Improvements to existing technology would be implemented, tactics and operations optimized. Still, he was haunted by the knowledge that on too many exercises his team had discovered the "enemy" sub after its fifteen minutes on the surface to fire its make-believe missile. And there were too many other exercises when the hull of his flagship absorbed the thud of an "enemy" submarine launched practice torpedo before his escorts knew of the submarine's presence or "destroyed" it.

Thach and Burke were not alone in their efforts. The Under-seas Warfare Advisory Panel to the subcommittee on military applications of the Joint Congressional Committee on Atomic Energy and several key congressional leaders were active in addressing the Soviet submarine threat. Several Navy bureaus prioritized research and development for the antisubmarine effort, and private industry was demonstrating more interest and investing their own money to a significant degree. And there were other encouraging events taking place in 1958 and 1959. On 1 July 1958, the first Fleet Ballistic Missile Submarine squadron, Atlantic Fleet was formed, and on 1 September an ASW laboratory was established at the Naval Air Development Center in Johnsville, Pennsylvania. On 15 April 1959, USS *Skipjack* (SSN-585), the first specially built nuclear submarine for attacking other submarines, was commissioned, and on 21 September 1959, the Regulus missile–armed *Barbero* (SSG-317) began deterrent patrol in the North Pacific, joining surface vessels there for the same purpose. Although only a stopgap effort, it nonetheless bought time for other deterrent technologies building and appearing on drawing boards. At the end of this assignment Thach viewed Task Group Alpha as a complex, difficult puzzle that had not been completely solved and would not be for some time. But he nonetheless thought that improvements in methods, tactics, and doctrine plus the better-identified equipment requirements would discourage the Soviets from using their submarines for blackmail or intimidation.

On 11 December 1959, Rear Admiral Thach stood on the flight deck of another of the Atlantic Fleet antisubmarine carriers (*Randolph*) for the change of command ceremonies detaching him from his command of Carrier Division 16 and Task Group Alpha. Rear Adm. John E. Clark would take his place, and in the near future Alpha would become Baker, then Charlie as it continued in its experimental mode. But Thach was not finished with his antisubmarine role. Selected for advancement to vice admiral effective January 1960, he left Norfolk and headed first to Washington, D.C., where his new three stars awaited him, and then west to Hawaii.

17

Commander, Antisubmarine Warfare Force, Pacific

By the time Vice Admiral Thach assumed duties as commander, Antisubmarine Warfare Forces, U.S. Pacific Fleet at NAS Ford Island on 1 March 1960, he was well aware that almost any technological breakthrough for U.S. Navy submarines would soon be followed by a similar occurrence in Soviet submarines.[1] In his continuing speeches to industrial and educational entities throughout this assignment, the admiral seldom failed to note that ten undetected Soviet submarines, each carrying twenty missiles, could devastate a major portion of the United States. Quickly acknowledging that he did not believe the Soviets had such capacity in 1960, he just as quickly stated that both sides were currently building submarines that would have that capability. Before he left his new command in mid-1963, he reminded listeners that change, technological progress, and new submarine construction were occurring at an exceedingly rapid pace, and that the Soviet Union did have ballistic missile submarines capable of destroying the United States.[2]

On 18 March 1960, the U.S. Navy enjoyed its first successful underwater launch of a rocket, and a week later, on the twenty-fifth, it accomplished the first launch of a guided missile from a nuclear powered sub (*Halibut* SSGN-587). In addition, the fleet ballistic missile submarine *George Washington* (SSBN-598) commissioned at Groton, Connecticut, on 30 December 1959, launched its first Polaris missile while submerged on 20 July 1960. The time from the Polaris missile's first test from Cape Canaveral on 20 April 1959 to being test fired at sea on 27 August 1959 was only four months. And just over a year later on 15 November 1960, sixteen A-1 Polaris missiles were aboard when the new SSBN-598 became operational. While U.S. intelligence agencies believed that the Soviets had not reached all these milestones, there was no solid estimate of how long it would be before they did. Always consistent, Thach let his listeners know it would be sooner rather than later.

When Thach received orders to form an antisubmarine defense force command for the Pacific similar to the one in the Atlantic for which he played such an important role, he reported first to CNO Burke at the Pentagon. Burke kept Thach in Washington for all of January and February 1960 for more reasons than one. Concerned about the growing Soviet presence in the Pacific, Burke wanted to upgrade the U.S. Navy's ASW capability there. One of the first impediments to doing so was his knowledge that the then commander in chief, U.S. Pacific Fleet, Adm. Herbert G. Hopwood, would not be receptive to having units of his command transferred to a new organization even though it would still be under his supervision. A second reason for retaining Thach in Washington was to give him time to meet with both civilian and military officials to share information pertinent to the new organization. These discussions assisted Thach in working on the charter for the new command and in determining the number of staff required and proficiencies desired. Equally important in Burke's mind was that he wanted a third star on Thach's uniform (approved by President Eisenhower) before he met with Admiral Hopwood.

Up to 1960, most eyes and attention in antisubmarine warfare emphasis had been on the Atlantic, primarily because the bulk of American industry and shipping was centered along and immediately inland from the East Coast. Too, the heart of Soviet submarine construction and operational bases faced the Atlantic, as the bulk of the Soviet Union's population and industrial capacity were located in the western region of that country. While the industrial heartland of the United States would remain in the East and Midwest for the near future, the nation had strategic interests in the Pacific, particularly Japan, the Philippines, and South Korea. The desire to keep these countries free from communism and open to trade required a response to the growing Soviet threat and potential growth of the Communist Chinese Navy.

Of course, the Pacific had not been without an antisubmarine force, and Thach's first experience in antisubmarine duty had been in the Pacific during the Korean War. The change in 1960 was greater emphasis on such a force and an organizational change that reflected the greater threat. As both Burke and Thach expected, Admiral Hopwood did not welcome the new slot for commander,

Antisubmarine Warfare Force, Pacific, believing that ASW should remain as a subdivision of his staff and that the new command did not warrant a vice admiral. Burke, however, believed that ASW was not receiving the attention required for a major response to Soviet activities and potential and that the Pacific Fleet ASW effort was neither as good nor as future oriented as the Atlantic effort in which Thach had been so instrumental. Thach was given to understand that studies on the subject of Pacific ASW readiness had resulted in acrimonious debates between Burke and Hopwood even though there was precedence in the Atlantic Fleet for the organization change. And he recalled that Burke's parting remarks to him in late February ended with a comment that he had seen Thach handle similar organizational disputes and he knew he could "handle it."

Upon reporting to Admiral Hopwood at Pearl Harbor, Thach immediately experienced what he and Burke expected. Indeed, Hopwood sent his deputy to preside over the 1 March commissioning ceremony for the new command. Hopwood was very polite and cordial, but he all but totally ignored Thach's oral and written recommendations for change. Unhappy with having the new independent command thrust upon him would have been enough to make Hopwood unhappy but he also had to give up some of his staff—particularly commander, Submarine Force, Pacific—and some surface ships to the new organization.[3] Thach's handpicked staff from outside the extant Pacific command was smaller than that of the Atlantic, but this gesture was too small to affect Hopwood's greater concerns. For the early days of this assignment, the only positive was that classmate and friend Rear Adm. Paul Ramsey was Hopwood's deputy. But even Ramsey's intercession resulted only in confirmation that Hopwood thought Thach was trying to take over his entire fleet. Still, the charter required action so Thach modified his tactical approach—a new spin of the weave to meet the challenge. Remembering that in aerial combat he could only fight one plane at a time even though others would have to be met, Thach approached Hopwood with the thought that all the provisions and functions of the new charter could not be assumed or implemented at one time. Consequently, Thach asked that only the first four be approved for implementation. That would allow him to get started working on those and the others could be approved when Hopwood thought the new command was ready and sufficiently organized to take them on. This step-by-step assumption of the new command's functions got Hopwood's attention and approval.

For the nearly six months Thach and Hopwood served together their relationship was polite but correct. That changed when Adm. John Sides relieved

Hopwood on 30 August 1960. Admiral Sides, who was well acquainted with Thach, thought that Thach was the expert in the ASW field and he should do whatever he thought best to accomplish the ends both men felt best for the Pacific Fleet, the Navy, and the nation. The assignment of Admiral Sides also smoothed relations between Thach and commander, Submarine Force, Pacific Rear Adm. Roy S. Benson, who served as ComSubPac from 12 March 1960 to June 1962. Benson provided Thach with all information requested, but his successor, Rear Adm. Bernard A. Clarey, was reluctant to share all the submariners' secrets. Statistical data were needed on the capabilities and limitations of submarines, destroyers, and aircraft that operated in the ASW role. Although Thach appreciated the outstanding war record of the highly decorated Clarey and felt sympathy for the desire of the ASW people to protect their professional secrets, he nonetheless needed to know what was happening, especially in submarine-against-submarine exercises. Given the technological advances in newer submarines, especially in speed, it was becoming apparent that the best antisubmarine defense was probably going to be another submarine. While not every statistic or tactical practice result he wanted made its way to his desk, he was satisfied that enough was available to prepare him for on-site inspection and evaluation of exercises. While witnessing exercises, secrets not previously revealed in classified memos were revealed.

TRAINING EVALUATION

Training at the complex on Ford Island never ended. Motivation, however, was never a problem, because the training exercises were for the most part actual operations against unfriendly submarines. Every day Thach's command attempted to detect, identify, and classify all underwater contacts. And everyone knew that an order could come at any moment for a detected, identified and classified contact to be destroyed. On occasion, however, the training centered on friendly ships in planned exercises.[4]

Thach and his staff developed an Operational Readiness Evaluation (ORE) for live exercises. When a hunter-killer task group deployed to the western Pacific, Thach scheduled them to operate off Hawaii long enough for he and his staff to conduct an ORE. The ultimate goal was to assess how well the team worked together and the grading process was used for the purpose of helping the task group understand weakness in their teamwork and how it could be improved. As in the Atlantic, nonparochialism and cross-pollination was at the heart of the teamwork emphasis. Even Thach's staff conducted the formal aspects of the ORE by having surface officers speak on aircraft procedures, submariners speak on surface ship concerns, and pilots discuss submarine and surface ship tactics.

Aboard the ships and aircraft of a new hunter-killer task group, Thach and his staff observed exercises off Hawaii. In remarks to a hunter-killer group (HUK), he told listeners the ORE would not be easy and that he did not want it to be. He noted it would be a test of stamina as well as skill as a team.[5] Exercises complete, the staff returned to the new operational control center to evaluate the exercise. Remembering the mountain of paperwork that accumulated when he first reported to Carrier Division 17 in 1951, Thach resolved to have his staff assemble all the important data from the ORE in two days. As the new task group usually departed for the western Pacific in less than a week after the ORE, Thach knew that the results of the evaluation were beneficial only if presented before the task group left Hawaii. With the ORE complete Thach invited the task group leaders to his operational control center for a meeting. With track charts already positioned discussions centered on the grades for each unit of the hunter-killer group and how well they had performed their respective responsibilities. Track charts helped the leaders see where they were at all times and what they were doing when a submarine was detected, or not, and how many false contacts were determined and how much time it took to downgrade or eliminate a contact. Armed with both the positive and negative results of Thach's ORE, hunter-killer task groups sailed to the western Pacific with more confidence in themselves and their weapons. And with potential problems in their procedures already known, members of the task group were better prepared to focus on the real threat.

THE INCREASED SOVIET PRESENCE IN THE PACIFIC

Helping prepare hunter-killer task groups occupied much of Thach's time, but at least once a year the Department of Defense conducted a worldwide nuclear war exercise. Due to Soviet introduction of a submarine ballistic missile capability in 1961, Thach's ASW command was included. To establish an operational control authority (OCA), Thach incorporated some Naval Districts and Sea Frontier commands around the Pacific Rim from the United States mainland through Alaska and to Japan. This incorporation allowed necessary functions to be handled without creating a new staff. The Canadian Navy was invited to participate, and they accepted, a situation warranted by rising fears that the cold war could turn hot. Competition between the two superpowers and distrust increased in 1960, and tensions in Germany led to a call-up of U.S. reserves on 1 October 1961. Given the unsettled state of international relations, the Department of Defense worldwide exercise took on even more significance.

Adding to Thach's worries was the knowledge that ballistic missiles were on the threshold of markedly increased range and ordnance capability. The Polaris A-1 had a range of twelve hundred miles, and the A-2, which would be ready in 1962, could fly fifteen hundred. Being prepared for tests in 1962 was the A-3, which would have a range of twenty-five hundred miles. Satisfaction with these advances in Thach's mind was offset by knowledge that the Soviets would undoubtedly match these capabilities and that a range of fifteen hundred miles was sufficient for Soviet submarines from both oceans to strike any target in the United States. And intelligence sources indicated that Soviet missile-launching submarines were operating more regularly close to the U.S. Pacific Coast. And while Thach often spoke of Soviet ballistic missiles, no mention is found in his papers of Soviet nuclear-tipped torpedoes. While danger was pervasive in ASW work, reminders that they might be the target (CVS aircraft carrier) of such a weapon in no manner enhanced morale. It was known that Soviet antiship missiles existed, but there was a good chance these could be effectively countered.

Tracking Soviet missile submarines was a major responsibility for Thach and his command. The process began by spotting a submarine when it left a Pacific port, usually either Vladivostok or Petropavlovsk, and then following it across the ocean. Often it passed first through Alaskan waters, where the Alaskan Operational Control Authority had responsibility, then into the Canadian OCA, and then to the Western Sea Frontier. Ideally, the contact would be followed by a plane or single ship until an ASW task group could pick it up. Cooperation of each district was essential to ensure that the contact was not lost, and if that meant continuing with the contact by crossing over into another jurisdiction, it would be done. Control of a tracking destroyer from the Alaskan command would come under the control of the Canadians once it entered their area, but the ship or plane was to stay with the contact until another ship or plane could relieve it. At one period during Thach's Pacific ASW duty, he had to pass orders to the commander, Western Sea Frontier, who was also a vice admiral and senior to him. Happily, that flag officer, Robert Goldthwaite, fully understood the purpose and requirements of the organizational structure, and he informed Thach that he would "get action" whenever needed.

Constant technological innovations improved U.S. capability for locating Soviet submarines, classifying, and tracking them. While the ability to determine probability whether or not a submerged Soviet submarine was carrying ballistic missiles was in place, there was very little reason for a Soviet submarine not carrying such weapons to operate off the U.S. Pacific coast. The idea of a conventional war with Soviet submarines attacking the U.S. merchant fleet was by 1961 rapidly growing more remote in U.S. thinking. In the Soviet Union, the concept and role of nuclear weapons varied from one political era to another, but in the early 1960s the pervasive fear on both sides was that any military confrontation would quickly evolve, if not start, with nuclear weapons.[6] Still, added to the strategic worries was knowledge that Soviet submarine strategy was rapidly moving from a defensive posture to offense. While ballistic missile submarines posed the major concern, the prospect of nuclear powered Soviet submarines with antiship missiles constituted a concern of nearly equal proportion.

Although Thach had long been an advocate of short, necessary-only radio communications throughout his career, he found that nearly all his duty stations experienced the same tendency to broadcast too much too often. While preparing for and during the Department of Defense worldwide exercise Thach sent word to all his people that he wanted to know everything that his command needed to know and to "try real hard" not to communicate anything else. Benefits from his pronouncement paid dividends quickly during the exercise as information arrived on Thach's desk before appearing on Admiral Sides's desk. Even though all routine communications traffic was being held in abeyance, there were still enough high-priority messages to jam circuits for commander, Pacific Fleet.

Always quick to recognize and adapt to new technology, Thach invited himself and his staff to make more use of computers. Tracking charts heretofore drawn exclusively by hand were programmed into the big computer available and submarine contacts were programmed. Inevitably the new technology proved a challenge on occasion particularly when one printout indicated the same submarine contact was in two widely separated places in the Pacific. The problem was that the programmer had forgotten to allow for travel around certain islands. More surprising to Thach was the fact that the error was not discovered sooner. But once corrected, the computer proved to be a major assist in keeping track of all submarines, those friendly and those not. And in short time the computer was also used to record performance data and help determine correlation between factors to assist in recommendations for change and modification to policy and procedures.

Although not as revolutionary as computers the briefing room in Thach's office building was constructed for optimum viewing and acoustics. No microphones were necessary and everyone in a natural pitch could hear every other voice in the room. The room had three large screens to help listeners see charts and was bugged with recorders to make a tape of all briefings, conversations and interviews. Although the tape recordings served a primary purpose of assisting administrative specialists to prepare exact transcripts of meetings, on one occasion they helped preserve Thach's standing with the chief of naval operations. An overzealous reporter misquoted Thach on the subject of congressional budget appropriations for new aircraft carriers versus money for his ASW program. In fact, Thach wanted more money for new carriers in addition to more money to meet the increased Soviet submarine threat. But when the article appeared in the *New York Times*, Thach was quoted as saying that money should be diverted from building new carriers in favor of the ASW program. In August 1961, classmate and friend Adm. George Anderson had taken over as chief of naval operations from retired Admiral Burke, and the article justifiably upset Anderson. Friends of both Anderson and Thach at the Pentagon sent word to Thach about the article and Anderson's reaction. As he had not been directly accused by Anderson, Thach had to bide his time until the issue surfaced. Again his combat experience and years of practicing discipline in the air paid dividends, as he remained patient waiting for the right moment to defend himself. A few weeks later, Admiral Sides told Thach that Anderson had mentioned the article in a letter to him and the door was then open for Thach to respond. A tape of the actual interview was provided to Sides, who forwarded it to Anderson. That Anderson was pleased with the tape was evidenced by his later acquiescence to Thach's request to remain at Pearl in his ASW command for a year longer than Anderson had planned.

THE CUBAN MISSILE CRISIS

On numerous occasions from 1958 to 1962, Thach called attention to the protracted Berlin crisis and repeated Eisenhower and Kennedy's view that the war of words over that contested city could quickly turn to general war. However, in his 1971 oral history interview, he stated that he did not consider the October 1962 Cuban missile crisis a "crisis." Not until just over ten years after his death was it revealed that the Soviets did indeed have missiles ready to launch from that Caribbean island and that Cuba's dictator, Fidel Castro, had agreed to support a Soviet decision to go to war if necessary. Still, Thach had to address the event in October 1962, even though he was nearly halfway around the world from the area of potential conflict. While some military bases in the southeastern United States went on alert for the first time, an occurrence not experienced during either World War II or the Korean War, Thach found himself as concerned with diplomatic matters as with his ASW operations.

In October 1962, Thach maintained his normal schedule and flew to Tokyo while world attention centered on Moscow, Havana, and Washington. One of Thach's missions was to visit Japan's Maritime Self Defense Force officials that had developed a post–World War II Navy designed primarily for coastal patrol and antisubmarine warfare. On his first visit, in 1960, he promised to work closely with the Japanese and visit often. By mid-1962, he had earned the respect of the former enemy by not only visiting but also cooperating with them as their Navy developed and modernized.[7] Upon his arrival in October, the U.S. ambassador to Japan requested Thach to visit him earlier than scheduled to consult on the seriousness of the Cuban crisis. Thach told him that he did not think the Soviets would launch a nuclear war over a country as peripheral to their interests as Cuba. Germany yes, Cuba no. Further Thach informed the ambassador that given the high state of readiness of the U.S. military that the Soviets would not have the element of surprise. Finally, Thach informed the ambassador that no unusual Soviet submarine activity had yet been detected in the Pacific. All these factors in conjunction seemed to point toward a resolution short of war. But Thach concluded by saying that either nothing would happen or there would be a nuclear war. Both men agreed, however, that a high state of alert should be in place.[8]

Leaving the ambassador, Thach then met with the head of the Japanese Navy, Admiral Sadayoshi Makayama. Thach was immediately impressed that the Japanese wanted to deal directly with the looming crisis in a meaningful manner. Thach suggested the Japanese increase surveillance in the Sea of Japan, particularly around the Soviet submarine base at Vladivostok to see what activity was occurring. Other sites to watch were identified and coordination details including communication frequencies were worked out between the respective staff's. A written agreement and plans made to meet the immediate threat, Thach was then invited to review and offer advice on the latest Japanese top secret research and development projects. Before leaving Japan, Thach offered to release some classified equipment to the Japanese and agreements were made to continue sharing information for emerging ASW technology. There was concern on the American side, however, pertaining to Japan's lack of an antiespionage act. Security within the Japanese Navy was excellent, but in 1962 there was insufficient security to prevent Japanese business interests from selling military secrets.

Happily, the world did not go to war in 1962, and both sides retreated from the confrontation off Cuba believing they had saved face. Both sides took credit for preserving the peace and both were justified in doing so. But both sides also became even more serious in preparing for war and for searching for a way to coexist without having to destroy each other.

ONE LAST YEAR IN THE PACIFIC

Soon after Admiral Anderson learned that Thach had been misquoted in the press, the two men visited. Neither the article nor the tape recording was mentioned. Anderson, who had earned his wings at Pensacola the same year as Thach, wanted his classmate to return in 1962 to Washington to take over the job as deputy chief of naval operations (air). Thach, however, felt he was too involved in the rapid developments then taking place in the relatively new ASW command and desired to stay one more year. Anderson understood his friend's logic and agreed to let him stay one more year, but he noted that in 1963 he expected to have Thach back at the Pentagon.

Having been directly associated with antisubmarine developments for the past four years, Thach was nowhere near being burned out on the assignment. For an aviator to turn down, actually delay, an assignment to take the top aviator job in the Navy was documented proof of his enthusiastic interest in remaining at Pearl for the additional year.[9] While in the Atlantic Thach took what existed and after two years reported that the ASW effort was woefully inadequate. In the Pacific, technology was beginning to catch up with the known needs and it was an exciting time to see equipment both ashore and at sea vastly improve the entire ASW endeavor.

Thach's experience in ASW paid dividends as the more he worked in the program the more ideas came to him to accelerate both technology and the

Navy's administrative response to the newer devices. Procedures and policies had to change and Thach had the satisfaction of knowing he was at the frontier of ASW. He fully appreciated that he had an opportunity to look across the frontier to what was previously thought impossible or improbable to what might be or could be done. And he could look back and see that many of his ideas and innovations were standard practice in 1962.

Although time ran out before he could bring it into being, Thach conceived of a three-dimensional tactical development range for the shore-aerial-sea components of a hunter-killer group to refine its art. Inspired by a laboratory experiment at the University of Washington, Thach pushed his idea only to be told that the Navy already had a research range in the Caribbean. For the longest time, he seemed unable to convince others that the difference he was recommending was not another research range but a tactical development range. After Thach moved on to other assignments, the tactical range was established.

Another of Thach's endeavors was to create a program for the control and protection of merchant shipping. NATO (North Atlantic Treaty Organization) had already developed such a program for the Atlantic and which countries would do what during a confrontation, but responsibility for NATO ships that steamed into the Pacific was not strictly defined. Both the United Kingdom and Australia looked at the problem and decided that the U.S. Navy should accept the responsibility. As no other U.S. Navy command seemed anxious to accept the new international responsibility, Thach had it written into his charter. Before friendly ships could be protected there first had to be a way of knowing where they were and a method of keeping track of them. In time, further utilization of the big computer at Pearl allowed Thach and his staff to know where all ships were at a given time and control was thereby established.

Protection was another matter. Naval ships had the speed, electronics, and ordnance means to give them a better than even chance to protect themselves. Slow merchant ships did not mount defensive electronics or ordnance. In case of limited war, merchant ships were to be directed either to port or to a position where a convoy could be formed. Still, in the early 1960s the U.S. Navy did not have enough ships to properly convoy or protect merchant ships. No merchant shipping meant no international trade and no importation of critically needed raw materials. The best hope was that such a war would not be protracted, but in the short run some plan was necessary. Somewhat surprisingly, merchant ships were not initially enthused about providing information as to their cargo, departure date, normal speed, and destination. When reminded that they could save money by not having to unnecessarily change course when SOS calls went out and other ships might be closer, companies agreed to provide the needed data. Even then, the information was not supplied directly to the U.S. Navy. Instead, the information noted above plus weather conditions were reported to the U.S. Coast Guard. In the end, everyone was satisfied.

At the time of his oral history interview there were many developments in ASW that Thach could not discuss because they were classified and highly sensitive. The underwater mapping of the oceans was progressing, underwater sound systems for surveillance were being developed and constantly improved, and the use and refinement of computer technology was ongoing. Computer stations analyzed acoustic data while scientists and technicians worked to construct instruments that would process information faster. In time computer technology in aircraft was greatly enhanced and soon after the admiral's retirement, it went into space. Known to the public were new developments such as new antisubmarine torpedoes, long-range, jet-powered patrol planes and vastly improved helicopters.[10] But with all the advances neither side during the cold war marched out in front of the other sufficiently to allow either to be confident that it could win a nuclear showdown with missiles being launched from land, sea and the air. Recent revelations, however, have indicated that late in the cold war the Soviets did obtain information via espionage that allowed them to track U.S. submarines, and the Soviets seem not to have known the full scope of the United States' ability to track their submarines. But in the late 1950s and early 1960s, it was Vice Admiral Thach's responsibility to move the frontier of antisubmarine defense as far as the state of the art could take it. Acknowledging that there never could be "absolutely airtight security from an under-seas counteroffense system," he went on to state that "the impossibility of achieving 100 percent success cannot, of course, be an excuse for ignoring the challenge."[11] Indeed, the search for optimum antisubmarine tactics had to continue.

By the time of his relief on 13 June 1963 by Vice Adm. John T. Hayward, Thach had organized and established the new ASW command to help ensure an integrated defense of the continental United States and of shipping at sea. Much of the new antisubmarine equipment in use and on the drawing boards plus proven new policies and procedures could be traced back to his insights and efforts, and he could acknowledge "tremendous strides."[12] And he had initiated the use of automated modeling techniques for Pacific Ocean anti-submarine warfare operations, training, and evaluations. Fully appreciating the power of the media, Thach offered constant reminders that research and development laboratories had to provide advanced concepts for needed antisubmarine equipment, that industry had to translate the products of science into tactical tools, and that Congress had to appropriate the necessary funds. In all these roles, Thach was successful and as proud of his responses to the antisubmarine challenges as any he had initiated during World War II or Korea.

18

Deputy Chief of Naval Operations (Air)

On 8 July 1963, Vice Admiral Thach assumed his duties as deputy chief of naval operations for air, naval aviation's top billet. The first officer to hold the title (1943–44) was Thach's late friend, Adm. John S. McCain, and Thach could not help but think that his old boss would have been pleased with his ascension to the post. From his office in the Pentagon, Thach would be responsible for the generation of requirements, research and development, design, procurement, and allocation of all naval aircraft, naval air weapons systems, and their associated support equipment. And he would be responsible for the preparation, submission, and justification before Congress of the naval budget for aircraft and associated systems to the tune of $3 to $4 billion annually.

Thach relieved Vice Adm. William A. Schoech and in the process inherited his aide, Capt. James B. Cain, a World War II fighter ace who had also flown F2H Banshees in combat during the Korean War. The two officers had first met at Jacksonville in 1943, when then Ensign Cain was present when then Lieutenant Commander Thach made his first flight in the new F6F Hellcat. Praising everything about the new fighter except for its heavy fuel consumption, Thach then followed Cain on a discovery tour of local refreshment fountains. Reunited at the Pentagon, Cain arranged for the band to play "Paddlin' Madalyn Home" during the change of command ceremony, a gesture immensely appreciated by both Vice Admiral and Mrs. Thach.[1]

Days at the Pentagon quickly proved long and demanding, and it seemed everyone in industry needed to see the DCNO (air) immediately and for urgent reasons. To get a handle on who really needed to visit, Thach and his aides asked last-minute callers if they could meet at 7:15 P.M. Those who showed at 7:15 really did need an appointment while those who did not decided the late hour did not meet their schedule.

THE COLD WAR AND VIETNAM

While Thach was in the Pacific from 1960 to mid-1963, the cold war threatened to get hot on occasions other than Cuba in October 1962. Soviet belligerency in Germany threatened to again cut off Berlin, and in August 1961 a tangible expression of the Soviet attitude was manifested as construction of the Berlin Wall began. As with all of the Iron Curtain, the Berlin Wall was intended by the Communists to keep their people in rather than keep others out.

While in the United States most attention centered on the Communist world and its two major powers, the Soviet Union and Red China, much of the rest of the world was consumed with problems of colonialism and nationalism. India had broken away from Great Britain in 1947 and the map of the world was soon thereafter in constant flux as colonial powers relinquished political control to local inhabitants—usually after armed conflict—mostly in South Asia, the Middle East, and Africa. In Southeast Asia, the French had been evicted from French Indochina (Vietnam) in 1954, but civil war ensued between Communists in the northern portion of the country and the semidemocratic south. Communist sympathizers in the south, the Vietcong, threatened the pro-American South Vietnamese government and President John F. Kennedy responded in 1961 by authorizing U.S. military advisers to assist. In November 1963, the sitting government of South Vietnam was overthrown, Kennedy was assassinated, and a new president, Lyndon B. Johnson, inherited the instability and conflicts in Asia. Only some fifteen thousand American advisers were in Vietnam when Johnson became president. But before Thach left the Pentagon, the gradual escalation of troops after the August 1964 Gulf of Tonkin incident drove the number from two hundred thousand (in 1965) to over half a million before the end of the decade. In his 1971 oral history, Thach's comments on Vietnam comprised only a few lines, but as earlier he thought that no war should be fought without an intention to win. His cold war view was that the Communists meant what they said when they spoke of worldwide domination.

And given the breadth of his naval experience, he viewed the war in Vietnam as a Communist move to block the nearby Straits of Malacca along with other maritime chokepoints—Panama Canal, Suez Canal—and would continue the effort "until someone steps on their fingers."[2]

In 1971, Thach did not think anyone was stepping hard enough on Communist fingers. He particularly faulted the continuing practice, initially championed by Secretary of Defense Robert S. McNamara, of gradualism ("Where if the enemy does something, then we do something, hoping to discourage him from doing anything like that again"), which led only to piecemeal mistakes. Such a policy, according to Thach, was one of the violations of the basic principles of warfare and went completely contrary to a careful study made by the Weapons System Evaluation Group. Thach doubted that McNamara ever read that study and again noted that studies were no more effective than the willingness of people in authority to pay attention to them.

Despite the tragic events in November 1963, both in the United States and Vietnam, there was one event that fall memorable for a pleasant occasion. Thach flew aboard the *Forrestal* to observe one of the tests to determine the big carrier's ability to land a plane never intended to operate from any carrier. In spite of repainting the designated landing strip on the angle deck, there would be only fifteen feet clearance between the plane's wing and carrier's island. The twenty-nine-year-old aviator at the controls of the C-130 Hercules transport was Lt. James H. Flatley III, the son of Thach's good friend and fellow fighter pilot who had coined the phrase "Thach Weave." The occasion had more meaning to Thach than it might otherwise have had, inasmuch as Vice Adm. James H. Flatley Jr. had died of cancer in July 1958 while Thach was at sea with Task Force Alpha. Watching the heavy, four-engine propjet with its 132-foot wingspan enter the landing pattern, there was good reason to worry because loading weight was purposely increased after each landing. But for all tests over four days young Flatley piloted the transport aboard for twenty-nine touch-and-go landings, twenty-six unarrested full-stop landings, and twenty-six unassisted takeoffs. Not satisfied with his aide's first draft of a Distinguished Flying Cross for Flatley, Thach sent him back to the typewriter. The second draft got the job done and the citation was approved. Over Vietnam soon after, Commander Flatley earned many of the same combat medals his father earned in World War II and later also rose to flag rank before his retirement in 1987.[3]

PROCUREMENT OF AIRCRAFT

Although he had spent the previous six years of his naval career in antisubmarine work, airplanes were nonetheless one of the significant tools in the ASW program. From late 1957 to mid-1963, Thach dealt daily with Navy planes and needed few briefings on the subject when he arrived at the Pentagon in July 1963. Navy fighter capability was much greater in 1963 than it had been during the Korean War when its best fighter was the F9F-5 Panther which was no match for a MiG-15. Born of the Korean War disparity between the Navy's best jet and their Russian-built adversaries, a 1952 requirement for a Navy fighter superior to the MiG-15 resulted in the Ling-Temco-Vought (LTV) F-8 Crusader that went aboard carriers in 1957. Faster than its Soviet opponents and capable of meeting them at high altitude with both Sidewinder missiles and 20-mm cannon, the Crusader demonstrated again how well the Navy and industry could work together to expand the state of the art to meet military needs. Also in the Navy inventory when Thach was DCNO (air) was the exceptional McDonnell Douglas F-4 Phantom II (on carriers from late 1960), which served in multiple roles as fighter, high-altitude interceptor, and attack plane. Notable Navy attack (strike) planes for the period included the McDonnell Douglas A-4 Skyhawk and the Grumman A-6 Intruder, aboard carriers from 1956 and 1963 respectively.

Visiting the several aircraft manufactures that were currently serving or hoping to sell future planes to the Navy consumed Thach's time when he was not in meetings or his office. When he reported to the Pentagon in 1963, Thach knew he was taking time off his life by arriving early, sitting through meetings for most of the day, and then starting paperwork at his desk after 5:00 P.M. As it had been during the Floberg years, dinner at home was served most often between 10:00 P.M. and midnight. But when visiting aircraft contractors, the hours were seldom better. Thach and a few members of his staff often flew in a converted Douglas AD Skyraider—hardly the epitome of quiet, comfortable air travel—for such conferences. Meetings were generally held in the morning with, hopefully, a round of golf in the afternoon, and dinner was usually an exercise in entertainment as much as nourishment. At age fifty-eight when he began this duty, Thach still had great enthusiasm for golf but a considerably diminished appetite for alcohol and entertainment. While visiting one contractor early during his tour, the after-dinner party raged on to the early morning hours at which time Thach announced he was going to the men's room. But before announcing his departure he had asked his aide, Capt. Cain, to "give him twenty minutes" before informing the company dignitaries that in fact he had called a cab and was heading back to the hotel to get some sleep. Thach appreciated the effort to show him a good time, and he did not want to break up the party, but he needed rest. On another, similar occasion when everyone around him was feeling little to no pain as the result of freely flowing liquor, Thach advised his host that he hoped their company would not make any important decisions that evening relevant to the Navy. And during another such party, Thach and his aide planned escape through a bathroom window. (Small wonder that Thach later suggested Captain Cain as CO of the ASW carrier *Yorktown* CVS-10.) As DCNO (air), Thach understood that contractors would do most anything to impress him. What they did not know is that all he was interested in was facts, statistics, honest projections, and an early end to the day.[4]

The procurement process began with the Navy setting the operational requirements for a new plane. Some new planes were low risk meaning that known technology could be utilized to obtain the desired aircraft. Others were high risk, pushing the state of the art with the goal of meeting stated performance standards with heretofore-unavailable technology. When planning for a new fighter requirements were that it first had to work, had to be able to get aboard a carrier, and could not be second best in combat. Tentative specific operational requirements were written and forwarded to the chief of naval materiel. This essentially made the chief of naval operations the customer and the chief of naval materiel with all his bureaus the producer. The chief of naval materiel then established a proposed technical approach stating how that office proposed to meet the various requirements, substantiated the technical feasibility of each system, and offered possible alternatives and tradeoffs listing advantages and disadvantages. Special considerations for any fighter included speed, performance in rate of climb, time required to produce a given number of planes, man-hours required for maintenance, the number of planes expected to be in repair, and the number of hours in commission. Paperwork then traveled back and forth until all relevant offices were in basic agreement. Once the Navy settled on its requirements and plan, the request for approval went up the line to civilian authorities and finally to Congress. In this manner, the Ling-Temco-Vought A-7 Corsair II was born and its progress was overseen by Thach during his tenure as DCNO (air). The design had been initiated on

17 May 1963 for a competition to produce a light attack aircraft (VAL) to replace the Skyhawk. A new strike plane was needed in a hurry because it had been too long since the Navy had been able to get a new one built and the Skyhawk dated back to 1952. Speed of production, therefore, became a major requirement with production needed no later than 1967. Rather than following precedent for most earlier planes, Thach changed the language of the contract bids for the incentive to be a penalty for not meeting requirements rather than being cost plus fixed fee or other positive incentives. Thach believed that would separate the men from the boys because only the people who really knew how to build planes would have the confidence to bid. And the amount of a penalty for a given characteristic put the spotlight on where a company's strengths and weaknesses were.

As with any other aircraft the Navy wanted to obtain, Thach had to defend this new strike plane before Congress. This task was never easy and was especially so when one of the congressmen on the committee represented a district with a contractor that had not won the bid. Like most other naval officers who had to testify before Congress or one of their committees, Thach knew he could not afford to make a congressman look foolish. The goal was to be careful, to not fight with him so that his vote could be won. And he had to remember not to lose his temper, even though he was not given to that type of a social lapse. Above all, he had to remember that many congressmen knew their stuff and could ask pertinent questions. Consequently, statistics for documentation of cost effectiveness had to be ready. Of course some congressmen spoke only to appease their constituents, and when they did, naval officers knew to sit quietly while the elected officials ensured that their comments made it into the *Congressional Record*.

LTV won the bid contest on 11 February 1964, signed the contract on 19 March, and basically designed the new plane around their earlier F-8 design. The new strike aircraft had a shorter fuselage than the fighter with no afterburner, less sweep back on the wing and no provision for varying the wing incidence. The strike version's wings were strengthened to carry 15,000 pounds of ordnance, all external except for the one 20-mm M61-A1 multibarrel gun. Manned by only a pilot, the plane was introduced aboard carriers on 1 February 1967 and was in combat before the year was over. LTV was particularly happy that Thach chose the name "Corsair II" to honor the Vought company that had built the highly successful gull-wing World War II/Korea F4U Corsair. As aircraft nicknames were not official before World War II, the prewar Corsairs were not counted. Additionally the company was highly pleased that the U.S. Air

Force also adopted yet another Navy-developed plane. And Thach was especially happy that the Corsair II proved excellent for close air support, a matter still close to his heart.

Procurement matters were always a priority for the DCNO (air), and it had its high moments. In addition to the Corsair II, Thach was also particularly pleased with his efforts to gain congressional funding for improvements in the Grumman E-2 airborne early warning aircraft that could see low-flying aircraft well beyond the horizon. Pushing the electronic state of the art, the new system was high risk, but Thach explained to the House Armed Services Committee that sought-for improvements would bring benefits no other country would have. There was justifiable concern that the project would cost more than announced and Thach confirmed that it might. But the rewards were too beneficial not to proceed. The committee agreed, the program went forward, and the Navy ended up with a highly successful system.

THACH, MCNAMARA, AND THE TFX CONTROVERSY

When Thach first got news of his assignment as DCNO (air) in May 1963, he was optimistic that he would be able to work well with his civilian superiors. He was doubtful that he would be as close to the secretary of the navy, Fred Korth, soon succeeded by Paul H. Nitze, or the secretary of defense, Robert McNamara, as he had been with Secretary John Floberg, but he was nonetheless upbeat. Thach's optimism, however, was not as evident in the newspaper offices of the *Honolulu Star-Bulletin* when the transfer became known. There had been major problems between the civilian leadership of the Kennedy administration and the military, and the newspaper writers openly wondered if Thach—"the Navy's undisputed authority in this field [aircraft]"—would be able to accept current policies. Noting Secretary of Defense McNamara's "distinctly unmilitary" background and "Father Knows Best" attitude, the writers noted that "it takes a considerable degree of devotion to civilian control to agree" that Adm. George Anderson and General Curtis LeMay were wrong "in the TFX plane controversy." Ending on a relatively more positive view, however, the article stated that Secretary McNamara, "who may not know a fan from an afterburner[,] . . . has picked the best man to be found."[5]

The TFX controversy the Hawaiian newspaper writers referred to centered on McNamara's desire for commonality in the procurement of some aircraft.

Particularly he wanted the Air Force and Navy to build an airplane that could be used by both services.[6] The idea of commonality had precedent as the AD Skyraider and Phantom II were highly successful for both the Navy and Air Force (although the avionics in the Phantom were different). And in the years since the TFX dispute the idea has reappeared and appears destined for success with the Joint Strike Fighter (JSF).[7] Still, the program championed by McNamara became one of the more infamous failures in U.S. military procurement history.

The problem was that the Navy wanted an air superiority carrier fighter and the Air Force wanted a bomber. After viewing the normal preliminary studies, CNO Anderson and General LeMay concluded that the TFX design did not meet the needs of their respective services. Anderson had defended the conclusions to such a degree that McNamara asked Kennedy to relieve him as CNO which did occur two months after Thach arrived back in Washington. Although concerned about Admiral Anderson's impending retirement Thach knew he could work with his successor, Adm. David L. McDonald, and initially thought he could work with McNamara. McNamara had announced his desire to apply cost effectiveness to the military in the same manner as used in Detroit when he was an executive with Ford Motor Company. Knowing that McNamara had been the first nonmember of the Henry Ford family to become president of the Ford Motor Company and one of the executives to have contributed to its recent success, Thach had reason for optimism. Indeed, he was "delighted" and thought that McNamara's announcement that he would make decisions "based on facts and cost-effectiveness instead of political pressure was the most encouraging thing he had heard in a long time."

Once at the Pentagon, however, Thach soon found that McNamara was far more concerned with cost than effectiveness, and his great expectations diminished. And the TFX controversy was not over because McNamara was determined to build the plane despite growing problems. Most alarming to Thach, however, was his discovery that McNamara and members of his staff brought from Detroit were violating the absolute rule utilized by the Weapons Study Evaluation Group that realistic input had to form the basis of any study. There could be no sins of omission as inadequate assumptions ensured a wrong answer and that only through a valid cost analysis could a procurement item be properly appraised. While with WSEG, Thach had learned that a more costly item could be far more cost effective over time if the proper assumptions had been made in the original study. Determined to streamline the military as he had at Ford, McNamara set up inflexible policy on design changes during research and development and remained adamant during the two years Thach was DCNO (air).

The Corsair II was low risk in 1963–64, given the state of the art and the performance parameters desired, but the TFX was high risk. First and foremost, the Air Force specifications were for a high-performance strategic bomber with low level capability while the Navy wanted an air superiority fighter that could launch the planned long-range Phoenix missile. Second, the Air Force requirement that the plane have long range necessarily required the design to allow for a heavy fuel load which when combined with ordnance meant considerable size and weight—attributes not desired for a carrier plane. Third, and most disturbing, was McNamara's intransigence in regard to studies demonstrating that the multimission aircraft project was technically infeasible. For this procurement battle, the Air Force and Navy were largely on the same side with both arguing against the design of a multipurpose, multiservice design.

Thach synthesized his view that the entire TFX effort was an attempt to combine the qualities of a moving van with that of a racer. And he knew that quantitative statistical analysis, greatly favored by McNamara and his staff, was not sufficient to determine a solution to a naval aviation problem without also employing qualitative analysis. But during his tenure Thach saw study after study document that the developing design was going to be too heavy for carrier landings and was not going to meet certain requirements such as acceleration to combat speed. In 1971, Thach recalled that he wrote a letter to Secretary Korth in 1963 that the F-111B (F-111A for the U.S. Air Force) would not meet Navy requirements. Somewhat angrily, Korth told Thach not to put such thoughts in writing because McNamara did not want congressional spies to know that the Navy was not happy. And in 1965, after seeing the negative results of wind tunnel tests, Thach wrote a letter to Secretary Nitze recommending that the concept of commonality be put aside and allow both services to develop their own planes. CNO McDonald approved the letter and agreed with Thach that the project should be derailed before the major expense of production began. But Thach was not present for the meeting between Nitze, for whom he developed great respect, and McNamara but surmised that McNamara declined to accept the letter even though a congressional subcommittee did see it. What he did know was that McNamara, in 1965 sometimes referred to in Washington as "the assistant President," took personal control of the project and met with contractors without Thach or the Navy project managers being present.[8]

After Thach's tour of duty as DCNO (air), Congress finally killed the project, the Air Force having taken some five hundred F-111A models that did not satisfactorily meet their needs and the Navy having six experimental versions.[9] A congressional report after Thach's retirement but for which he was consulted concluded that the TFX program was "a series of management blunders . . . [with] poor decisions at the highest levels of the Department of Defense, which compounded error upon error." Continuing, it added, "appointees . . . during the McNamara era overrode expert advice to impose personal judgments on complex matters beyond their expertise." Finally the report noted that "this sorry record has done nothing to enhance public confidence in the integrity and competence of the people who are charged with preserving the national security."[10] In looking back on this period many have voiced the opinion that the same attitude displayed by the civilian leadership on the TFX issue was in part responsible for widening the war in Vietnam.[11]

While significant sums of money were wasted, there was one blessing that did emerge from the mess after the project was cancelled in 1968. Having worked on the Navy version of the TFX/F-111B project with General Dynamics, the Grumman Corporation won a new design contest that incorporated the F-111's engines and other innovations with their experience with variable swept wings into the F-14 Tomcat. Capable of launching the still developing Phoenix missile, the gap for a new air superiority fighter in the Navy's inventory was filled when the new fighter went aboard carriers in 1972 and is still operational into the twenty-first century.

Although the TFX/F-111 issue was the most highly visible difference between the thinking of Thach and McNamara from 1963 to 1965, there were other areas of disagreement. Five weeks after arriving at the Pentagon, Thach put in writing his disagreement with McNamara on the number of carriers to be maintained in active status. Thach believed the number of attack carriers should continue to be at least fifteen; McNamara wanted fewer. And there was the question of replacing the five aging *Essex*-class carriers still operating with the fleet. Building new carriers was necessary in order to both replace older carriers that could not accommodate all of the newer Navy planes and maintain the necessary skills and specialized industrial complex that could build modern carriers.[12]

THACH AND RICKOVER

As DCNO (air), Thach had to communicate with other naval leaders. With the new nuclear-powered *Enterprise* (CVN-65) at sea since its commissioning on 25 November 1961, Thach had to do business with Vice Adm. Hyman G. Rickover, the brightest star of nuclear propulsion. Thach and Rickover had a long history before Thach reported for duty as DCNO (air). Although Rickover (Naval Academy class of 1922) had been in the Navy longer than Thach, both were still captains when the two officers began to have frequent discussions. While Thach was working with Secretary Floberg, Rickover contacted Thach regularly concerning appropriations for the nuclear program. For most of that time Thach believed he was a big help to Rickover in ensuring things got done when the secretary's office could help; both officers strongly believed in the potential of nuclear propulsion for the Navy. Thach was serving with Floberg on 14 June 1952, when the nuclear submarine *Nautilus* was laid down, and he felt justifiable pride in knowing his association and cooperation with Rickover had resulted in such an occasion. He was exceedingly pleased that the "development of the atomic power plant for the submarine . . . involved radical new engineering procedures, development of completely new metals, ingenious systems of lubrication and protection from radiation for the crew."[13] That a nuclear propelled aircraft carrier could soon be built was a belief, aspiration, and goal for both Thach and Rickover.

While the Thach-Rickover relationship was fruitful and cordial during Thach's days with Floberg in 1952–53, by 1963 the relationship was not on the same high plateau. Both officers had been promoted to vice admiral before Thach reported back to the Pentagon and Rickover's contributions had lifted him to legendary status. Unfortunately for Thach, Rickover's well-known propensity for arrogant behavior was more apparent in the mid-1960s, when Thach was no longer a conduit to help further his primary interests. The two officers still needed to communicate with each other on matters of mutual national interest, but Thach did not enjoy the relationship then as much as he had in the early 1950s. Usually Rickover was not as rough with Thach as with others, but from 1963 into 1965 some telephone conversations found both men fluent in profanity.[14] In sum, Thach strongly applauded Rickover's vision and work, and knew him to be an excellent engineer. Thach was pleased to have assisted Rickover throughout their Navy careers, but personality-wise, Thach thought Rickover to be "the most insulting human being I've ever known." Thach also noted that Rickover had a propensity to make pronouncements in fields other than engineering as if he were also an expert in those endeavors and this did not sit well with him or many others.

By January 1965, Thach was beginning to think of retirement. The mandatory retirement age was sixty-two, and he was then just shy of his sixtieth birthday. Madalyn had not been in good health and openly longed to retire and return to the moderate climate of San Diego and to family and lifelong friends. Vice Admiral Thach did not discourage her longings for retirement and home, and in the cold winter of Washington in 1965, he too longed for the warm weather of southern California and the well-groomed golf courses that were open twelve months of the year. With the end of his thirty-eighth year on active duty less than six months away, both expected to trade their Navy life for civilian status soon.

19

Four Stars, Last Call, and Honors

As the holiday season approached in the late fall of 1964, Vice Admiral Thach and his wife talked more frequently about retirement. In January 1965, he would complete five years in his three-star rank, and mandatory retirement at age sixty-two was then only a little over two years away. The subject of his possible retirement was also on CNO David L. McDonald's mind. McDonald and Thach often shared lunch together at the Pentagon in the presence of others, and after one meal McDonald invited Thach to his office for a private conversation. There, McDonald asked him, "Are you considering retiring any time soon?" Assuming the CNO wanted a younger man as DCNO (air), Thach, in a spirit of cooperation and willingness to do what was best for the team, offered to retire within two weeks. Smiling, McDonald said he had asked the question only to determine if Thach had any plans or commitments that might entice him into retirement. He said that he had talked with the secretary of the navy and both wanted to offer Thach the four-star rank of full admiral and send him to London as commander in chief, U.S. Naval Forces, Europe.

COMMANDER IN CHIEF, U.S. NAVAL FORCES, EUROPE

Thach was so surprised and pleased by McDonald's offer that he left the CNO's office without thanking him. Returning later to do so, he was reminded again that Secretary of Defense McNamara and the U.S. Senate had to approve the recommendation before President Johnson could announce it. Thach was allowed to share the possible promotion and move only with his wife, and while he was excited about the offer, Madalyn was considerably less taken with it. Although the admiral would be responsible for the operations of naval forces in Europe and the Mediterranean, including the U.S. Sixth Fleet, Madalyn knew that this job would require a nonstop schedule of diplomatic gatherings and she would be nearly as busy as her husband. Not in the best of health, Madalyn was also concerned that the climate in Europe might further exacerbate her problems. Too, the San Diego–Coronado area had been home for most of her life, and she was anxious to return there. But she was also a dutiful Navy wife, and she began to pack.[1]

Orders were cut on 27 January 1965, Thach was detached from his job at the Pentagon in February, and the journey to Europe began. On 25 March, Thach relieved Adm. Charles D. Griffin and soon met with his new boss, Gen. Lyman L. Lemnitzer, supreme allied commander, Europe, and commander in chief, U.S. Forces. Lemnitzer was no stranger to Europe, having recorded a distinguished record there during World War II and later in Korea. Although past the mandatory retirement age (born 1899), Lemnitzer, formerly chairman of the Joint Chiefs of Staff, was retained on active duty because of his extraordinary competence and extensive experience in Europe. Signing Thach's guest book in May 1965, the general recorded his full title, but on subsequent visits he signed only as "Lem." Thach found the general "a very fantastic person . . . so highly qualified." Particularly impressive, he thought, was the general's ability to never lose his temper and to remain calm and stable in difficult situations. Even in tense moments he was able to carefully consider multiple factors. Madalyn was equally impressed with the general ("very nice") and his wife Katherine, finding her "friendly and most cordial with no put-on airs." She did, however, have to remember, and discretely remind others, not to attempt to assist Mrs. Lemnitzer, who walked with a cane. And the Thaches found the Lemnitzers easy to entertain, as both preferred a quiet evening rather than parties of any size.[2]

Headquarters for U.S. Naval Forces, Europe was located in historic Grosvenor Square in the heart of London, very close to the American Embassy. Jimmie and Madalyn had living quarters (a flat) in the building above the offices and also a country house (Romany) next to a golf course, all which went to the incumbent in the job. The job, of course, was the reason for their being in

Europe, and on his first day at work, Thach insisted on walking around headquarters to meet his staff. Coming upon one enlisted man who had just completed working on a copying machine, the admiral was introduced and extended his hand. The young man, whose hands were covered with ink, declined to take the admiral's hand, explaining the obvious problem that would result from a handshake at that moment. With his left hand, Admiral Thach grasped the young man's right arm and then grasped his right hand, ink and all, completing the introduction. Thach, who always smiled when meeting someone new, moved on, leaving a smile on the face of the ink-covered enlisted man and smiles on the faces of all within the building when news of the event spread. Admiral Griffin was a well-respected, outstanding naval officer but had been more formal in interactions with his staff.[3] One of Thach's staff officers, Capt. Joseph C. Wylie Jr., (later Rear Admiral) commented, "I just loved Jimmy Thach."[4] And soon many others in the headquarters building came to share the same thought. His flag secretary, Capt. E. T. Wooldridge, recalled that Thach was "a great guy to work for."[5]

Following procedure set by Admiral Griffin, Captain Wooldridge continued staff and office procedure as it had been for the first several days Thach sat in his new office. After a few days, however, Wooldridge could not help but notice that the stack of papers in Thach's in basket was growing rather than diminishing and he felt compelled to ask if the new CIC wanted to institute different procedures. Thach replied that he did. If a matter was urgent and particularly significant, he wanted to know about it. Otherwise, the staff should handle the mountain of paperwork.[6] Thach "was a good, commonsense, gentle, sensible man [that] . . . never crossed up his staff once he learned who they were and what their capabilities were."[7]

Even though he learned early that he had a good staff to which he could delegate both official responsibilities and social activities, Thach did not expect the job to be easy. While he did not have to wear a NATO hat, he did have a second hat to wear as commander, Eastern Atlantic. With the two titles the area of his responsibility was from the eastern tip of Greenland to the northwest coast of Africa, and he was one of only three U.S. Navy fleet commanders, the other two being for the Atlantic and Pacific. Sixth Fleet at the time was comprised of fifty ships built around two (usually) attack carriers with 165 planes. That meant that any problem between Greenland and the Near East or aboard the ships was his problem. To friends and family, Madalyn wrote, "Jimmy is finding that he has a bit more of a job than he had anticipated but it's an important one and he's digging in. As long as he has Romany for the weekends and a bit of golf his spirit stays sanguine and he's well worth living with."[8]

Madalyn Thach soon learned that there would be many days when she would not be able to live with her husband at all due to his need to travel extensively. She could not accompany him when he went aboard ships, and there were other occasions when her health, particularly continuing eye problems, did not permit her to travel. Thach did not feel he needed to be in France unless General Lemnitzer called for him, and he felt that would not be often as the general had a firm hand on the difficult matters in Paris. But he was firmly convinced he needed to visit the smaller countries throughout the region, especially in critical places like Morocco. His ultimate objective was to get to know his counterparts and political leaders so that in an emergency everything could move faster. He also desired to know exactly who and what he would be dealing with should war occur.

Early into his new duties, Thach traveled to Morocco, its location at the entrance to the Mediterranean making it a potential chokepoint. Like many other countries on the African continent, Morocco had been under colonial control and only recently (1956) had gained independence. Without a tradition of democracy, the country was under internal stress from tribal rivalries, French and Spanish loyalists, and religious differences. Added to the unsettling equation was an effort by a small minority to introduce communism. And if all these problems were not enough, Thach found that some of Ambassador Henry J. Tosca's subordinates were less than supportive or helpful to Thach's desire to meet Moroccan officials. Nonetheless, he persevered, Tosca intervened, and the admiral met several high-ranking officials and developed a particularly close relationship with several. Two bonds glued the men together: all wanted democracy to work and all worried about Communist takeover efforts. That the men developed good personal relations was a result of mutual admiration.

Thach became especially fond of General Mohamed Oufkir, who had been awarded a United States Army Silver Star while fighting with the Allies during World War II. Oufkir led Thach on tours in Morocco that few if any tourists would ever see, and when the admiral made subsequent visits Oufkir demonstrated his esteem for Thach by discarding all his medals except for the Silver Star. A few years later, however, Thach was saddened to learn that Oufkir and most other ranking officials he had befriended, and who had befriended him, had lost their lives in the continuing political turmoil attendant to the nationalist fervor and furor of the era. And he was disappointed that his work to better relations between the United States and Malta was largely lost when the political winds there changed. Thach had scheduled carrier visits to bolster the government in power during his duty in Europe, but after he left, the country moved more to an independent stance with the consequent loss to the West of an important strategic base. Disappointed though he was, that occurrence became yet one more example why the flexibility and mobility of aircraft carriers was paramount in both military and diplomatic thinking.

In this tour of duty, as in so many previous assignments, Thach constantly found the need to preach to Washington on the need for aircraft carriers. The essence of his message was the same as in the past but current events allowed him to cite new examples. The U.S. Air Force was not the enemy in the mid-1960s, but the continuing loss of overseas bases was accelerating and those that remained were often overcrowded. The loss of bases and overcrowding was being compounded by alliances between new national leaders even though they were at odds on nearly everything except to expel foreign influence.

Wars of national liberation presented yet another consideration for Thach in Europe and planners in Washington. While one country in turmoil might require a fixed land base for air operations while the issue was in doubt, that place on the globe might soon prove to be of little or no value for an air base once stability was regained. And even economics was being seen in a new light. It was becoming apparent that "land-based air in the long run costs more, can only do part of the job, cannot protect itself or its logistic pipeline, and is politically vulnerable" . . . in addition, it cannot protect the high seas commerce.[9] And documentation of the salient role of carriers came from an unexpected source very soon after Thach left his European assignment but not before he was involved in one last push for them.

Given the special relationship between the United States and United Kingdom, Thach always hastened to acknowledge that the British had led the way in significant carrier developments such as the canted flight deck, mirror-landing technique, hurricane bow, and steam catapults. And he often noted that the U.S. Navy shared its top-secret submarine and Polaris missile technology only with the British. The British appreciated Thach's tip of the hat to them, but the issue at hand was whether or not to fund aircraft carriers. Not unexpectedly, Thach tried to help the British keep the few carriers they still had and to invest in future development. Budget reductions were as much a consideration for the

Royal Navy as for the United States Navy in the mid-1960s. Still, Thach pointed out to his British friends, and to others in Europe and Washington, that carriers were just as relevant in the 1960s as they had been and were becoming even more so. Even in the atomic age carriers had low vulnerability to surprise attack due to their mobility, could cover all the ocean area of the world and 90 percent of world land areas. Without political restraint to their movements, carriers could quickly respond and even skirt bad weather that often closed land bases. Emphasizing economy, Thach stated that advanced-area airfields did not have logistics facilities, and that the investment of building a land base could not be retrieved if lost in the fortunes of war or through political fluctuations. Another consideration along economic lines then being proved in Vietnam was that a hand grenade could close runways. And finally, along economic lines, Thach reminded listeners that a carrier could have many lives or incarnations just as U.S. Navy carriers that began life as attack carriers were converted to antisubmarine work, amphibious ships, plus communications and command platforms.

Particularly relevant in Thach's presentations to the British was that more strikes could be made from a carrier close to a target than from a long-range bomber that would have to use a "sledge hammer to pound a thumb tack."[10] At the time he shared these thoughts with the British, they were under considerable pressure from Secretary McNamara to buy some of the F-111A bombers that had just been forced on the U.S. Air Force. McNamara's major selling point was that the long-range bomber could fly from Britain to the Indian Ocean, still a major area of British concern. In the end, the British did buy some F-111s, but Thach recalled in his oral history that they felt "sandbagged" when McNamara authorized three supercarriers just prior to his departure as secretary of defense in 1968. McNamara, who finally saw the economic value of carriers, did not acknowledge it in time to help the British champions of aircraft carriers, and Thach remembered their feeling was that "he [McNamara] really sold us down the river."

While there was no shortage of problems in London and elsewhere demanding the admiral's attention, there were also pleasures, and near the top was an invitation to meet Queen Elizabeth. On 22 November 1965, Admiral and Mrs. Thach were invited by David K. E. Bruce, U.S. ambassador to the Court of St. James (1961–69), to attend Queen Elizabeth's Evening Reception for the Diplomatic Corps at Buckingham Palace. Following the formal reception, all guests were to join in the Throne Room for champagne and hor d'oeuvres. Specific instructions in the ambassador's 22 November 1965 letter, including a standing diagram, accompanied the invitation for the event scheduled for the twenty-fourth. The Thaches were to be at the palace no later than 2115, were advised to leave their coats in their car, and were to present a green-colored card to an official who would then direct them to the Picture Gallery, where the group was to assemble. The ambassador and his wife would stand on the front row, and Admiral and Mrs. Thach would stand on the second row alongside Maj. General John L. Hardy, USAF, and his wife. The third and fourth rows were comprised of twenty lower ranking U.S. civilian and military officials. Altogether some three hundred guests representing various nations would join with British diplomats. The letter of invitation continued on to instruct invitees on the protocol for the evening, noting that diplomatic missions would be greeted in order of mission's precedence (Nigeria and Cyprus being greeted immediately before the U.S. and followed by New Zealand and Burma). Finally, the letter of invitation and instruction advised that no one was to take the queen's hand unless she offered and no one from the third or fourth rows was to do anything unless requested by one of the royals. Should the queen or queen mother engage anyone in conversation, the first acknowledgment was to be "Your Majesty," followed by "Ma'am." In response to a male member of the royals, the first response was to be "Your Royal Highness," followed by "Sir."

At the reception, the marshal of the diplomatic corps presented Ambassador and Mrs. Bruce to the queen, who then walked together along the second row. All that Thach expected was to be presented, bow, watch Madalyn curtsy, and then do the same for the remainder of the royal procession, including the queen's husband, Queen Mother, and five others. As the queen approached, Thach was greatly impressed by her beauty, especially her sparkling eyes, later recalling that her eyes and vivacious personality overshadowed all the jewels that adorned her. Pleasantly surprised, the queen asked Admiral Thach about his duties, and before he could complete his answer, she interjected and asked that her husband, Prince Phillip, who had a naval background, be present to hear the description of his CIC duties. Impressed before he arrived, Thach left the palace with an even greater estimation of the queen and her husband.

For his first nine months in Europe, Thach made twenty-eight official trips that totaled seventy-two days and carried him to over a dozen countries throughout his region of responsibility. Repeat visits for 1965 included only Paris, the headquarters of his boss, plus Morocco and two trips back to Washington. Some trips became necessary on short notice, one being a visit to the

Netherlands. It came to the admiral's attention that the Dutch were upset that the rear admiral with whom they had been working had been replaced by a captain with the move being interpreted as a loss of respect and concern. So Thach polished all four of his stars and set out for the Netherlands. The visit was a success according to the ambassador and naval attaché stationed there and egos were soothed.[11]

For most months throughout his tenure, only Sundays were usually free of official meetings or social obligations. Each month usually saw the Thaches attending one to six dinners, hosting an average of two dinners, and attending or hosting an average of seven luncheons or receptions. Guests included a wide range of backgrounds. Fellow military officers were seldom absent while foreign diplomats, captains of commerce (J. Paul Getty), and media notables (Dan Rather) were often attendees. Even old friends such as Douglas Fairbanks Jr. and Thacher Longstreth occasionally were present. A totally free weekend, defined as one without guests, was such a rare occasion that it merited mention in Madalyn's letters to family.[12]

Movie star Douglas Fairbanks Jr.'s association with Thach went beyond being a frequent guest. A reserve U.S. Navy captain, Fairbanks, who had earlier become acquainted with Thach in Hollywood, asked to be recalled for temporary duty to work with Thach on his staff. As he had traveled extensively in Europe he had met and could visit with people Thach could not approach in his official position. Still, their opinions were important to Thach, and Fairbanks was occasionally tasked to make certain visits and seek opinions and advice. Being a good speechwriter and having a good feel for the British, Thach turned to Fairbanks to write a Fourth of July speech. While seldom needing help for any speech he wanted to make, for such a speech before an audience comprised of both American and English guests, the choice of words was especially important. Fairbanks prepared the text, injecting a generous dose of humor, and that, plus Thach's innate sensitivity for such occasions, made for both a successful speech and appreciative audience.

Another matter was equally sensitive and Thach went out of his way not to initiate the subject. However, the war in Vietnam was everyday news and questions pertaining to America's role in that war brought a constant stream of questions from diplomats, foreign military officials, and the media. In response when directly confronted, Thach drew his questioners' attention to the fact that the United States had no territorial designs in Vietnam or any nearby country. He emphasized that North Vietnam was the aggressor with the objective of total conquest and that the United States was trying to help the Vietnamese shape their own destiny. No apology could be offered for the United States honoring a pledge to defend South Vietnam's independence and defend the principles of freedom. And he usually concluded his line of thinking by stating, "How can we negotiate if the other side prefers to fight and refuses to come to the conference table?"[13] In time, Thach's Moroccan friend, General Oufkir, proved to be prophetic when he told Thach that the United States would lose the war in Washington rather than Vietnam. Oufkir, who noted that the French had lost Vietnam in Paris, saw all the same mistakes of the mid-1950s being repeated again in the 1960s.

Thach had a somewhat standardized speech for his European audiences in regard to communism. The speech was designed to strengthen both those who already championed democracy and those who were tempted to look toward communism and socialism. The subject of Communist China was usually held in abeyance unless attached to discussions pertaining to Vietnam. For Europeans the emphasis was on the Soviet Union, its presence in Eastern Europe a daily reminder of its potential peril to still-free areas of Western Europe. Thach believed the Soviets had not mellowed as some thought based on the contemporary talk of coexistence. He believed they no longer rattled rockets because they knew they had no missile gap in their favor. In speeches on the subject he noted that the Soviets continued the propaganda theme that wars of national liberation were just wars, popularly supported and therefore should be fully and militarily supported. Thach stated the Soviets' real goal was to impose rule by force and that they would export weapons anywhere to take advantage of instability, pursue a policy of divide and conquer, and use their veto to block UN peace keeping efforts. According to Thach, in the end the only thing stopping the Soviets was the effective deterrence built by the United States and NATO. With both the United States and the European countries in NATO dependent on control of the seas and sea lines of communication, there was no choice but to remain strong.[14]

In addition to the constant threat of communism and the war in Vietnam, there were other unpleasant matters Thach had to deal with from 1965 to 1967. During a training exercise, an Air Force B-52 accidentally dropped an atomic bomb in the ocean just off Spain. Relations between Spain and the United States were already strained, and fears of radiation contamination exacerbated matters. Happily the bomb was recovered, but at no small expense in time and money. When a subordinate naval flag officer informed Thach that much of his operating budget had been reduced in the recovery effort, Thach forwarded the recovery bill on to the Air Force.

Another problem might seem to have been less demanding, but Thach knew from the beginning it would not be. Having already served on a number of selection boards, in the late summer of 1965 he was asked to serve again. Although unhappy to leave his responsibilities in Europe and not happy to make a judgment on a number of officers who were friends as well as professional colleagues, he accepted the charge. On 19 September 1965, he flew back to Washington, put in very long days, shoveled through thirteen hundred files of captains, and, just before returning to London on 4 October, presented a list of names for advancement to rear admiral. The nine-member panel was allowed to select only seventeen, a task as professionally challenging as any other duty.

In the early weeks of 1967, the daily challenges lessened as Admiral Thach and his wife began the forthcoming transition to retirement. Some of the social obligations were delegated to subordinates, much to their delight, and Thach spent more time at his country house and the adjacent golf course.[15] In late March, the expected letter from CNO McDonald arrived. Dated 24 March, it read in part, "Orders relieving you of all active duty incident to your retirement, delivered with regret, but with my best wishes and congratulations for a job 'Well Done.'" Also to Thach's great satisfaction was the announcement that Adm. John S. McCain Jr., son of the World War II admiral for whom Thach served as operations officer, would succeed him.

With his retirement effective 30 April, Thach held his retirement ceremony in the office lobby at his London headquarters. Present were the members of his staff, all the British officials he had worked with, and Ambassador Bruce. All were happy, but no smile was any larger or more heartfelt than Madalyn's. Finally, she too could retire.

RETIREMENT AND LAST CALL

With a Navy career that had consumed two-thirds of his life now over, Thach had no burning desire to find another occupation that would be as demanding. For the first months as a civilian, he and Madalyn took time to visit friends and family in New England and the mid-Atlantic states. Although both knew they would end up in Coronado, they investigated real estate throughout the eastern

United States. While looking, Thach also had a few job interviews, but nothing appealed to him until after he arrived back in southern California.

Aware of Thach's Navy career and recent retirement, Dan Houghton of Lockheed wrote the admiral and asked him to visit before committing to another job. Happily, both men were thinking consultant status rather than full-time employee. On 26 September 1967, Thach signed a consultant agreement with Lockheed Aircraft Corporation for its Lockheed-California Company Division, which specialized in the design, improvement, and development of Lockheed products and research efforts. His services were to be performed at times and places designated by the company with forty-eight hours notice of dates and places. Remuneration would be $875 a month for the retainer fee plus expenses and an additional $175 a day for each full day he worked after the first five full days in a calendar month. Meal allowance was $10.50 a day and 10 cents a mile for travel. For the next three years, Thach was very active in assisting Lockheed with major business ventures.

Few people were as knowledgeable about aviation and in a better position to be a consultant than Thach. When he first began working with Lockheed, the company was in competition to obtain the F-14 contract and was in contention until December 1968, when Grumman and McDonnell Douglas were named the finalists. Although disappointed to lose the F-14 contract, Lockheed had numerous other projects that required Thach's services and contributions. Among the most important was the company's effort to develop aircraft that could fill the void left by the impending possibility of the loss of the CVS carriers. The old *Essex*-class carriers were at or near the end of their service lives, and newer ways of conducting antisubmarine warfare were being sought. In discussions with Vice Admirals Connolly and Turner Caldwell, Thach found documentation of his own view that design proposals quite similar would place more emphasis on the low bid.[16]

Lockheed had produced a number of successful aircraft for use during World War II and thereafter. The PV Ventura land-based patrol bomber was followed by the highly successful P2V Neptune that was used for patrol and antisubmarine search from 1947 into the late 1960s. The company had also produced the USAF F-80 Shooting Star jet fighter, the large airborne early-warning Constellation, the superb C-130 Hercules transport and tanker, and the P-3 Orion antisubmarine patrol bomber. Development of the S-3A carrier-based, antisubmarine search-and-strike aircraft to replace the Grumman S-2 was under way when Thach joined the company, and much of his work was liaison between the Navy and Lockheed. Known as the VSX project, Lockheed submitted proposals in 1968, was selected in August 1969, tested in 1971, and produced the S-3A Viking. Fast, high flying, and with long-range and sophisticated avionics, the Viking made it possible to retire the *Essex*-class antisubmarine carriers (CVS). Throughout the 1967–71 period, however, there were still major reservations pertaining to the potential and performance of the Viking. Vice Adm. Evan P. Aurand, commander of Antisubmarine Force Pacific, confided to Thach that two A-7 Corsairs would have to be offloaded from a carrier to make room for one Viking and that space was lacking for its equipment.[17] Sharing some of Aurand's concerns over the possible loss of the CVS carriers was Vice Adm. William F. Bringle, who was looking for ways and means of maintaining adequate force levels of aircraft carriers. To Bringle, Thach suggested that some carrier names be changed to "General Purpose Support Ship" and that they could combine the ASW function with other missions in an effort to counter criticism by those who were continuously attempting to reduce carrier force levels.[18] It was apparent, however, that some ASW fixed-wing aircraft might be placed aboard attack carriers. Consequently, Thach advised his company that this eventuality "could also enhance the feasibility of procurement of a tanker derivative of the S3A."[19]

Before retiring from Lockheed in 1971, Thach advised the company on antisubmarine developments, fighter aircraft development, advances in missiles, and advanced fighter tactics. Too, he advised the company on pilot problems and opinions, and met with instructors and students of the Advanced Weapons School at NAS Miramar to discuss effective methods of combating new Soviet-built high performance fighters.[20] But as much as he enjoyed being with fighter pilots and discussing fighter tactics, his greatest contributions as a consultant revolved around the antisubmarine effort. With the ASW program in transition, his questions pertaining to the contemporary and future issues were as significant as his recommendations on the parts of the world that could best be served by patrol rather than scouting. And as always, he was quick to defend the Navy, often noting that any reduction in force throughout the military would hurt the Navy most because it could not rebuild as fast as the other services.[21]

After concluding his contractual obligations with Lockheed, Thach still found people seeking his advice, particularly historians. This was not a new experience, as in 1956 he had reviewed the widely distributed book *Zero* by Masatake Okumiya, Jiro Harikoshi, and Martin Caidin. Thach noted in his review, "The story written by Masatake Okumiya deserves far better treatment than it received in the hands of the collaborator, Mr. Caidin and the publisher." The "editing job is poor," he said, but he concluded that "the story Okumiya wanted to tell does come through and I wonder, had our country been defeated and occupied, how many Americans could have written about the war in so objective a manner." Also in 1956, Thach assisted Gallu Productions for a Navy Log presentation entitled *Thach Weaves a Trap*.[22]

In 1971, Cdr. Etta Belle Kitchen visited frequently on behalf of the Naval Institute's oral history program, and in 1974 John Lundstrom spent the better part of a week as a guest in the Thach home while researching material for his several books on the Pacific theater. Also in 1974, Thach read scripts for Jack V. Fogarty and Associates for television's *Flight Deck* presentation.[23] Speaking engagements were not as frequent as they had been while he was on active duty, but invitations were still extended. One of the more notable of these was on 20 September 1979 at the National Air and Space Museum, Smithsonian Institution, where Thach joined George Gay and Rear Adm. Max Leslie for a seminar on the Battle of Midway.

By 1977 Thach's oral history was in print. The very last question he responded to during the 1971 interview was whether or not he might write a book on his life. His response was, "I am not going to write a book, maybe." In 1975, Thach worked with Paul Stillwell, then managing editor of the Naval Institute *Proceedings* on an article, and the two men remained in contact, finally deciding to collaborate on a book. However, in December 1980, Thach began to experience serious health problems, and although doctors "did everything possible . . . he . . . had a miserable time."[24] Suffering from internal bleeding, Thach developed septicemia and finally pneumonia, which resulted in his death at Balboa Naval Hospital on 15 April 1981, only four days short of his seventy-sixth birthday.[25] Memorial services were held on 20 April at North Island Chapel, and interment followed at Fort Rosecrans National Cemetery, Point Loma, San Diego County. On 9 August 1987, Madalyn died and was buried next to her husband.

HONORS

Honors came to Admiral Thach long before his death. His many combat decorations from World War II and Korea were daily reminders of success. Madalyn Thach also received a number of awards, most for her work with Navy wives and their charitable organizations. Retirement from the Navy, however, did not end her association with the Navy as she was called upon to sponsor the USS *Charleston* (LKA-113), an attack cargo ship. On 2 December 1967, Madalyn was at Newport News Shipbuilding and Dry Dock Company in Virginia for the ceremony. Alongside Charleston, South Carolina, mayor J. Palmer Gaillard, the two watched the new ship slide down the ways bearing a large sign on the bow that read "Sink Army." Later in the day, the Navy football team did just that (19 to 14) in Philadelphia.[26]

Even before retirement from the Navy, Admiral Thach was active with the American Fighter Aces. Serving as president and a national director, he remained active until his death. After his death, honors were bestowed in a continuing stream. In 1981, he was inducted into the Arkansas Aviation Hall of Fame, and on 3 October 1982 into the *Yorktown* Carrier Aviation Hall of Fame. Aboard the retired aircraft carrier in Charleston Harbor, South Carolina, Thach's induction speaker was old friend Thacher Longstreth, then president of the Philadelphia Chamber of Commerce. Inducted during the same ceremony were other old friends no longer living, including Vice Adm. Jimmy Flatley, Lt. Cdr. Edward H. O'Hare (Noel A. M. Gayler speaking), Rear Adm. C. Wade McClusky, Adm. Marc A. Mitscher, Adm. Arthur W. Radford, and Lt. Cdr. John C. Waldron. Also inducted and present that day was Capt. David McCampbell, who, with thirty-four kills, was the Navy's top World War II ace. At the conclusion of the ceremony, Captain McCampbell and the speakers for the deceased inductees were invited to throw wreaths into the harbor as the band played the "Navy Hymn." By coincidence the wreaths thrown by Mrs. Flatley and Thacher Longstreth fell as one onto the water and Mrs. Flatley commented that given the close relationship between the two men that it was appropriate that they again share a moment together.[27]

The next honor came only two months later, when Todd Shipyards Corporation at Long Beach, California, launched the Guided Missile Frigate USS *Thach* (FFG 43) on 18 December 1982. Secretary of the Navy John Lehman had invited Madalyn to be the sponsor on 3 June 1982. William H. Webster, director of the FBI, was the speaker at the launching and christening ceremony. Admiral Thach would have been pleased with Webster's comments regarding the Navy's role in meeting domestic security needs and the relationship between effective domestic law enforcement and national defense. Commissioned on 17 March 1984 at Long Beach Naval Station, the ship displaced thirty-eight hundred tons, was 453 feet long with a 47-foot beam, and had a speed of twenty-eight knots and crew of two hundred. Carrying an assortment of missiles for protection against both ships and planes, ASW torpedo tubes, guided missiles, and one MK 75 76-mm/62-caliber rapid-firing gun, Thach was the thirty-seventh ship of the Oliver Hazard Perry class of guided-missile frigates. With a mission to provide antiair, antisub, and antisurface protection for underway replenishment groups, convoys, and amphibious forces, the frigate had ASW computer technology long sought and championed by its namesake. While Admiral Thach may or may not have approved of the minimum-manning concept, no doubt he would have been pleased with the ship's motto, "Ready and Able."

In recognition of Admiral Thach's antisubmarine work and contributions, the Navy annually awarded the best ASW squadron the Admiral Thach Award in the 1970s and 1980s. On 24 May 1979, Thach had been present to present the large silver bowl award to VS-38, the Red Griffins. Sponsored by Grumman Aircraft Engineering Corporation the award was "symbolic of meritorious achievement by a carrier ASW squadron and recognizes the dynamic traits and inspiring leadership exemplified by Admiral Thach during his naval career."[28]

On 4 May 1984, Thach was inducted into the Hall of Honor at the National Museum of Naval Aviation, NAS Pensacola. Also inducted on the same day was Jimmy Flatley. They were the first two fighter pilots to be inducted. On 8 May 1992, Butch O'Hare was inducted, and the three were once again together.

During World War II, Jimmie Thach, Butch O'Hare, and Jimmy Flatley worked out carrier fighter tactics with little or no thought as to how history might view their efforts. They were only trying to find a way to keep themselves and the members of their squadrons alive in combat against a formidable opponent. Because they succeeded, a grateful nation has rendered and will continue to render honorable remembrance of their names and contributions.

Despite his innovations and outstanding work in antisubmarine warfare and avocation of a strong Navy, Jimmie Thach will most likely continue to be remembered primarily for his development of the Thach Weave. How he would be remembered, however, was not a great concern. Important to him were his Navy, his country, and the many friends and peers he treasured and to whom he was especially considerate and gracious. And if he could have chosen between historical standing or an outstanding score on the golf course, aviation enthusiasts should not be too disappointed to know that he would have spent more time pondering club selection than the special place in history he earned.

Appendix

Combat and Command Decorations

NAVY CROSS

"For distinguished service in the line of his profession, as Commander, Fighting Squadron THREE, when, on February 20, 1942, in enemy waters, he led his squadron in repeated attacks against two nine plane formations of Japanese twin-engined heavy bombers which resulted in the destruction of sixteen of the eighteen enemy aircraft engaged. Through his courage and skill he shot down one enemy bomber and, with the assistance of his teammates, shot down a second bomber."

NAVY CROSS (SECOND AWARD)

"For extraordinary heroism and distinguished service as Commanding Officer of Fighting Squadron THREE in action against enemy Japanese forces in the Battle of Midway on June 4, 1942. Pursuing the bold and fearless tactics of a great fighter and a skillful airman, Lieutenant Commander Thach led a division of his squadron on a mission providing protection for our own attacking torpedo squadron. Facing intense anti-aircraft fire, the squadron under his efficient command, attacked an overwhelming number of enemy Japanese fighters, shooting down three of them. Again, in the afternoon, he led a determined and effective attack against enemy torpedo planes which were attacking his carrier, shooting down one of them in this engagement. His great courage, inspiring example and his complete disregard for his own personal safety were in keeping with the finest traditions of the United States Naval Service."

DISTINGUISHED SERVICE MEDAL

"For exceptionally meritorious service to the Government of the United States as Commander of a carrier Fighting Squadron during the first six months of the war. The remarkable quality of leadership displayed by Lieutenant Commander Thach was exemplified in the thorough and comprehensive training of his pilots in both their brilliant combat tactics and excellent gunnery. The resultant high state of combat efficiency attained by his squadron enabled it to play a decisive and major part in the destruction of nineteen of the twenty enemy Japanese bombers which attacked an aircraft carrier on February 20, 1942, and to make an essential contribution to the success of the air attack on Salamaua and Lae, New Guinea, on March 10, 1942. The effectiveness of his unique system of fighting plane combat teams, evolved from a detailed study of action reports of the Coral Sea Battle and taught not only to his own pilots but to all of the fighting squadrons in the Hawaiian area, was demonstrated north of Midway Island on June 4, 1942. In this engagement, Lieutenant Commander Thach led a four plane division of fighter planes from his squadron against twenty enemy "zero" fighters during the successful attacks by our carrier based planes against enemy carriers, destroyed six enemy fighters and repulsed the others with the loss of only one of his four planes."

SILVER STAR

"For distinguishing himself conspicuously by gallantry, and intrepidity in action while serving as Operations Officer on the Staff of the Commander, Task Group Thirty Eight Point One during the period 13 through 15 October 1944, off Formosa. While the Task Group was covering the withdrawal of two crippled cruisers against repeated and persistent enemy air attacks, his skillful handling of our air forces, effective strategy and unerring judgment forced the enemy to break off his attacks, and in a large measure contributed to the salvaging of the crippled ships. His courage and disregard for his own safety were at all times in keeping with the highest traditions of the United States Naval Service."

LEGION OF MERIT

"For exceptionally meritorious conduct in the performance of outstanding services to the Government of the United States as Operations Officer on the Staff of Commander SECOND Carrier Task Force, Pacific, during operations against enemy Japanese-held Palau, Philippine Islands, Formosa, the Nansei Shoto Group and the Coast of Indo China, from August 18, 1944, to January 25, 1945. Displaying exceptional ingenuity as a tactician and utilizing previously unexploited aircraft potentialities in the conduct of aerial operations, Captain (then Commander) Thach contributed materially to the infliction of serious damage on hostile aircraft, shipping and ground installations and made possible the recapture of important territories from the Japanese. His professional skill and devotion to duty were in keeping with the highest traditions of the United States Naval Service."

LEGION OF MERIT (SECOND AWARD)

"For exceptionally meritorious conduct as Commanding Officer of the USS SICILY in operations against the enemy in Korea from August 3, 1950 to January 15, 1951. With outstanding ability, energy and high technical skill, he operated his ship and its embarked air group as a unit of the United Nations Naval Forces, furnishing invaluable support to our fighting forces in Korea by the destruction of enemy air opposition, troop concentrations and the interception of the enemy lines of communications, supplies and bases. During this period the ship maintained a superior performance in every phase of operations and the immediate and effective response to calls for extra effort reflect the highest caliber of leadership."

BRONZE STAR

"For meritorious achievement while serving as Operations Officer on the Staff of Commander SECOND Carrier Task Force, Pacific, during operations against enemy Japanese forces from April 10 to September 2, 1945. Planning and brilliantly executing the sledge-hammer attacks of Task Force THIRTY EIGHT, Captain Thach paralyzed the enemy sea and air forces, disrupted their communications and leveled their industries, thereby contributing to the final defeat and surrender of Japan. His professional skill and devotion to duty were in keeping with the highest traditions of the United States Naval Service." (Captain Thach is authorized to wear the Combat "V.")

NAVY COMMENDATION MEDAL WITH COMBAT "V"

"For distinguished service in the line of his profession as Commander Fighting Squadron Three, when, on March 10, 1942, he led his squadron during the attack by carrier's air group on enemy surface vessels in a distant enemy area (Salamaua and Lae, New Guinea). His squadron harassed the enemy ships by repeated strafing despite heavy anti-aircraft fire. These attacks resulted in enemy casualties and seriously disrupted the anti-aircraft fire directed at our dive bomber and torpedo planes."

Notes

Preface

1. Robert J. Cressman, Ships Names and Sponsors; Naval Historical Center to Mrs. John S. Thach, 19 May 1982.

Chapter 1. A Razorback Goes to Sea

1. During an informal conversation in 1974 with John Lundstrom, Thach indicated he preferred "Jimmie" rather than "Jimmy." Correspondence to and from Thach throughout his naval career shows both spellings, but from before World War II to retirement, nearly all official papers were signed either John S. or J. S. Thach, with personal letters usually ending with just "Jim." The admiral's wife, Madalyn Thach, nearly always used "Jimmy" in her correspondence.
2. *Lucky Bag*, 187.
3. James Harmon Thach III, telephone interviews and correspondence with the author from 19 June 2002. Thach contributed much of the family information presented in chapter 1.
4. *Lucky Bag*, 187.
5. Ibid.
6. John S. Thach training records, National Museum of Naval Aviation, Pensacola, Fla.
7. Thach's oral history reverses the order of Bogan and Radford as squadron commanding officers. Also not mentioned is his three-month duty on an aircraft tender.
8. Frank Albert, conversations with author from 1982.
9. Wildenberg, "Helldivers," 60–61.
10. Original in Thach Papers, Emil Buehler Naval Aviation Library, National Museum of Naval Aviation, Naval Air Station Pensacola, Florida (hereafter cited as TP).
11. Capt. James B. Cain, interview with author, 27 April 1997, and conversations with author from 1984.

12. David Lister, conversations with author from 1982.

13. James H. Thach III, Captain Cain, and Dorothy Flatley, conversations with author from 1994.

Chapter 2. Test Pilot, Patrol Planes, and Scout Planes

1. Experimental flight testing formerly performed at both Norfolk and Anacostia is currently concentrated at the Naval Air Test Center at Patuxent River, Maryland.

2. See Thach with Wooldridge, "Checking Out This Year's Models," 36–45.

3. Swanborough and Bowers, *United States Navy Aircraft since 1911*, 195–204. Also see Gunston, *Grumman*, 9–15.

4. Cdr. Perry W. Ustick, interview with author, November 1984.

Chapter 3. Fighting Squadron 3 Readies for War

1. In July 1937, carrier squadrons were renumbered according to the hull number of their carrier.

2. Thach remembered the engines as Pratt and Whitneys in his oral history.

3. Clyde E. Baur (VF-3 mechanic) to author, 3 March 1994.

4. TP.

5. Ibid.

6. Thach recalled in his oral history that the information appeared in the 1941 spring bulletin.

7. Lundstrom, *First Team*, 477–78.

8. Points of interest pertaining to this memo are that the six-plane division is implied. The major thrust of the memo was simply that small British fighter forces were having considerable success against large numbers of German aircraft (mostly bombers). Given that the British fighter formation in July 1940 was the six-plane division, Browning and other American aviators presumed that this division was part of the reason for the British success. It also should be remembered that the Battle of Britain was still in the incipient stage in July 1940, and tactics changed before the battle concluded three months later. NA, copy in Flatley Papers, currently at Patriots Point Naval and Maritime Museum, Charleston Harbor, South Carolina (hereafter cited as FP).

9. Lundstrom, *First Team*, 478.

10. *Essex* (CV-9) was also authorized in the 17 May 1938 act.

Chapter 4. World War II and VF-3's First Battles

1. Memo from Cdr. Paul Ramsey to VF-2, c. spring 1941, FP.

2. Lundstrom, *First Team*, 40–41.

3. Thach's comments were included in his 26 August 1942 interview at the Bureau of Aeronautics and with the exception of one word (machine) in the cited text was copied verbatim from one of his little black books.

4. Thach commented in his oral history that he opened up "with all six of those guns. I hadn't fired six guns at once before. It was really a good blast of tracers." However, the F4F-3 (BuNo. 3976) had only four guns.

5. Thach recalls the bombers as flying at 8,500 feet; other sources record 11,500 feet.

6. Thach, Bureau of Aeronautics interview, 26 August 1942, 3.

7. Some sources, including *Fateful Rendezvous*, the first book in this trilogy, reverse the order of destruction of the bomber responsible for Ensign Wilson. This account presents Thach's remembrance from his oral history. Details of the battle were omitted in Thach's twenty-seven-page Bureau of Aeronautics interview on 26 August 1942 ("I won't bore you with the details because you've no doubt read it all").

8. To avoid confusion with other sources, this narrative uses the terms "first" and "second" to mean only the order in which the Japanese attacked. Actually, the first Japanese formation to attack was numbered "Second" division, and the second wave to approach Lexington was numbered "First."

9. In a March 1994 telephone interview, Admiral Gayler expressed admiration for the pilot and gunner of the slow floatplane for rising to oppose such overwhelming odds. He added that he was already in a dive to strafe an enemy ship when the floatplane appeared before him, and that he did not have to interrupt his dive to shoot at the plane before descending farther to strafe an enemy ship.

10. KGU-NBC radio broadcast, 30 March 1942, O'Hare Papers, Enterprise Collection, Patriots Point Naval and Maritime Museum, Charleston Harbor, South Carolina (hereafter cited as O'Hare Papers).

11. In his 26 August 1942 interview at the Bureau of Aeronautics, Thach commented on his new pilots observing that "they could land aboard a carrier but they couldn't shoot." In his oral history, he was more detailed, noting that the new pilots carrier qualifications had been made in training planes that had a different (more stable) configuration of landing gear than the Wildcat.

Chapter 5. Battle of Midway

1. Capt. Robert M. Elder, telephone interview with author, 27 March 2002, and conversations with author from 1987.

2. Thach, Bureau of Aeronautics interview, 6.

3. Thach's oral history remembrance of Massey's formation is not in concert with remembrances from other sources, including Cdr. Tom Cheek, USN (Ret.), who recalled Massey's twelve TBDs in right echelon of two divisions. See Lundstrom, *First Team*, 348, and Cressman et al., *Glorious Page*, 97.

4. The VT-6 Action Report estimated two hits but none certain. Action Report in Enterprise Collection, Patriots Point Naval and Maritime Museum, Charleston Harbor, South Carolina (hereafter cited as EC), donated by Rear Adm. Robert E. Laub, VT-6's senior officer to survive, plus conversations and correspondence with Laub from June 1984. Also, George Gay, conversations and correspondence with author, from July 1987; Capt. Albert K. "Bert" Earnest, conversations with author from 11 October 1987; and Cdr. Wilhelm G. "Bill" Esders, conversations and correspondence with author from 11 October 1987.

5. This first kill is not described in Thach's oral history but is included in VF-3's Action Report and in subsequent articles, documents, and speeches in his papers.

6. In his oral history, Thach recalled Massey being the first to go down and praised the enemy for going first for the flight leader, but other accounts indicate that at least one other TBD was lost prior to Massey.

7. It is my understanding that the first time Commander Esders formally offered this story was during the dedication of the Battle of Midway Torpedo Squadrons Memorial aboard *Yorktown* (CV-10) at Patriots Point Museum on 11 October 1987. Having overseen the construction of the large exhibit, the master of ceremonies asked me to be the main speaker. Feeling quite inadequate to the task, given that the only four then-living pilot survivors (all retired) of the three carrier torpedo squadrons were present, I requested they come and stand beside me during my brief remarks. Comments finished, I then invited each to speak (Esders, VT-3; Rear Adm. Robert E. Laub, VT-6; George Gay, VT-8; plus Capt. Albert "Bert" Earnest, VT-8 Detached, who flew from Midway). Gay and Earnest declined; Laub spoke very briefly, and then Esders, resplendent in white uniform, held the large audience on *Yorktown*'s flight deck spellbound as he detailed how Jimmie Thach had saved his life on 4 June 1942. Even though privileged to hear Commander Esders make similar speeches later, the eloquence and emotion evoked on the morning of 11 October 1987 was never surpassed.

At a later program on *Yorktown*, the fiftieth anniversary of the Battle of Midway in June 1992, Esders, Gay, and Earnest returned along with other veterans of the battle. All spoke on that occasion, and happily that event too was taped. Esders and Laub's Navy Crosses are on permanent display in Patriots Point's Midway exhibit. There are several published accounts of this incident, one being "Torpedo Three and the Devastator—A Pilot's Recollection," by Cdr. Wilhelm G. Esders, USN (Ret.), *Hook*, August 1990, 35–36. There was an occasional variance in Esders speeches and articles regarding the turn toward the Zero at exactly the right moment. In some, he also seemed to allow that the TBD pilot might turn away rather than toward the attacker depending on the angle of attack.

8. Thach's logbook lists 3.0 hours for the morning flight and 2.0 for the afternoon, but both entries appear somewhat inflated when cross-referenced with other sources.

He also recorded the same bureau number for both flights (BuNo 5171), but other sources indicate the F4F-4 he flew that morning (BuNo 5171) required too many repairs, thereby placing him in another Wildcat (BuNo 5174) for his afternoon mission. His other log book entries for the remainder of June placed him in BuNo 5127 on 6 June and the seemingly popular—and apparently repaired—5171 on 12 June (twice) and 18 June. Obviously, Thach had more important things on his mind in June than administrative matters.

9. Some accounts record a meeting between Thach and Vice Admiral Fletcher. In a letter of 11 April 1949 to Samuel E. Morison, who was then writing his *History of United States Naval Operations in World War II* volume on Midway, Thach wrote that he only reported to Commander Arnold. Thach and Arnold's intent to directly report to Fletcher was preempted by radar indications of the first enemy attack. TP.

10. CO USS *Yorktown* to CinCPac, "Report of Action for June 4, 1942 and June 6, 1942," 18 June 1942, 15, Patriots Point.

11. Capt. John Adams, conversations with author from 1977.

12. Copy of message to VF-8 in TP.

13. Walter Lord, author of *Incredible Victory*, who had direct contact with many Japanese principals associated with the Battle of Midway, expressed surprise that reviewers of his book did not emphasize his revelation that the Aleutians campaign was not a diversion. Text dated 16 November 1968, EC, and conversation with author, 5 May 1988.

14. Thach, Bureau of Aeronautics interview, 6.

15. Adams, conversations with author; Capt. E. Scott McCuskey, conversations with author from June 1992; and Rear Adm. William N. Leonard, telephone interviews with author, 12 March 1996 and 13 June 2002.

Chapter 6. The Weave Validated and Named

1. Memo from Miles Browning to Rear Admiral Spruance, 13 June 1942, NA.

2. VF-3 Action Report, Lt. Cdr. J. S. Thach, 4 June 1942, NA and PPM. Words underlined are those in the report, and Thach was consistent in spelling "zero" without a capital Z.

3. Thach, Bureau of Aeronautics interview, 6.

4. Ibid., 10.

5. Ibid., 14.

6. Then enlisted pilot Stephen Smith (later commander), one of the four VT-6 pilots to return to the *Enterprise* on 4 June 1942, accosted Gray after the battle and, according to his widow during a conversation in October 1984, never fully forgave Gray for not coming to VT-6's aid. Conversations with Lt. Cdr. Richard "Dick" Best of VB-6 revealed he understood Gray's dilemma. But other conversations and correspondence with his

1942 enlisted back-seat gunner during the battle, Cdr. James Murray, indicate he was not as understanding. Rear Adm. Robert E. Laub, USN (Ret.), was the senior surviving VT-6 pilot who attacked the Japanese carriers at Midway. In an October 1984 interview, correspondence, and later conversations, Laub totally absolved Gray. He also stated that given the Wildcat's performance at low altitude against a Zero, Gray and his pilots would have been little help to VT-6 and would probably have been decimated had they engaged anywhere near VT-6's level of attack.

7. United Press stories, 22 and 24 April 1942.

8. Lt. Cdr. Dale Harris letter to Lt. Cdr. James H. Flatley, 4 July 1942, FP.

9. Nimitz message to King, 21 June 1942, courtesy of John Lundstrom.

10. CO VF-10 to Commander, Carriers, U.S. Pacific Fleet, Subject: The Navy Fighter, 25 June 1942, FP. On page 2, Flatley wrote that his letter "is written for carrier-based fighters, but the principles apply in general to any fighter squadron."

11. The Navy Fighter, 1.

12. Capt. Stanley W. Vejtasa to author, 20 April 1996. Vejtasa also recalled that Flatley passed along Thach's concerns in regard to jumping planes with counsel not to overdo it—accompanied with a wry smile.

13. Dorothy Flatley, interview with author, 19 December 2002.

14. Thach's original small black notebooks are with the TP. Flatley's reference pertaining to Thach was in a letter to Cdr. Frank W. "Spig" Wead, 24 June 1943, FP.

15. Capt. J. M. Shoemaker, ComCarPac, to Thach, 15 July 1942, FP.

16. Thach to ComCarPac, n.d. but c. 25 July 1942, FP.

17. Thach, Bureau of Aeronautics interview, 27.

18. Harris to Lt. Cdr. James H. Flatley Jr., 4 July 1942, and Cdr. J. B. Pearson to Flatley, 15 July 1942, both in FP.

19. Rear Adm. Edward L. Feightner, interview with author, 14 March 1996.

20. Vejtasa to author.

21. Capt. T. Hugh Winters, interview with author, 1 November 1996.

22. Vice Adm. Bernard H. Strean to Rear Adm. John G. Crommelin, 9 April 1984, and Crommelin to Strean, 22 April 1984, EC.

23. John Lundstrom to Capt. Gordon Firebaugh, 13 November 1987, and Firebaugh to Lundstrom, 10 November 1987 and 1 December 1987, courtesy of John Lundstrom.

24. In Crommelin's same 22 April 1984 letter to Vice Admiral Strean he also commented on a 23 June 1942 meeting at Pearl Harbor with Thach and Jake Swirbul of Grumman Aircraft. Although not fully understanding some of Thach's comments during the meeting, Crommelin credited Thach's recommendation that climb be the most significant factor in the design of a new fighter for the creation of the F8F Bearcat. The Bearcat arrived too late for combat but was the fastest climbing propeller driven Navy fighter.

It may be instructive here to note that Crommelin, a longtime acquaintance, was not overly happy with the early inclusion of Thach and Flatley into NAS Pensacola's

National Museum of Naval Aviation's Hall of Honor and the *Yorktown* Hall of Fame. After some of his other favorites, including Admiral Halsey, were inducted, he mellowed somewhat. Too, once provided with documentation counter to some of his earlier views, his usual response was, "I was not aware of that," and he then promulgated his new understanding as ardently as he had previously argued against it. At his own induction aboard the *Yorktown* in 1987, he spoke glowingly only of others—including Thach and Flatley.

25. Lundstrom to Firebaugh, 8 December 1987.
26. Capt. J. P. Monroe, Commander Air Force, Pacific Fleet to Thach, 7 August 1947, TP.
27. Thach to Monroe, 29 August 1947, TP.
28. Clyde H. Tuomela, Deputy Program Manager F-14 Test and Evaluation, to Thach, 12 March 1973, TP.
29. Rear Adm. Frederick C. Sherman letter to Judge E. F. Hunsicker, 9 October 1942, TP.

Chapter 7. The Thach Weave Gets Help from Hollywood

1. Interoffice communication from Carl Nater, Walt Disney Productions to Thach, 21 July 1943, TP.
2. "Air Navy Chennault," *Time*, 14 June 1943, typescript copy in TP.
3. In Thach's oral history, he stated "chief of basic training." He did not record Rear Admiral Murray's name, but Murray was at Pensacola during Thach's tour at Jacksonville.
4. See "Performance and Characteristics Trials Japanese Fighter," Technical Aviation Intelligence Brief no. 3, Bureau of Aeronautics, 4 November 1942, NA; and Lundstrom, *First Team*, 533–36.
5. Check off List: Demonstration-Lecture of the Overhead Gunnery Run, n.d., TP.
6. WLW Cincinnati radio broadcast, 27 October 1942, TP.
7. Jack G., Simon and Schuster Inc. to Thach, telegram, 27 November 1942, TP. The *Collier's* articles were published 5 and 12 December 1942, those dates being the last days the magazine was to appear on shelves.
8. Roland "Jerry" Gask in *Newsweek*, 28 May 1943, copy in TP.
9. Transcripts of Naval Aircrewmen Graduation, NBC, 14 March 1944, TP. A number of radio broadcasts from various radio networks are in the TP. Scripts often did not lend themselves to extemporaneous participation, and it is apparent that at least a few minutes of formal or informal rehearsal were necessary. In Thach's first radio interviews in early 1942 his participation was mainly question and answer, but by late 1942 he had written scripts handed to him to follow and was as much an actor as others on many programs.

Chapter 8. Battle of the Philippine Sea

1. Even after modernization in the 1950s, the *Lexington* still amazed visitors aboard her while making the extremely sharp turn from Pensacola Bay into the Gulf of Mexico during her days as a training carrier.

2. Eyerman wrote commendable accounts of his experience on *Lexington*, but the stories about him recorded in Potter's *Burke* and Taylor's *Mitscher*, plus Thach's oral history, treat different aspects of his 15 June experience on the bridge. Only Thach's recollection is included herein.

3. Lt. Cdr. Joseph R. Eggert Jr., USNR, telephone interview with author, 26 February 1996.

4. Vice Admiral Strean, interview with author, November 1994; Rear Adm. James Ramage, interview with author, 27 April 1996; and Cdr. Robert L. "Les" Blyth, USNR, interview with author, 6 October 1994.

Chapter 9. Operations Officer: Destination Leyte

1. My 2002 *Reaper Leader* biography of Flatley is incorrect on page 203, wherein Thach was placed on Spruance's flagship, *Indianapolis* (CA-35). McCain was aboard the cruiser during part of the Marianas campaign, but conversations between Thach and Burke were on *Lexington*.

2. Lt. Charles D. Ridgway III, USNR, conversations with author, 4–6 October 1994. Also see Clark and Reynolds, *Carrier Admiral*, 194–95.

3. Winters interview.

4. Leonard interview, 13 June 2002.

5. Cdr. James H. Flatley Jr. to Thach, 10 December 1944, FP.

6. Rear Adm. R. Emmett Riera, interview with author, 28 February 1996, and conversations with author from 1980.

7. Ibid.

8. Thach's oral history account of the Battle of Leyte Gulf is difficult to reconcile with other sources, as he placed some events of the twenty-fifth on the twenty-fourth. He also credits McCain for turning back west before orders arrived, but this too is arguable.

9. Winters interview. Also see Winters, *Skipper*, 124.

Chapter 10. Operations Officer: Destination Tokyo Bay

Note: For this chapter, the major source in addition to Thach's oral history was the twenty-four-page War Diary, Commander Task Force 38, 1 January–31 January 1945, which provides the daily chronology and describes actions and units involved complete with latitude and longitude positions (NA with a copy in TP). Also especially helpful was the Commander Third Fleet War Diary, 22 July 1945 and Commander Task Force 38 Action Report, 31 August 1945 (NA with copies in the FP).

1. Capt. D. M. Bradford Williams, conversations with author, October 1984. Also see Inoguchi, Nakajima, and Pineau, *Divine Wind*.
2. Ibid.
3. Leonard interview, 12 March 1996.
4. Message from Commander Third Fleet to Commander Task Force 38, 3 November 1944, TP.
5. Halsey to Nimitz, 6 October 1944, NA.
6. Lt. Cdr. W. N. Leonard to Vice Adm. McCain, memo, n.d. but c. November 1944, TP.
7. All terms coined by Thach, McCain, or others on the staff are presented exactly as used, except for capitalization on the original 11 November 1944 four-page memo from Commander Task Force to Task Groups 38.1, 38.2, 38.3, and 38.4, and a separate, undated "Glossary of Current Operational Terms," both in TP. Some of the terms are presented differently in Thach's oral history and other sources. Also, the description of Moose-trap in Thach's oral history is paraphrased.
8. Pocket notebooks in TP.
9. Operations plans, NA, TP, EC, and FP.
10. All locations, times, and direct quotes for events on 21 January are from War Diary, CTF 38, 1 January–31 January 1945, NA and TP.
11. Richard Johnson, wounded aboard *Ticonderoga*, conversations with author, from 1989.
12. Original scripts in TP.
13. See McCain, "So We Hit Them," 14 July 1945, 12–13, 40, 42, 44, and 21 July 1945, 22–23, 37, 39.
14. Cinpac Intelligence Report, "Japanese Air Force."
15. McCain to TF 38, "Selection of Japanese Targets for Carrier Based Attack," 26 June 1945, in response to Nimitz's 16 May 1945 letter, TP.
16. Ibid.
17. Ibid.
18. Ibid.
19. TF 38, 1 July–15 August 1945, two-page untitled damage total assessment, TP.
20. "Comments on Bombardment versus bombing of Japan Steel Works," 17 July 1945, TP.
21. Associated Press news release, 30 July 1945, TP.
22. Untitled list, 1 July–15 August 1945, TP.
23. Commander Third Fleet message to CinCPac, 15 August 1945, NA.

Chapter 11. Perspectives on Task Force Personalities

1. In his oral history, Thach described in detail how Lt. Cdr. (later Capt.) Arthur L. "Art" Downing assisted in one particularly delicate rescue. Thach recalled Downing as a fighter pilot and ace. Downing was neither a fighter pilot nor ace, but he is generally acknowledged as one of the most outstanding dive-bomber pilots of the war. He was CO of Bombing Squadron 14 aboard *Wasp* while McCain and Thach were aboard.
2. Leonard interview, 13 June 2002.
3. See Sherman, *Combat Command*.
4. Clark and Reynolds, *Carrier Admiral*, 237.
5. Leonard interview, 13 June 2002.
6. McCain, "So We Hit Them," 14 July 1945, 12–13, 40, 42, 44, and 21 July 1945, 22–23, 37, 39.
7. Longstreth comments to Bryan; and see Longstreth with Rottenberg, *Main Line Wasp*.
8. Leonard interview, 13 June 2002.
9. Correspondence in TP.

Chapter 12. Hollywood Again, Pensacola, and Unification

1. Thach to Daves, 28 September 1945, TP.
2. Thach to Roger W. Kahn, Grumman Aircraft, 14 February 1950, TP.
3. Thach to L. A. "Jake" Swirbul, 8 March 1950, TP.
4. Ibid.
5. Thach speech, c. 1950, TP.
6. Thach to Capt. J. H. Flatley, 27 May 1949, FP and TP. Also Thach to William Faralla, U.S. Naval Photo Center at Anacostia, 29 July 1949, and Capt. R. W. Denbo to Thach, 3, 10, 18 August 1949, TP. Jimmy Flatley would have appreciated Captain Denbo's compliments to Thach pertaining to Flatley's assistance in helping with the color film of jet operations needed for *The Naval Aviator*. However, he would have wondered where Denbo got the idea he was of Italian rather than Irish ancestry. The matter was important enough to Flatley that Thach most likely advised Flatley of Denbo's compliment and ancestral faux pas. Just as surely, Flatley would have immediately penned a letter to Denbo to confirm his Irish roots. Thach was not as interested in his ancestral heritage; his roots were in Arkansas.
7. NA and FP.
8. Vice Adm. J. D. Price to Chief of Naval Personnel, 5 October 1948, TP.
9. Melville B. Grosvenor, *National Geographic Magazine* to Thach, 29 September 1942, TP.
10. Thach to Hanson W. Baldwin, 10 May 1949, TP.
11. Thach to William Randolph Hearst Jr., 18 January 1949, TP.

12. Thach to Lloyd Wendt, 30 November 1948, 17 January and 17 March 1949, TP and O'Hare Papers; Thach to Lt. Warren A. Skon, 21 December 1948, TP.

13. Thach to Thacher Longstreth, 11 February 1949 and 10 February 1950, and Longstreth to JST, 22 November 1948, 14 December 1948, and 28 December 1949, TP.

14. Madalyn Thach to *Life-Time*, 15 July 1946, TP. In referring to USAF Gen. Carl A. Spaatz's article "If We Should Have to Fight Again," Mrs. Thach wrote, "His so-called analysis of the air lessons learned in the last war is immature, incomplete, and criminally misleading."

15. Rear Adm. E. C. Ewen to Thach, 2 May 1949, TP.

16. Statement by Thach delivered before the House Armed Services Committee, n.d., NA.

17. *Pensacola Journal*, 28 October 1949, p. 1.

18. Ibid.

19. Thach to Jack Leahy, 14 December 1949, TP.

Chapter 13. Korean War Combat on USS Sicily

1. Thach to George W. Bradham, 20 February 1950, TP.

2. Thach to Davis Merwin, 21 March 1950, TP.

3. Thach to Eugene E. Wilson, 21 March 1950, TP.

4. Thach to Merwin, 21 March 1950, TP.

5. Thach to Capt. B. C. McCaffree, CNO Op-54, 21 February 1950, and McCaffree to Thach, 24 February 1950, TP.

6. Thach to Capt. W. R. Caruthers, CO, Fleet Sonar School, NOB Key West, Florida, 9 March 1950, TP.

7. Capt. W. R. Caruthers to Thach, 11 March 1950, TP.

8. Thach to Rear Adm. H. S. Duckworth, Op-50, Navy Department, Washington D.C., 27 April 1950, TP.

9. Ibid.

10. *San Diego Evening Tribune*, 16 June 1950, p. B-20.

11. Thach insisted in his oral history that he entered Kobe on 2 August and left on the third. However, from this place forward in the narrative on Korea dates, times, and location (derived from latitude and longitude positions) are from the War Diary, USS *Sicily* (CVE-118) for the months of August, September, October, November, and December 1950 and January 1951. Also, organizational terminology utilized is that of the War Diary rather than of Thach's oral history (i.e., Joint Operations Center rather than Joint Communications Center, as Thach recalled).

12. Lt. Gen. Robert P. Keller, interview with author, 27 August 2002, and subsequent communications with author. Also see Keller, "Korean War Perspective," 66–72. Also,

General Keller identified Maj. Ken Reusser for the low-level attacks on the disguised buildings.

13. Denson, "Captain Thach's Phantom," 19. This article mentions only "the Russians," but Thach's worry about Chinese submarines is quoted in numerous other sources.

14. A commanding officer inviting himself to a squadron ready room was not common during either World War II or the Korean War. For both his carrier commands with Block Island (CVE-106 in 1952–53) and Lake Champlain (CV-39 in 1955–56), Jimmy Flatley did not enter a squadron ready room to talk with pilots unless he first requested permission from the air officer and squadron commander, according to Capt. John Lacouture, USN (Ret.) (conversations dating from 1993 and correspondence from 1996). Lacouture, who served with Flatley when he was a carrier commanding officer, recalled that Flatley was always around to observe, and his pervasive interest was apparent. But everyone knew he would offer suggestions, recommendations, changes, and orders only through the established chain of command. Equally effective in one-on-one conversations or communicating with groups, Flatley and Thach were both liked, but each stayed within their own personality to accomplish their objectives.

15. War Diary, USS Sicily, August 1950.
16. Ibid.
17. Associated Press, 14 August 1950, TP.

Chapter 14. Inchon, Chosin Reservoir, and Final Korean Operations

1. All times, dates, locations, and actions in this chapter are derived from War Diary, USS Sicily, September to December 1950, and January 1951.
2. Ibid., September 1950.
3. Ibid.
4. Ibid. In his oral history, Thach credited his planes with only two of five, assigning the others to marine artillery. Other sources also mention only five tanks.
5. Ibid.
6. Ibid., August and September 1950.
7. Keller interview.
8. War Diary, USS Sicily, November 1950.
9. Ibid.
10. Ibid.
11. Maj. Robert P. Keller, USMC, CO VMF-214 to Thach, 13 November 1950.
12. War Diary, USS Sicily, November 1950.
13. Message from COMNAVFE to Sicily, 29 November 1950.
14. San Diego Union, 2 March 1951, p. A-6.
15. Ibid.

16. Ibid.
17. Ibid.
18. Secret Intelligence estimate, Major General Harris, 5 December 1950, TP.
19. War Diary, USS Sicily, January 1951.
20. *San Diego Union*, 6 February 1951, p. 1.
21. *Los Angeles Times*, 9 February 1951.
22. Rear Adm. Arleigh A. Burke, personal and confidential letter to Capt. Alexander McDill, Office of the Secretary of the Navy, 9 October 1950, NA.

Chapter 15. Commanding Officer to Flag Rank

1. Thach to Capt. A. W. McKechnie, 15 November 1951, TP.
2. Ibid.
3. Eugene Zuckert oral history, Truman Presidential Museum and Library, 27 September 1971, 53.
4. Thach to Wesley Price, letter in *Saturday Evening Post*, 7 July 1952, TP.
5. Thach in *Aviation Week*, 10 March 1952, TP.
6. Thach to Rear Adm. O. B. Hardison, CO Fleet Air, Jacksonville, 29 March 1952, TP.
7. Thach to Capt. James H. Flatley Jr., 28 May 1952, FP.
8. Thach to Davis Merwin, 22 April 1952, TP.
9. Speech by Adm. William M. Fechteler, CNO, 25 April 1952, TP.
10. Thach to Charlotte Knight, letter in *Collier's*, 19 June 1952, TP.
11. Thach to John F. Floberg, 3 July 1952, TP.
12. Thach to Harriet C. Ainsworth, 4 July 1952, and Thach to J. Gordon Ainsworth, 28 January 1953, TP.
13. Thach to Capt. George W. Anderson, 11 October 1953, TP.
14. CO, Puget Sound Naval Shipyard to Thach, 23 April 1954, TP.
15. Madalyn Thach to John Thach, 13 March 1954, TP.
16. Thach to Capt. William S. Harris, 31 August 1955, TP.
17. Rear Adm. Rufus Rose, CO Amphibious Training Command, Atlantic Fleet to Thach, 29 July 1955, and Thach to Rose, 2 August 1955, TP.
18. Thach to Rear Adm. Frank Akers, 24 August 1955, and Thach to Capt. Thurston B. Clark, 25 August 1955, TP.
19. Lt. Gen. Sam E. Anderson, Director Weapons Systems Evaluation Group to Thach, 25 August 1955, TP.
20. See Wilson, "Roles and Missions Muddle," 37–39, 43–48; "Admiral and the Atom," 25–26, 31–32; Burke, "H-bomb Cannot Wipe Out U.S. Navy," 82–88, 90, 92, 94; Berry, "Bombers and Aircraft Carriers," 40–42; Hittle, "Why Russian Seapower?" and "Scientific Navy for Peace."
21. Thach to CNO Burke, 5 August 1955, TP.

Chapter 16. Task Group Alpha

Note: The most significant source for this chapter was a number of Rear Admiral Thach's speeches and notes from 1957 into 1959 archived in the TP. Many are undated, but their content provides indications for the period. Also in the TP is an undated paper titled "Antisubmarine Defense Group Alfa," by Capt. Thomas D. McGrath, USN, which provides an excellent overview of Task Group Alpha. Thach's speeches and notes provide considerably greater detail than his oral history.

1. Roskill, *War at Sea*, Appendix Q.
2. Polmar, *Guide to the Soviet Navy*, 2.
3. *Chicago Sun-Times*, 30 September 1958, p. 16. The same thought was constantly repeated in speeches and other articles during the period.
4. James Harmon Thach III, interview with author, 19 June 2002.
5. Perfall, "Submarine Menace," 11c.
6. See Brownlow, "Navy Asks Industry Aid," 32–34.
7. Original in TP.
8. Thach's immediate superior, Vice Adm. Frank Watkins, wrote in Thach's guest book on 10 April 1958, "Do not forget that case of Jack Daniels." Several correspondents also wrote similar comments in his guest book. Liquid public relations had prominent place throughout Thach's naval career. TP.
9. Original in TP.

Chapter 17. Commander, Antisubmarine Warfare Force, Pacific

1. Vice Admiral Thach's official title changed three times during this assignment. From December 1959 to July 1961, it was commander, Antisubmarine Defense Force, U.S. Pacific Fleet. From July 1961 to September 1961, it was commander, Antisubmarine Force, U.S. Pacific Fleet, and from September 1961 to his detachment in June 1963, it was as presented in the chapter heading.
2. "Soviet Sub Force Looms as Top Maritime Threat," *San Diego Union*, 26 May 1963, TP.
3. "Thach will draw his forces from existing air, sea and submarine unites in the Pacific without a significant build-up of the Pacific Fleet." *Honolulu Advertiser*, n.d. but c. March 1960, TP.
4. "Pacific Sub Experts Play Deadly Game," *Honolulu Star-Bulletin and Honolulu Advertiser*, 5 May 1963, TP.
5. Remarks aboard USS *Yorktown* CVS-10, August 1961, Patriots Point.
6. Not everyone agreed with nuclear war being inevitable, including at least one writer who was acquainted with Thach and his ASW work. The basis of this theory was that the Communists believed they had been most successful in limited war and that

they knew that nuclear war spelled the end to both sides. C. W. Parcher, "In My Opinion," *Glendale (Calif.) News-Press*, 4 October 1962, TP. A later perspective from the Soviets was that a conventional war might not escalate to nuclear use. See Friedman, *Submarine Design and Development*, 101.

7. Admiral Sadayoshi Makayama, Chief of Japan's Maritime Self Defense Force, acknowledged Thach as a close associate due to Japan's emphasis on antisubmarine operations. He also noted in an MSDF newspaper article that Thach was America's equivalent of Japan's Pearl Harbor planner and famous pilot Minoru Genda, a general in 1962 and later a statesman. The same article highlighted many of Thach's wartime and prewar accomplishments after stating Thach "was an average boy in the Naval Academy but became very famous as pilot." MSDF *Newspaper*, 20 July 1962, TP.

8. In March 1979, former secretary of state Dean Rusk spent a week at the University of Charleston (West Virginia) to lecture both on campus and in the community. Then a professor at UC, this writer's assignment was to be his escort to the various classes and meetings. While sharing some moments together between appearances, Rusk revealed that the U.S. military would have taken action in Cuba "on Tuesday had the matter not been resolved on Sunday." Thirteen years later, both Russia and Cuba revealed the full story of Soviet missile capability and resolve to use them in 1962. Admiral Thach would no doubt have felt vindicated that his "all or nothing" projection was correct.

It was particularly interesting in 1979 that Secretary Rusk declined to either discuss before the groups he addressed the Cuban "Tuesday" invasion or speculate on the course of history had President John F. Kennedy lived beyond 1963. Later he did openly discuss both subjects.

9. U.S. Naval Station, Ford Island rather than NAS in 1963.

10. "Pacific's Antisubmarine Warfare Force Observes Third Birthday Today," *Honolulu Star-Bulletin and Honolulu Advertiser*, 1 March 1963, TP.

11. Ibid.

12. "Adm. Thach Hails Allies' Joint Antisub Exercises," *San Diego Union*, 27 June 1963, TP.

Chapter 18. Deputy Chief of Naval Operations (Air)

1. Cain interview. Cain also recalled that both he and Thach were greatly impressed with the ruggedness of the Hellcat's R-2800 engine. During a night exercise at Jacksonville in 1943, Cain's wingman attempted to join on the red light on his wing but instead joined on the red light of a water tower. Though shattered, and with two pistons faithfully pumping nothing but air, the damaged Hellcat landed without incident.

2. *San Diego Union*, c. May 1967, TP.

3. Rear Adm. James H. Flatley, conversations with author from 1986; and Cain interview. Dates of the C-130 tests were 28 October, 8, 21, and 22 November 1963. Number of touch and go, full stop and launchings were derived from Flatley's log books. Many published sources list twenty-one stops and launchings.
4. Cain interview and conversations.
5. *Honolulu Star-Bulletin*, 22 May 1964, TP.
6. See "Hybrid," *Newsweek*, 10 December 1962, 60, and "Contracts," 78.
7. See Miller, "JSF Sets the Standard," 38–40, 42.
8. Martin, "Power in the Pentagon," 30–34. In this same article praising McNamara as "the assistant President," a top Vietnam planner addressed McNamara's propensity to want to quantify every element. The assessment by the Vietnamese official concluded with, "It's just as important—perhaps sometimes more important—how people feel." While there is no way to know whether or not Vice Admiral Thach ever saw this article, undoubtedly he would have been pleased to know he was not alone with his thoughts on quantitative versus qualitative value.
9. For details and thoughts on McNamara and the TFX, see Miller, "Crash of TFX," 32–35. Also see Shifley, "Commonality," 28, and Outlaw, "McNamara's Band of Ignorant Analysts," 29–30.
10. TFX Contract Investigation Report, 91st Cong., 2d sess., Senate No. 91-1496, 1970.
11. See McNamara with VanDeMark, *In Retrospect*.
12. Thach speech, 15 August 1963, TP.
13. Thach to Charlotte Knight, *Collier's*, 25 June 1952, TP.
14. Cain interview. Cain stated that Thach had him listen on a silent phone to many conversations and take notes so Thach would not have to depend entirely on memory for the matters discussed and agreements made.

Chapter 19. Four Stars, Last Call, and Honors

1. Mrs. Dorothy Flatley, conversation with author, October 2002.
2. Madalyn Thach to family, 23 May 1965, TP.
3. Capt. E. T. Wooldridge, interview with author, 21 June 2002.
4. Rear Adm. Joseph C. Wylie oral history, 78, courtesy of Paul Stillwell.
5. Capt. E. T. Wooldridge, correspondence with author, 20 June 2002.
6. Wooldridge interview.
7. Wylie oral history, 77–78.
8. Madalyn Thach to family, 23 May 1965, TP.
9. Thach speech, n.d. but c. 1965, TP.
10. Ibid.
11. Madalyn Thach to family, 6 September 1965, TP.

12. Ibid.
13. Thach speech, 4 June 1965, TP.
14. Ibid.
15. Wylie oral history, 77 and 79–81.
16. Thach to D. M. Wilder, Executive Vice President for Lockheed California Company, 3 April 1968 and 4 October 1968, TP.
17. Thach to Wilder, n.d. but c. May 1970, TP.
18. Thach to Wilder, 9 December 1970, TP.
19. Thach to Wilder, 13 May 1970, TP.
20. Thach to Wilder, 2 March 1970 and 2 June 1970, TP.
21. Thach to Wilder, 13 July 1970, TP.
22. Originals for both Zero and Gallu in TP.
23. J. V. Fogarty to Thach, 1 July 1974, TP.
24. Hamilton, *Shipmate*, 57.
25. Thach autopsy report, 16 April 1981, TP.
26. *Charleston News and Courier*, 3 and 16 December 1967.
27. Rear Adm. James H. Flatley III, interview with author, 2 June 2003, and conversations with author from 1986. Also related in Longstreth with Rottenberg, *Main Line Wasp*, 127.
28. CNO OPNAV Instruction 1650, 20 January 1971, TP.

Bibliography

The two major sources for this book were the oral history interview with Adm. John S. Thach for the U.S. Naval Institute and the Thach Papers (TP), currently archived at the Emil Buehler Naval Aviation Library, National Museum of Naval Aviation, Naval Air Station Pensacola, Florida. Although an invaluable source, the Thach oral history was not necessarily intended to address all the relevant historical and geopolitical background surrounding the admiral's career. And as Admiral Thach acknowledged during the oral history interviews, many dates—especially when some people interacted with him and certain events occurred—plus details, exact names, and the spelling of some people's and ships' names escaped his memory. Fortuitously, his surviving professional and personal papers at Pensacola and other sources enabled a reconciliation of his oral history reminiscences. These sources include correspondence and interviews with Admiral Thach's contemporaries and his nephew James Harmon Thach III as well as material in the National Archives (NA) and Naval Historical Center (NHC), plus the Vice Adm. James H. Flatley Jr. Papers (FP), the O'Hare Papers in the Enterprise Collection (EC), and other holdings within the Patriots Point Naval and Maritime Museum, Charleston Harbor, South Carolina. Reconciliation of selected, more salient variances between Thach's oral history and other sources are cited in the notes to preclude misinterpretation by future researchers. Finally, quotations in the text not otherwise cited are from *The Reminiscences of Admiral John S. Thach*, the two-volume oral history Thach recorded in 1971, which was published by the Naval Institute Press in 1977.

Books

Alden, John D. *The Fleet Submarine in the U.S. Navy*. Annapolis: Naval Institute Press, 1979.
Barlow, Jeffrey G. *Revolt of the Admirals: The Fight for Naval Aviation, 1945–1950*. Washington, D.C.: NHC, GPO, 1994.
Buell, Cdr. Harold L., USN. *Dauntless Helldivers*. New York: Orion Books, 1986.
Cagle, Malcolm W., and Frank A. Manson. *The Sea War in Korea*. Annapolis: Naval Institute Press, 1957.
Clark, Adm. J. J., and Clark Reynolds. *Carrier Admiral*. New York: David McKay, 1967.

Cressman, Robert J. *That Gallant Ship USS Yorktown (CV-5)*. Missoula, Mont.: Pictorial Histories Publishing, 1985.

Cressman, Robert J., Steve Ewing, Barrett Tillman, Mark Horan, Clark Reynolds, and Stan Cohen. *A Glorious Page in Our History: The Battle of Midway, 4–6 June 1942*. Missoula, Mont.: Pictorial Histories Publishing, 1990.

Ewing, Steve. *American Cruisers of World War II*. Missoula, Mont.: Pictorial Histories Publishing, 1986.

———. *Reaper Leader: The Life of Jimmy Flatley*. Annapolis: Naval Institute Press, 2002.

Ewing, Steve, and John B. Lundstrom. *Fateful Rendezvous: The Life of Butch O'Hare*. Annapolis: Naval Institute Press, 1997.

Field, James A., Jr. *History of United States Naval Operations: Korea*. Washington, D.C.: GPO, 1962.

Friedman, Norman. *The Postwar Naval Revolution*. Annapolis: Naval Institute Press, 1986.

———. *Submarine Design and Development*. Annapolis: Naval Institute Press, 1984.

———. *U.S. Aircraft Carriers: An Illustrated Design History*. Annapolis: Naval Institute Press, 1983.

Garrett, Richard. *Submarines*. Boston: Little, Brown, 1977.

Gay, George. *Sole Survivor*. 1979. Revised, Naples, Fla.: Privately printed, 1986.

Grossnick, Roy A. *United States Naval Aviation, 1910–1995*. Washington, D.C.: NHC, GPO, 1997.

Gunston, Bill. *Grumman: Sixty Years of Excellence*. New York: Orion Books, 1998.

Inoguchi, Capt. Rikihei, Cdr. Tadashi Nakajima, and Roger Pineau. *The Divine Wind*. Annapolis: Naval Institute Press, 1958.

Jurika, Stephen, Jr., ed. *From Pearl Harbor to Vietnam: The Memoirs of Admiral Arthur W. Radford*. Stanford, Calif.: Hoover Institution Press, 1980.

Karig, Walter, Malcolm W. Cagle, and Frank A. Manson. *Battle Report*. Vol. 6, *The War in Korea*. New York: Rinehart, 1952.

Lawson, Robert L. *The History of U.S. Naval Air Power*. New York: Military Press, 1985.

Leckie, Robert. *Conflict: The History of the Korean War*. New York: G. P. Putnam's Sons, 1962.

Longstreth, W. Thacher, with Dan Rottenberg. *Main Line Wasp: The Education of Thacher Longstreth*. New York: W. W. Norton, 1990.

Lord, Walter. *Incredible Victory*. New York: Harper and Row, 1967.

Lundstrom, John. *The First Team and the Guadalcanal Campaign*. Annapolis: Naval Institute Press, 1994.

———. *The First Team: Pacific Naval Air Combat from Pearl Harbor to Midway*. Annapolis: Naval Institute Press, 1984.

MacArthur, Douglas. *Reminiscences*. New York: McGraw-Hill, 1964.

McNamara, Robert, with Brian VanDeMark. *In Retrospect: The Tragedy and Lessons of Vietnam*. New York: Times Books/Random House, 1995.

Marolda, Edward J. *Carrier Operations: The Vietnam War*. New York: Bantam, 1987.
McClendon, Dennis E., and Wallace F. Richards. *The Legend of Colin Kelly, America's First Hero of World War II*. Missoula, Mont.: Pictorial Histories Publishing, 1994.
Morison, Rear Adm. Samuel E. *History of United States Naval Operations of World War II*. Vol. 3, *The Rising Sun in the Pacific, 1931–April 1942*. Boston: Little, Brown, 1948.
———. *History of United States Naval Operations of World War II*. Vol. 12, *Leyte: June 1944–January 1945*. Boston: Little, Brown, 1958.
———. *History of United States Naval Operations of World War II*. Vol. 13, *The Liberation of the Philippines: Luzon, Mindanao, the Visayas, 1944–1945*. Boston: Little, Brown, 1963.
———. *History of United States Naval Operations of World War II*. Vol. 14, *Victory in the Pacific, 1945*. Boston: Little, Brown, 1960.
Olynyk, Frank J. *USN Credits for the Destruction of Enemy Aircraft in Air-to-Air Combat, World War II*. Aurora, Ohio: Privately printed, 1982.
Polmar, Norman. *Guide to the Soviet Navy*. Annapolis: Naval Institute Press, 1983.
Polmar, Norman, Eric Wertheim, Andrew Bahjat, and Bruce Watson. *Chronology of the Cold War at Sea, 1945–1991*. Annapolis: Naval Institute Press, 1998.
Potter, E. B. *Admiral Arleigh Burke*. New York: Random House, 1990.
———. *Bull Halsey*. Annapolis: Naval Institute Press, 1985.
Rearden, Jim. *Koga's Zero: The Fighter that Changed World War II*. Missoula, Mont: Pictorial Histories Publishing, 1995.
Reynolds, Clark G. *Admiral John H. Towers: The Struggle for Naval Air Supremacy*. Annapolis: Naval Institute Press, 1991.
———. *Famous American Admirals*. New York: Van Nostrand Reinhold, 1978.
———. *The Fast Carriers: The Forging of an Air Navy*. New York: McGraw-Hill, 1968.
Roskill, Capt. S. W., RN. *The War at Sea 1939–45*, vol. 1. London: Her Majesty's Stationary Office, 1956.
Sherman, Adm. Frederick C. *Combat Command: The American Aircraft Carriers in the Pacific War*. New York: Dutton, 1950.
Solberg, Carl. *Decision and Dissent: With Halsey at Leyte Gulf*. Annapolis: Naval Institute Press, 1995.
Stafford, Cdr. Edward P., USN. *The Big E: The Story of the USS Enterprise*. New York: Random House, 1962.
Stern, Robert C. *U-Boats in Action*. Carrollton, Tex.: Squadron/Signal Publications, 1977.
Stumpf, David K. *Regulus: The Forgotten Weapon*. Paducah, Ky.: Turner, 1996.
Swanborough, Gordon, and Peter M. Bowers. *United States Navy Aircraft since 1911*. Annapolis: Naval Institute Press, 1982, 195–204.
Taylor, Theodore. *The Magnificent Mitscher*. Annapolis: Naval Institute Press, 1991.
Tillman, Barrett. *Corsair: The F4U in World War II and Korea*. Annapolis: Naval Institute Press, 1979.
———. *The Wildcat in World War II*. Annapolis: Nautical and Aviation Publishing, 1983.

U.S. Naval Academy. *The Lucky Bag*. Annual of the Regiment of Midshipmen. Annapolis: United States Naval Academy, 1927.
U.S. Navy, NHC. *Dictionary of American Naval Fighting Ships*. Washington, D.C.: GPO, 1959–70.
Winters, T. Hugh. *Skipper*. Mesa, Ariz.: Champlin Fighter Museum Press, 1985.
Wooldridge, E. T., ed. *Carrier Warfare in the Pacific: An Oral History Collection*. Washington, D.C.: Smithsonian Institution Press, 1993.

Articles

Adams, Capt. John P., USN. "The First Use of the 'Thach Weave' in Combat." *Foundation*, Spring 1988, 40–41.
"The Admiral and the Atom." *Time*, 21 May 1956, 25–26, 31–32.
"Air Navy Chennault." *Time*, 14 June 1943, typescript copy in TP.
Berry, Robert W. "Bombers and Aircraft Carriers—Secretaries and Senators." *Air Force*, July 1956, 40–42.
Brownlow, Cecil. "Navy Asks Industry Aid in ASW Battle." *Aviation Week*, 14 July 1958, 32–34.
Burke, Arleigh A. "H-bomb Cannot Wipe Out U.S. Navy." *U.S. News & World Report*, 4 May 1956, 82–88, 90, 92, 94.
Connaughton, Lt. Cdr. Sean T. "Revolt of the Admirals: Part Deux." *Proceedings*, February 1999, 77.
"Contracts." *Newsweek*, 10 December 1962, 78.
Denson, John. "Captain Thach's Phantom." *Collier's*, 14 October 1950, 18–19.
Esders, Cdr. Wilhelm G., USN (Ret.). "Torpedo Three and the Devastator—A Pilot's Recollection." *Hook*, August 1990, 35–36.
Giangreco, D. M. "The Truth about Kamikazes." *Naval History*, May/June 1997, 25–29.
Hamilton, Rear Adm. Thomas J., USN (Ret.). *Shipmate*, July–August 1981, 57.
Hittle, Col. J. D. "Why Russian Seapower?" *Marine Corps Gazette*, November 1956. Reprinted in *Congressional Record*, 7 January 1957.
"Hybrid." *Newsweek*, 10 December 1962, 60.
Keller, Lt. Gen. Robert P., USMC (Ret.). "A Korean War Perspective." *Foundation*, Fall 1987, 66–72.
Martin, Dwight. "The Power in the Pentagon." *Newsweek*, 6 December 1965, 30–34.
McCain, Vice Adm. John Sidney. "So We Hit Them in the Belly." *Saturday Evening Post*, 14 July 1945, 12–13, 40, 42, 44; 21 July 1945, 22–23, 37, 39.
McCuskey, Capt. Scott E., USN, and Bruce Gamble. "Time Flies: The Oral History of Captain E. Scott McCuskey, USN (Ret.)." *Foundation*, Spring 1992, 56–61.
Miller, Vice Adm. G. E., USN (Ret.). "The Crash of TFX." *Foundation*, Fall 1988, 32–35.

———. "JSF Sets the Standard for Aircraft Acquisition." *Proceedings*, June 2003, 38–40, 42.
Outlaw, Rear Adm. E. C., USN (Ret.). "McNamara's Band of Ignorant Analysts." *Foundation*, Fall 1988, 29–30.
Perfall, Art. "The Submarine Menace: America's Open Cellar Door." *Newsday*, 6 May 1959, 11C.
"A Scientific Navy for Peace." *Newsweek*, 9 April 1947, copy in TP.
Shifley, Vice Adm. Ralph L., USN (Ret.). "Commonality." *Foundation*, Fall 1988, 28.
Stillwell, Paul. "John Smith Thach, Adm., USN, 1906 [sic]–1981." *Navy Times*, 25 May 1981, 21.
Thach, John S. *Aviation Week*, 10 March 1952.
———. "The Red Rain of Battle: The Story of Fighter Squadron Three." *Collier's*, 5 December 1942, 14ff; 12 December 1942, 16ff.
Wendt, Lloyd. "Five Jap Planes Measure the Heroism of Butch O'Hare." *Chicago Sunday Tribune Graphic Magazine*, 25 September 1949, 10ff.
Wildenberg, Thomas. "Helldivers." *Foundation*, Fall 1998, 60–61.
Wilson, Gill Robb. "The Roles and Missions Muddle." *Air Force*, July 1956, 37–39, 43–48.
Wooldridge, Capt. E. T. USN (Ret.). "Checking Out This Year's Models: Flight Testing in the Golden Age." *Foundation*, Spring 2001, 36–45.

Oral Histories

Thach, John S. *The Reminiscences of Admiral John S. Thach*. Vols. 1 and 2. Annapolis: Naval Institute Press, 1977.
Wylie, Joseph C. *The Reminiscences of Rear Adm. Joseph C. Wylie, Jr.* Annapolis: Naval Institute Press, 2003.
Zuckert, Eugene. Oral history. Truman Presidential Museum and Library, Independence, Mo., 1971.

Interviews, Correspondence, and Conversations

Asterisks mark those interviewees and correspondents known deceased in 2003.

Adams, Capt. John, USN (Ret.). Conversations with author from July 1977.
Albert, Frank. Conversations with author from 1982.
Baur, Clyde E. Letter to author, 3 March 1994.
Best, Lt. Cdr. Richard, USN (Ret.).* Conversations with author from October 1988, and correspondence with author from 1989.

Blyth, Cdr. Robert L., USNR. Interview with author, 6 October 1994, and subsequent conversations with author.
Brown, Capt. Carl, USN (Ret.). Conversation with author, 6 October 1994.
Buell, Cdr. Hal, USN (Ret.). Interviews with author, 29 February 1996, 20 April 1996; letter to author, 15 October 1996; and conversations with author from 1984.
Cain, Capt. James B., USN (Ret.).* Interview with author, 27 April 1997, and conversations with author from 1984.
Carmichael, Lt. Robert, USNR. Conversation with author, 6 October 1994.
Condit, Rear Adm. James W., USN (Ret.).* Conversations with author from 1987.
Crommelin, Rear Adm. John G., USN (Ret.).* Conversations with author, 1982–96.
Duncan, Capt. George, USN (Ret.).* Conversation with author, 6 October 1994.
Earnest, Capt. Albert K., USN (Ret.). Conversations with author from 11 October 1987.
Eggert, Lt. Cdr. Joseph R., Jr., USNR. Telephone interview with author, 26 February 1996.
Elder, Capt. Robert M., USN (Ret.). Telephone interviews with author, 25 March 1996 and 27 March 2002; and conversations with author from 1987.
Esders, Cdr. Wilhelm G., USN (Ret.).* Conversations with author from 11 October 1987.
Feightner, Rear Adm. Edward L., USN (Ret.). Interviews with author, 23 April 1994 and 14 March 1996; and conversations with author from 1993.
Flatley, Dorothy. Interview with author, 19 December 2002; and conversations with author from 1987.
Flatley, Rear Adm. James H., III, USN (Ret.). Interview with author, 2 June 2003, and conversations with author from 1986.
Gallaher, Rear Adm. Wilmer Earl, USN (Ret.).* Correspondence and telephone conversation with author, 1982.
Galt, Cdr. Dwight, USN (Ret.). Conversation with author, 6 October 1994.
Gay, George.* Conversations and correspondence with author from July 1987.
Gayler, Adm. Noel A. M., USN (Ret.). Telephone interview with author, March 1994.
Gray, Capt. James S., Jr., USN (Ret.).* Conversation with author following Gray's Battle of Midway remarks at a National Museum of Naval Aviation symposium, Pensacola, Fla., May 1987.
Keller, Lt. Gen. Robert P. Interview with author, 27 August 2002, and subsequent correspondence with author.
Lacouture, Capt. John, USN (Ret.) Correspondence with author from 1996, and conversations with author from 1993.
Laub, Rear Adm. Robert E., USN (Ret.).* Conversation and correspondence with author from 1984.
Leonard, Rear Adm. William N., USN (Ret.). Telephone interviews with author, 12 March 1996 and 13 June 2002.
Lister, David.* Conversations with author from 1982.

Lord, Walter.* Conversation with author, 5 May 1986.
Lovelace, Cdr. Don, Jr., USN (Ret.). Telephone interview with author, March 1994.
Lundstrom, John B. Conversations with author from 1988.
McCuskey, Capt. E. Scott, USN (Ret.).* Conversations with author from June 1992.
Murray, Cdr. James, USN (Ret.).* Conversations and correspondence with author from August 1982.
Outlaw, Rear Adm. Edward, USN (Ret.).* Conversation with author, 6 October 1994.
Ramage, Rear Adm. James, USN (Ret.). Interview with author, 27 April 1996, and conversations with author from 1988.
Reynolds, Dr. Clark. Conversations with author from 1984.
Ridgway, Lt. Charles D., USNR. Conversation with author, 6 October 1994.
Riera, Rear Adm. Emmett R., USN (Ret.).* Interview with author, 28 February 1996, and conversations with author from 1980.
Smith, Mrs. Stephen. Conversation with author, October 1984, and subsequent correspondence with author.
Stillwell, Paul. Director, History Division, U.S. Naval Institute. Correspondence with author, 13 February 2003, and conversations with author from June 2002.
Strean, Vice Adm. Bernard M., USN (Ret.).* Interview with author, November 1994.
Tate, Capt. Benjamin, USN (Ret.). Conversation with author, 6 October 1994.
Thach, James Harmon III. Interview with author, 19 June 2003, and subsequent correspondence.
Tillman, Barrett. Conversations with author from 1990.
Ustick, Cdr. Perry, USN (Ret.). Interview with author, November 1984, and conversations with author from November 1987.
Vejtasa, Capt. Stanley W., USN (Ret.). Correspondence with author, 20 April 1996 and 30 March 1998.
Vraciu, Cdr. Alexander, USN (Ret.). Conversations with author, from October 1993.
Winters, Capt. T. Hugh, USN (Ret.). Interview with author, 1 November 1996.
Wooldridge, Capt. E. T. Interview with author, 21 June 2002; correspondence with author, 20 June 2002; and subsequent conversations with author.

Action Reports and Official Records

Action Report 4 June–6 June 1942. From CO *Yorktown* (CV-5) to Commander-in-Chief, U.S. Pacific Fleet. NA and FP.
Action Report VF-3. Lt. Cdr. John S. Thach. Wording essentially the same as for BuAer interview, 26 August 1942.
Air Raids on Japan. 1944–45. NA and FP.
Battle of Leyte Gulf. 1944. NA and FP.
Cinpac Intelligence Report. "The Japanese Air Force." 4 June 1945. TP.

Commander Task Force 38 Action Report. 31 August 1945. TP.
Commander Third Fleet War Diary. 22 July 1945. TP.
Summary of War Damage. 8 December 1944–9 October 1945. TP.
War Diary, Commander Task Force 38. 1 January–31 January 1945. TP.
War Diary, USS Sicily (CVE-118). August, September, October, November, and December 1950, and January 1951. TP.

Newspapers

Charleston News and Courier. 3 and 16 December 1967.
Chicago Sun-Times. 30 September 1958.
Glendale (Calif.) News-Press. 4 October 1962.
Honolulu Star-Bulletin and Honolulu Advertiser. 1 March and 5 May 1963, 22 May 1964.
Los Angeles Times. 9 February 1951.
Maritime Self Defense Force (Japan). 20 July 1962.
Pensacola Journal. 28 October 1949.
San Diego Evening Tribune. 16 June 1950.
San Diego Union. 6 February 1951, 26 May and 27 June 1963, May 1967.

INDEX

Acheson, Dean, 197–98
Adams, John, 77, 82
Admiralty Islands, 131
aircraft. *See under specific manufacturer*
Air Groups: Fifteen, 93; Nineteen, 125
Akagi, 65, 73, 79
Akers, Frank, 8
Alaska, 102, 263
Albacore (SS-218), 117
Aleutian Islands, 20–22, 79
Anderson, George W., Jr., 228, 265, 267, 277–78
Anderson, Sam E., 235–36
Anti-Comintern Pact, 44
Anti-submarine Squadron Thirty-six, 249
Apra Harbor (Guam), 209, 216
Argentina, 233
Arizona (BB-39), 46
Arkansas Aviation Hall of Fame, 296
Arnold, Murr, 68, 70, 76
Aroostook (CM-3), 9, 19
Ashia (Japan), 210
at sea logistics service group, 131
Attu, 80
Aurand, Evan P., 294
Australia, 50, 58, 268
Aviation Week, 226

Badoeng Strait (CVE-116), 197–98, 202, 205, 207, 210–11, 219
Baker, Wilder D., 124–25, 129, 171
Baldwin, Hanson W., 187
Balikpapan, Borneo, 58
Barbero (SSG-317), 257
Bassett, Edgar R., 70–71, 73, 75–76, 88
Battle of Britain, 32, 36
Bauer, Louis H., 94
Beery, Wallace, 13
Belgium, 42
Belleau Wood (CVL-24), 117, 130, 143
Bell P-39 Airacobra, 62, 87
Bennington (CV-20), 158–59

Benson, Roy S., 261
Berlin, 240, 265, 272
Berlin Airlift, 185
Berner, Warren, 25–26
big blue blanket (also "thatched roof," "three strike system"), 148–49, 160, 164, 169–70, 174, 178
Bismarck Sea (CVE-95), 157
Bivin (DE-536), 224
Blue Angels, 181
Boeing aircraft: B-17, 18, 80, 105–6; B-29 Superfortress, 18, 156, 160, 163–64, in Korea, 203, 221, 226; B-36, 185, 189; F2B-1, 8–9; F4B-1, 17, 92; P-12, 17; XF6B, 17–18
Bogan, Gerald F. ("Gerry"), 5, 8–9, 10, 135
bombing squadrons: VB-2, 59; VB-3, 66–79; VB-5 59, 66
Bon Homme Richard (CV-31), 158
Bonin Islands, 114
Borie (DD-704), 164
Boston Naval Shipyard, 196
Bougainville, 58
Boxer (CV-21), 207
Boyington, Gregory ("Pappy"), 202
Brassfield, Arthur J. ("Art"), 68
Brazier, Robert B., 74
Brazil, 232
Bremerton, 32, 46, 49, 232–33
Brennen, Walter, 180
Brett, James H., 59
Brewster aircraft: F2A-1 Buffalo, 30, 32; F2A-2 Buffalo, 30, 32, 35; F2A-3 Buffalo, 32, 46, 48–49
Bringle, William F., 294
Brooks, George W. ("Red"), 93–94
Brown, Wilson, 50, 53, 58, 60
Browning, Miles R., 36, 83
Bruce, David K. E., 288–89, 292
Buckmaster, Elliott, 70
Bulgaria, 231
Bunker Hill (CV-17), 134, 155, 157–58

Bureau of Aeronautics (BuAer), 18–19, 24, 33, 81, 84–87, 91, 97, 101–3, 108, 170
Bureau of Ordnance, 195
Bureau of Ships, 195
Burke, Arleigh A.: with Thach at Philippine Sea, 111–19; at Leyte Gulf, 122, 126–27, 134, 136; mentioned 155, 169, 171; unification 189, 191; as CNO, 236, 244–45, 253–54, 256–57, 259–60, 265

Cabot (CVL-28), 144, 186
Cady, Gordon, 122
Caidin, Martin, 294–95
Cain, James B., 271, 275
Caldwell, Turner, 293
California (BB-44), 7
Canada, 263
Canberra (CV-70), 132–33, 174, 178
Canton, 152
Cape Canaveral, 258
Cape Engano, 136, 140, 152
Cape Hatteras, 253
Caroline Islands, 111
Carrier Division Fifteen, 197, 213, 220
Carrier Division Sixteen, 240, 243, 257
Carrier Division Seventeen, 220, 262
Castro, Fidel, 266
Cavella (SS-244), 117
Charleston, South Carolina, 241, 296
Charleston (LKA-113), 296
Chennault, Claire, 100
Cheek, Tom F., 70–73, 75–76
Chesapeake Bay, 18
Chicago Tribune, 187
Chile, 232
China: 34, 41, 44–45, 152, 162–63, 189; during Korean War, 197, 199, 202, 215–16, 218, 220, 259, 272, 291
Chitose, 11
Chiyoda, 117
Chosin Reservoir, 217–19
Cincinnati (CL-6), 22–24, 31
Clarey, Bernard A., 261
Clark, John E., 257
Clark, Joseph J., 96, 122, 130, 145, 159, 172–73, 175
Coco Solo. *See* Panama
Collier's, 106, 227
"Combat Doctrine: Fighting Squadron Ten", 86
Commencement Bay–class, 196, 199
Connolly, Thomas, 293

Consolidated aircraft: PBY-2 Catalina, 18, 20, 24–25, 69, 163–64; P2Y-3, 24–25; XP2H, 18–20
Cook, A. B., 97, 99–101, 106
Cooper, Gary, 180
Coral Sea, Battle of, 61–64, 66–68, 86, 88, 90, 97, 104, 117*Coral Sea* (CV-43), 231
Corl, Harry L., 73, 77
Cowpens (CVL-25), 130
Crommelin, John G., 8, 93–94, 188
Crommelin, Richard ("Dick"), 68, 188
Cuba, 26, 234, 238–40; missile crisis, 265–67, 272
Cubera (SS-347), 246
Curtiss aircraft: F8C-4 Helldiver, 9–10, 15; H-16, 6; P-40, 47; SOC-1 and SOC-3 Seagull, 22–24; SB2C Helldiver 109, 116–18, 130, 137, 152
Cusk (SS-348), 242
Cyprus, 289
Czechoslovakia, 41–42

Dace (SS-247), 134
Darter (SS-227), 134
Daves, Delmar, 180
Davis, William O., 8
Davison, Ralph E., 135, 145
Denfield, Louis E., 188, 190–91
Denmark, 42
Destroyer Division Twenty-eight, 246, 250
Destroyer Division Thirty-six, 250
Dibb, Robert ("Ram"), 67, 70–73, 77, 90
Disney, Walt: and Disney studios, 77–101, 104, 124
Douglas aircraft: AD Skyraider, 204, 229, 250, 274, 278; SBD Dauntless, 51, 59, 69–79, 86, 109, 116–18, 130; TBD-1 Devastator, 50, 59, 66–81, 86, 109
Doyle, James H., 215
Duckworth, Herbert S. ("Ducky"), 10–13, 124
Duerfeldt, Clifford H., 196
Dufiliho, Marion W. ("Duff"), 53–54, 56
Dutch East Indies. *See* Indonesia

Eastern Solomons, Battle of, 94, 109
Eder, W. E., 56
Edwards, Cliff, 13
Eggert, Joseph R., Jr., 116–17
Eisenhower, Dwight D., 259, 265
Elder, Robert M. ("Bob"), 69
Elizabeth, Queen, 288–89
English, James, 207

INDEX 331

Eniwetok, 119, 130
Enterprise (CV-6): Thach aboard prewar, 32, 37; Wake Island expedition, 34, 46; at Midway, 68–81, 83, 85, 169; mentioned, 39, 62, 88, 92–94, 157–58
Enterprise (CVN-65), 281
Esders, Wilhelm G. ("Bill"), 74–75, 77
Essex (CV-9), 93, 144, 152
Essex-class carriers, 104, 113, 130, 145, 148, 160, 192, 207, 221, 229, 280, 293–94
Ethiopia, 41
Evans, Robert C., 67
Ewen, E. C., 188
Export Control Act (1940), 45
Eyerman, J. R. ("Ed"), 114–15

Fairbanks, Douglas, Jr., 290
Felt, Don, 29
fighting squadrons: VF-1B, 9–10, 12–16, 28; VF-1, 93; VF-2, 34, 36, 49, 61, 66; VF-3, 28–40, 45–84, 92, 123, 178, 187; VF-5, 36; VF-6, 79; VF-8, 79, 82; VF-42, 59, 62, 86, 90; VF-10, 62, 86, 88–94, 181; VF-17, 202; VF-19, 125
Firebaugh, Gordon E., 93–94
Fitch, Aubrey W. 47, 61
Flatley, Dorothy, 88, 89, 296
Flatley, James H., Jr.: with Thach in Hawaii, 61–62; "Navy Fighter," 86–91; adopts Thach weave, 91–92; names weave, 93–94; at NAS Jacksonville, 96; with Thach and Disney, 97; Zero secrets, 102; recommended for operations by Thach, 122–23; TF-58 operations officer, 126–27, 133–34, 136, 155–56, 169, 179; in training command with Thach, 181, 183; at Olathe, 226; death, 273; honored with Thach, 296–97
Flatley, James H. III, 273
Fletcher, Frank Jack, 47, 49, 58, 60; at Midway 68–81
Floberg, John F. ("Jack"), 223–29, 274, 277, 281
Floeck, Robert, 213
Fogarty, Jack V., 295
Fordyce, Arkansas, 2–4
Formosa (Taiwan), 132, 140, 151–52, 169, 174, 178, 189, 197, 199
Forrestal, James V., 184, 186, 191
Forrestal (CV-59), 191, 226, 273
Forrestal-class, 226–28, 234
France, 42, 286

Franco, Francisco, 41
Franklin (CV-13), 140, 143, 157–58
Franklin D. Roosevelt (CV-42), 228, 233
Fuso, 136

Gable, Clark, 13
Gadrow, Victor M, 48
Gaillard, J. Palmer, 296
Gallaher, W. Earl, 79
Gallu Productions, 295
Gambier Bay (CVE-73), 136
Gannet (AVP-8), 19
Gardner, Matthias B., 228
Gask, Roland ("Jerry"), 106
Gates, Thomas, S., Jr., 254
Gay, George, 295
Gayler, Noel A. M., 53–54, 59, 123, 296
George Washington (SSBN-598), 259–60
Germany, 41, 43–45, 106, 141, 157, 241–42, 263, 266, 272
Getty, J. Paul, 290
Gilbert Islands, 110–11, 114
Gill, Frank F. ("Red"), 51, 56
Goldthwaite, Robert, 263
Goodwin, Hugh H., 231
Goss (DE-444), 224
Grafmueller, Alfred M., 124
Gray James S., Jr., 79, 85, 87
Great Britain, 42–45, 185, 242, 268, 272, 287
Greater East Asia Co-prosperity Sphere, 44
Greece, 231
Greenland, 285
Greer (DD-145), 43
Griffin, Charles D., 284–85
Grumman aircraft: A-6 Intruder, 274; F3F-1, 30, 32; F4F-3 Wildcat, 30, 33, 35, 37, 67, 85–86; F4F-4 Wildcat, 61, 66–82; "Navy Fighter" controversy, 83–91; comparison with Zero, 84–86, 102; F6F 3 Hellcat, 102, 109, 117–18, 132; final operations, 142, 152, 154, 161, 181, 186, 271; F6F-5 VFB-VFS introduction of, 182; utilization, 146–49; 162; F8F Bearcat, 182; F9F-5 Panther, 181–82, 274; F9F-6 Cougar, 229; F-14 Tomcat, 280, 293; Guardians, 196; S-2F Tracker, 247, 249, 251, 293; TBF-1/TBM Avenger, 109, 116–18, 130, 136–38; ordnance change 148; at Hong Kong, 152; explosion, 154; as trainer, 186; ASW role, 216
Guadalcanal, 62, 91–94, 108–9, 111, 181, 196

Guam. *See* Mariana Islands
Gulf of Tonkin incident, 152, 272

Hainan Island, 152
Halibut (SSGN-587), 258
Halsey, William F., Jr.: with Thach on *Enterprise*, 33; prewar fighter formations, 36–37; with TF-8, 50; misses Midway, 62; 81, 83; commands Third Fleet, 120, 126, 130, 132, 168–69, 171, 174; at Leyte Gulf, 133–7, 176–77; commands TF-38, 143, 145, 152–53, 155, 158–59, 163; surrender ceremony, 165–66, 178; unification 185, 189
Hamilton, Thomas J., 5
Hammann (DD-412), 80
Hancock (CV-19), 144, 152; explosion on, 154; at Ulithi, 155; damaged, 157
Hardy, John L., 289
Harikoshi, Jiro, 294
Harrill, Keen, 124
Harris, Dale, 86, 91
Haruna, 105–6, 163
Harvey, Warren W. ("Sid"), 23, 31–32
Hawaii: VF-3 training, 32; attack on and aftermath, 46–49; VF-3 trains before Midway, 61–62; Midway preparations, 64, 66; VF-3 returns from Midway, 80; VF-10 arrives, 91; Thach meets Swirbul at, 182; Thach's ASW headquarters, 257, 262
Haynes, Leon W. ("Lee"), 53
Hayward, John T., 269
Hean, J. H., 123
Hearst, William Randolph, Jr., 187
Hedding, Truman, 112–13
Helicopter Antisubmarine Squadron Seven, 250
Hell Divers (movie), 12–13
Hiroshima, 164
Hiryu, 47, 65, 78–79
Hitler, Adolph, 29, 41–43, 252
Hiyo, 117
Hokkaido, 158, 162–63
Holland, 42
Hollandia, 111
Hollywood, 12–14, 180, 290
Hong Kong, 152
Honolulu Star-Bulletin, 277
Honshu, 158, 162–64
Hoover, John H., 29, 175
Hope, Bob, 107
Hopwood, Herbert G., 259–60

Hornet (CV-8), 39; at Midway, 68–82; sunk, 109, 112, 169
Hornet (CV-12), 130, 137, 159
Houghton, Dan, 293
Housatonic, 241
Houston (CL-81), 132–33, 174, 178
Hungnam, 217–18
Hunley, 241
Hyuga, 152, 163

I-6, 49
I-168, 80
IBP (individual battle practice), 11, 30
IFF (identification signals), 146
immigration law (1924), 44
Inchon, 205, 209, 210–11, 213–15, 218
Independence (CVL-22), 135
Independence-class carriers, 160
Indochina, 152
Indonesia, 44–45
Intrepid (CV-11), 143, 157
Iowa (BB-61), 162
Ise, 152, 163
Italy, 41, 45, 185
Itami Air Force Base, 210
Iwo Jima, 126, 130–31, 155–57, 169, 171, 224

Japan: history of disputes leading to war, 20–22, 35, 41, 43–45; Korean War era, 197, 199, 200–203, 206, 210, 214, 224; submarine rockets fired at, 242; U.S. strategic interests in, 259, 263; during Cuban missile crisis, 266–67
Java. *See* Indonesia
Johnson, Louis A., 188, 190–91
Johnson, Lyndon B., 272, 284
Joint Task Force Seven, 210, 213
Jones, Dr. Leland D., 13
Joy, C. Turner, 201–2, 220
Junyo, 118

Kaga, 65, 73, 79
Kahn, Genghis, 141
Kalinin Bay (CVE-68), 140
Kamaishi, 162
kamikaze attacks, 141–53, 157–58, 162–64, 169, 171, 176
Kavieng, New Ireland, 58
Kawanishi N1K1 Shiden ("George"), 141
Kearny (DD-432), 43
Keller, Robert P., 202, 213, 216
Kelly, Colin P., Jr., 105

INDEX

Kennedy, John F., 233, 239, 265, 272, 278
Key West, 194–95, 240
Kimpo airfield, 204, 212–13
King, Ernest J.: with Thach at Norfolk, 18–20; during World War II, 87, 111, 126, 130, 133, 169; with Thach and McCain, 155–56; Thach visit on behalf of McCain, 166–67; Thach's opinion of, 177; unification, 189
Kincaid, Thomas C., 136, 189
Kiska, 80
Kitchen, Etta Belle, 172, 295
Kitkun Bay (CVE-71), 140
Knox, Frank, 91
Kobe, 163, 200, 210
Korea, 22
Korean War, 192–223, 226, 229–30, 240, 244, 246, 256, 266, 270–71, 274
Korth, Fred, 277, 279
Kumano, 138
Kunsan, 205
Kure, 163
Kurita, Takeo, 136, 138, 140–41
Kyushu, 157–60, 201

Lae, 58–60, 105
Langley (CV-1), 12
Langley (CVL-27), 153
League of Nations, 44
Leary, Herbert F., 49
Lehman, John, 296
Lemay, Curtis, 185, 277–78
Lemnitzer, Lyman L., 284, 286
Lend-Lease Act (1941), 43
Leonard, William N. ("Bill"), 68, 77, 82, 122, 126, 141, 174, 178, 189
Leslie, Maxwell F., 68, 71, 73, 76–77, 295
Lexington (CV-2): 12, 39, 46, 48–50; off Rabaul, Lae, and Salamaua, 50–60; at Pearl Harbor, 61–62; at Coral Sea, 64, 86, 104
Lexington (CV-16), 104: at Philippine Sea, 109–19, 170; at Leyte, 122, 134–36, 142–44
Leyte: prelude to invasion, 130–31, 171; battle of Leyte Gulf, 133–40, 176; securing of and continuing operations, 142–44, anchorage, 157
Life, 187
Lingayen Gulf, 143, 151–52
Lischeid, Walter E., 202, 213
Lockheed aircraft: C-130 Hercules, 273, 290; F-80 Shooting Star, 204, 290; P2V-3C Neptune, 188, 204, 250–51; P-3 Orion, 290; PV Ventura, 293; S-3A Viking, 290–91
Lockheed Aircraft Company, California Division, 293
Loening OL-8, 9
London Treaty (1930), 44
Longstreth, William T., 123, 132, 162, 166, 174, 179, 187, 290, 296
Lovelace, Donald A., 32, 45, 53–54, 56–57, 60, 66–69
LTV (Ling-Temco-Vought) aircraft: F-8 Crusader, 274, 276; A-7 Corsair II, 275–77, 279, 294
Lucky Bag, 5
Lundstrom, John B., 295
Luzon, 131–35, 143, 151–52, 171

MacArthur, Douglas, 143, 151–53, 177–78, 209, 211, 214
MacInnes, J. N., 123
Macomber, Brainard T., 70–73
Maddox (DD-731), 153
Magda, John, 181
Makayama, Sadayoshi, 266
Malta, 287
Manchuria, 44
Manila, 131, 134, 151, 177
Mankin, Lee P. Jr., 94
maps: Pacific theater, 42; Coral Sea and Solomon Islands, 51; Western Pacific, 117; Philippine area, 121; Korea, 193
Mariana Islands, 111; battle for, 113–19, 131, 134, 156, 163, 170
Marine fighting squadrons: VMF-211, 49; VMF-214, 201–2, 205, 207, 210, 212, 215, 217–19; VMF-221, 46; VMF-223, 219; VMF-323, 219
Marshall Islands, 109–11, 114
Martin, Charles, 125
Martin aircraft: B-26, 204; PM-1, 21
Mason, Charles P., 96
Massey, Lance E. ("Lem"), 66–74, 88
Mathews, Francis P., 188, 191
Mayport, 234
McCain, John S. ("Slew"): with Thach at NAS Jacksonville, 107–9; at Philippine Sea, 111, 113–14, 119; commands Task Group 38.1, 120–39; commands Task Force 38, 142–66; death of, 167; analysis of, 168–80; mentioned, 240, 271
McCain, John S., Jr., 292

McCampbell, David, 296
McClusky, Clarence Wade, 82
McCormick, Robert R., 187, 198
McCuskey, E. Scott, 82
McDonald, David L., 278, 283, 292
McDonnell Douglas aircraft: A-4 Skyhawk, 274–75; F2H-4 Banshee, 229, 271: F-4 Phantom, 274, 278
McFall, Andrew C., 106–8
McNamara, Robert S., 273, 277, 280, 283, 288
Merwin, Davis, 227
Midway: battle of, 49, 64–82; references to, 83–85, 90–91, 102, 104–6, 109, 113, 115, 122, 169, 178, 180, 187–88; seminar on, 295
Midway-class carriers, 183, 187, 227–29
Midway (CVB-41), 229–30
Mikuma, 79
Mindanao, 130, 136
Mindoro (CVE-120), 151, 171, 194
Mississippi (BB-41), 67
Missouri (BB-63), 162, 165–66
Mitchell, Samuel G., 79
Mitscher, Marc A.: at Midway, 81–82; at Philippine Sea, 109, 111–18; First Fast Carrier Task Force Pacific, 120, 122; at Leyte, 130–42; resumes command, 155; analysis of, 168–72; honored, 296; mentioned 123, 143–44, 157, 180
Mitsubishi aircraft: A6M2 Type 0 carrier fighter (Zeke), 33–38, 47–48, 58, 67, 70–78, 84, 86–87, 94, 101–3, 123; poor pilot performance, 11; Blue Angels show, 181; G4M1 Betty bomber, 54–58, 161
Monterey (CVL-26), 130
Montgomery, Alfred E., 142, 145
moose-trap exercises, 146–47, 151
Morocco, 286, 289
Morotai, 131
Morrow Board and Lampert Committee, 183
Moss, John, 125
Murray, George D., 7, 101
Musashi, 133, 135
Mussolini, Benito, 41
mutual support beam tactic. *See* Thach weave

Nagasaki, 164
Nagel, Conrad, 13
Nakajima "Myrt," 142
Naktong River, 203, 206, 208

National Air and Space Museum, 295
National Broadcasting Company, 107
National Geographic, 185
National Museum of Naval Aviation, 297
National Security Act, 184
NATO, 268, 291
Nautilus (SS-571), 241, 281
Naval Air Gunnery School, 104
Naval Air Operation Training Command, 96, 106
Naval Air Stations: Corpus Christi, 181; Ford Island, 33, 258; Jacksonville, 83, 88, 96–107, 234–35, 240; Norfolk, 14–16, 18–20, 183, 229, 232, 240, 247, 249, 252, 257; Olathe, 226; Pensacola, 7–10, 57, 85, 92, 95–96, 101, 166–67, 180–87, 192, 225, 267; San Diego (North Island), 7–9, 11, 14–15, 28–29, 32–33, 39, 41, 80, 86, 89, 91, 96–97, 102, 123, 156, 167, 195–96, 219
Naval Expansion Act (1938), 39
"Navy Fighter," 83–93
Netherlands, 290
Neutrality Act (1939), 42–43
Newfoundland, 43
New Mexico–class, 6
Newsweek, 106
New York Times, 187, 265
New Zealand, 289
Nigeria, 289
Nimitz, Chester W. (CinCPac), 60, 65, 79, 81, 87, 105, 126, 133, 143, 159, 169, 181, 189
Nitze, Paul H., 277, 279
Nixon, Richard M., 239
Normandy, 112
North American aircraft: AJ Savage, 188; F-86 Sabre, 220, 226; P-51 Mustang, 157, 203–4; SNJ, 181, 186
North Korea, 191, 197–99, 203, 206, 208–12, 214, 216–17, 221, 226
Norway, 42
Noshiro, 138

O'Hare, Edward Henry ("Butch,"): VF-3 engineering officer, 33; role in developing Thach weave, 38–39; Pearl Harbor news, 45; Medal of Honor action, 51–58; at Lae and Salamaua, 59; at Pearl Harbor with Thach, 60–61; assigned command of VF-3, 64; assumes command of VF-3, 80; at

Grumman factory, 86; converts Flatley to weave, 92, 94; death of, 170; airport named for, 187; honors, 296–97
Okinawa, 126, 131–32, 151, 155–60, 163, 169, 171, 197, 224
Okumiya, Masatake, 294
Oldendorf, Jesse B., 136, 138
Oliver Hazard Perry–class, 296
Omaha-class cruisers, 22
Ommaney Bay (CVE-79), 151
Operational Readiness Evaluation (ORE), 261–62
Operation Forager, 111
Operation Pulverize, 143
operations officer, duties of, 125–29
Orders for Gunnery Exercises (OGE), 11
Oufkir, Mohamed, 286, 291

Packard, Howard S., 94
Panama, 19, 20, 24–26, 177, 232, 273
Palau Islands, 111, 130–31, 134–35, 139, 142, 144, 146, 151, 153, 155
Panay (PR-5), 44
patrol squadrons: VP-5F, 24; VP-8, 250; VP-9, 19–24; VP-21, 250
Pearl Harbor: attack on and aftermath, 39, 45–49, 65, 136, 180; first operations from, 50; enemy bomber identification at, 58; breakup of VF-3 at, 60–61; fleet returns from Midway, 80; VF tactics discussed at, 94, 102; Thach to Mitscher's flagship, 109; Thach interviews staff at, 119–20, 124; Flatley at, 126; Thach respite at, 155–57; Thach return to 1945, 166–67; Korean War, 198–200, 219; Thach ASW headquarters, 260
Pearson, J. B., 91
Peleliu. *See* Palau Islands
Peron, Juan, 233
Perry, Matthew, 43
Pescadores, 152
Petropavlovsk, 263
Philippines. *See* Leyte, Lingayen Gulf, Luzon, Mindanao
Philippine Sea, battle of, 113–19, 123, 130, 132
Philippine Sea (CV-47), 207
phoenix missile, 279–80
Pine Bluff, Arkansas, 1
Pledge (AM-277), 211
Point Loma, 195, 295

Poland, 29, 42
polaris missile, 259, 263, 287
Port Moresby, 58, 60
Potsdam Declaration, 157
Pratt, William V., 7
Price, J. D., 184, 192, 194
Princeton (CVL-23), 135
Proceedings, 295
Puget Sound Navy Yard. *See* Bremerton
Pusan, 201, 203, 205–7, 209–10, 218
Pyongyang, 210, 214

Quinn, Charles H., 31

Rabaul, 50–60, 105, 111
Radford, Arthur C.: with Thach in VF-1B, 9–10, 14–16; prewar, 31–32; assists Thach with Disney films, 97, 99–100; in Pacific theater, 145, 159, 189, 191, 175, 186; during Korean War, 199, 219; with JCS, 236, 296
Ramsey, Paul H., 8, 34, 49, 61–62, 260
Rand, Sally, 13–14
Randolph (CV-15), 157–58, 250, 257
Ranger (CV-4), 39, 110
Rather, Dan, 290
Reeves, J. B. L., 125
Reeves, J. M., 183
Reeves, John W. ("Black Jack"), 184, 186–88, 192, 194
Regan, Herbert E., 220–21
regulus missile, 243, 257
Reuben James (DD-245), 43
Reusser, Ken, 204
"Revolt of the admirals." *See* unification
Reynolds, Clark G., 172
Rhee, Syngman, 214
Rhodes, Thomas W., 94
Rickover, Hyman G., 281–82
Ridgeway, Charles D. III, 123
Robin Moor, 43
Robinson, Joe T., 4
Rockwell Field, 9, 28
Romany, 284, 286
Rome-Berlin Axis, 41
Roosevelt, Franklin D. 42–43, 60, 86, 157
Roosevelt, Theodore, 44
Rota. *See* Mariana Islands
Royal Hawaiian Hotel, 60
Ruble, Richard W., 197–98, 202, 214
Russo-Japanese War, 44, 197
Ryukyu, 152

Saipan. *See* Mariana Islands
Saipan (CVL-48), 186
Sakishimas, 152
Salamaua, 58–60, 105
Salisbury Sound (AV-13), 184, 192, 194
Samar, Battle off, 135–40, 176
San Bernardino Strait, 133, 135, 137–38, 176
Sandpiper (AVP-9), 19
San Francisco, 44, 195
Santa Cruz Islands, 39, 92, 109
Santee (CVE-29), 140
Saratoga (CV-3), 10, 12, 28, 32, 39, 45–49, 97, 109, 157–58
Saratoga (CV-60), 226
Sasebo, 205, 207, 210, 214, 216, 219
Saturday Evening Post, 144, 156, 174
Scamp (SS-277), 80
Schoech, William A., 220, 271
scouting squadrons: VS-2, 56, 59; VS-5, 59, 66–79; VS-6B, 22–24; VS-6, 79; VS-21, 216–17; VS-38, 297
Sea Leopard (SS-483), 246
Sea of Japan, 163–64
Sellstrom, Edward R., Jr., ("Doc"), 51–54, 56–57
Seoul, 204, 209, 212, 214, 219
Shangri-La (CV-38), 158, 185
Sheedy, Daniel C., 70–73, 75–76, 81, 169–70
Sherman, Frederick C. ("Ted"), 50, 54, 57, 59–60, 144–45, 172–73
Shikoku, 158
Shoemaker, J. M., 89
Shoho, 64, 86
Shokaku ("Sho"), 64–65, 117
Sibuyan Sea, 133–34, 140
Sicily (CVE-118), 196, 198–99, 200–202, 204–8, 210–20, 223, 230, 246
Sides, John H. ("Savvy"), 236, 260–61, 264–65
Simon and Schuster, 106
Simpson, William F., 212
Sino-Japanese War, 197
Sisson, Charles A., 125
Sixth Fleet, 284–85
Sikorsky HSS helicopter, 250
Skipjack (SSN-585), 257
Smith, Capt. John, 2
Smith, Etta Bocage, 2
Soryu, 47, 65, 73, 79
South Korea, 191, 197, 199, 203, 209, 212, 214, 259

South Vietnam, 272
Southwick, Edward P. ("Bud"), 10–11, 13
Soviet Union: World War II, 42–44, 157, 165; explodes atomic bomb, 189; Korean war, 197, 199, 220; iron curtain, 183, 272; space race, 239; submarine threat, 241, 243–45, 248, 251–54, 256; 258–60, 262–64, 266, 269; Thach states USSR Cold war objectives, 291
Soviet Union MiG-15, 95, 204, 220, 226, 274
Spain, 291
Spanish-American War (1898), 41, 43
Spanish Civil War, 41
Sprague, Thomas L., 197–98
Spruance, Raymond: at Midway, 68–81, 83; at Philippine Sea, 114–19; final Pacific operations, 120, 126, 130, 155, 158, 166; Thach comparison of with Halsey, 176; mentioned 169, 171–72, 189
sputnik, 301
Stalin, Joseph, 239
Stanley, O. B. ("Burt"), 52–53
Stillwell, Paul, 295
St. Lo (CVE-63), 140
Strean, Bernard M., 93–94
Struble, A. D., 214
Sullivan, John L. 188
Sullivan, W. A., 177
Surigao Strait, 133, 136, 140
Suwannee (CVE-27), 140
Swirbul, Jake, 181–82

Tacloban airfield, 138, 142
Taegu, 203–4, 206
Taiho, 117
Task Force (movie), 180–81
Task Elements, TE-96.23, 202; TE-96.82, 202
Task Forces, USN: TF-8, 50; TF-11, 50–60; TF-16, 68–81; TF-17, 58–60, 68–81; TF-38, 130, 133–34, 143; securing Philippines, 151, 153, 155; final operations 158–59, 161, 163, 165, 206; typhoon inquiry 175; British carriers, 185; TF-58, 109, 113–19, 155–56; TF-90, 210, 215
Task Groups, USN: TG-Alpha ("Alfa"), 240, 244–57, 273; TG-38.1, 130–38, 142–43, 159–60, 176; TG-38.2, 135, 144; TG-38.3, 134, 160; TG-38.4, 135, 143, 159–60; TG-58.1, 130; TG-58.4, 124;

INDEX

TG-90.5, 210; TG-96.2, 202; TG-96.8, 202, 210; TG-96.9, 214
Task Units, U.S. Navy: Taffy One, 138; Taffy Two, 138; Taffy Three, 136, 138
Tatom, John, 123
TFX controversy, 277–80
Thach, John S. ("Jimmie"): youth through U.S. Naval Academy and nickname, 1–6; battleship duty, 6–7; flight training, 7–8; with VF-1B, 9–14; test pilot, 15–19; patrol planes, 19–22, 24–26; scout planes, 22–24; with VF-3 prewar, 28–40; with VF-3 WWII, 41–82; develops "Thach weave," 34–39; Wake Island expedition, 45–49; promotion to lieutenant commander, 49; operations off Rabaul and Bougainville, 50–58; attack on Lae and Salamaua, 58–60; training in Hawaii, 60–63; Battle of Midway, 64–82; response to "The Navy Fighter," 89–91; at Jacksonville 1942–44, 97, 100–101, 106; Disney-produced training films, 103–7; promotion to commander, 106; meets McCain, 107–8; at Battle of the Philippine Sea, 109–19; assembles operations staff for McCain, 122–25; duties as operations officer; 125–29; off Philippines and Formosa fall 1944, 130–33; at Leyte Gulf, 133–40; first kamikaze attacks, 140–42; defense for carriers, 145–50; Task Force operations officer November 1944–January 1945, 142–55; promotion to captain, 157; final Pacific operations, 157–65; at surrender ceremony, 166; death of McCain, 167; perspectives on personalities, 168–70; Pensacola training command, 181–94; CO of Sicily, 194–220; Chief of Staff CD-17, 220–21; with Secretary Floberg, 223–28; CO of FDR, 228–34; CO NAS Jacksonville, 234–35; promotion to rear admiral, 235; with WSEG, 235–38; with TG-Alpha, 239, 257; promotion to vice admiral, 259; CO ASW Pacific, 258–70; Deputy CNO (Air), 271–82; TFX/F-111 controversy, 277–80; relationship with Rickover, 281–82; promotion to admiral, 283; CIC Europe, 283–92; retirement, 292; Lockheed consultant, 293–94; death of, 295; honors, 295–97

Thach, John S. ("Jack") Jr. 26, 148
Thach, Madalyn Jones: marriage, 14; to Norfolk, 15; weave discussion, 37; San Diego 1938, 41, 45; to Pensacola 1942, 91; sends comic strips to JST, 146; with McCain's, 167; in Pensacola 1946, 181; writes articles, 187; in San Diego 1950, 195; in Washington, 224; in Europe 1953, 230–31; in Jacksonville, 234–35; at Pentagon, 271; poor health, 282; in Europe 1965–67, 284–86; meets Queen Elizabeth, 288–89; retirement, 292; death, 295; christens cargo ship and sponsors frigate, 296
Thach, Frances, 2
Thach, James Harmon, 1, 3
Thach, James Harmon, Jr., 1, 4–7
Thach, Jo Bocage Smith, 1
Thach, Josephine, 2–3
Thach, William Leland, 148
Thach (FFG-43), 296
Thach weave (mutual support beam tactic): development of, 34–39, 297; discussed with Flatley, 62; use at Midway, 72–73, 90; named, 92–93; challenges to, 93–95; in training, 103; in Blue Angels shows, 181; press release, 195; promotion, 235; Navy Log show, 295
Thorburn, Don B., 124
Ticonderoga (CV-14), 153, 155
Time, 254
Tinian. *See* Mariana Islands
"Tiny Tim" rockets, 126
Tokyo, 45, 156, 158, 161, 163, 165, 175, 200, 209, 266
Tomonaga, Joichi, 78
torpedo squadrons: VT-2, 59; VT-3, 66–82; VT-5, 59, 65; VT-6, 73–74, 85; VT-8, 73, 85; VT-8 (detached), 74
Tosca, Henry J., 286
Towers, John H., 123, 165–66, 173
Tripartite Pact (1940), 45
Truk, 111
Truman, Harry S., 190, 224
Two Ocean Navy Act (1940), 43

Ulithi. *See* Palau Islands
unification, 183–91, 194, 203
Uruguay, 232
United States Naval Academy ("Annapolis"), 4–6, 12

United States Military Academy ("West Point"), 4
United States (CV-58), 188–89

Valley Forge (CV-45), 192, 203, 207, 246, 249, 252
Vejtasa, Stanley ("Swede"), 88
Verity, Erwin, 98–99
Versailles, Treaty of, 41
Vietcong, 272
Vietnam War, 95, 272–73, 288, 290–91
Vinson-Trammell Act (1933), 39
Volcano Islands, 114
Voris, Roy M. ("Butch"), 181
Vought aircraft: F4U-1 Corsair, 83–85, 103, 145, 185–86, 202; F4U-4 (in Korea) 201–2, 204–8, 210, 212, 215, 217–19, 242–43. *See also* LTV

Wagner, Frank, 181
Wake Island, 34, 45–47
Walker, Walton, 203
Walton, Bill, 3
Warner Brothers, 98–100, 104
Washington Naval Conference (1922), 12, 44
Washington Post, 106
Wasp (CV-7), 39, 109–10
Wasp (CV-18), 130, 136–38, 142–43, 157–58
Wead, Frank W. ("Spig"), 13
Weapons System Evaluation Group, 235–38, 240, 273, 278

Webster, William H., 296
Wendt, Lloyd, 187
Widhelm, William J. ("Gus"), 112–13, 122, 170
Wilson, Jack, 55–56
Winters, Hugh, 93, 125
Wisconsin (BB-64), 102
Wolmi-do, 211
Wonsan, 214–17
Wooldridge, E. T., 285
Wright, Jerauld, 240, 254
Wright (AV-1), 19
Wright (CVL-49), 186
Wylie, Joseph C., Jr., 285

Yalu River, 216
Yamamoto, Isoroku, 65, 79–80
Yamashiro, 136
Yamato, 133, 136
Yap. *See* Palau Islands
Yellow Sea, 204, 206, 218, 220
Yokosuka, 163, 166, 196, 200, 216–17, 219
Yorktown (CV-5): attack on Lae and Salamaua 58–60; at Coral Sea, 62, 64; at Midway, 65–84, 86, 169; mentioned, 39
Yorktown (CV-10), 93, 157, 275
Yorktown Carrier Aviation Hall of Fame, 296

Zero fighter. *See* Mitsubishi A6M2 Type 0 carrier fighter
Zuikaku ("Zui"), 65, 117, 136

About the Author

Steve Ewing is the author of several books on naval history, including *Memories and Memorials: The World War II U.S. Navy 40 Years after Victory*; *American Cruisers of World War II: A Pictorial Encyclopedia*; *USS Enterprise (CV-6): The Most Decorated Ship of World War II—A Pictorial History*; *The Lady Lex and the Blue Ghost: A Pictorial History of the USS Lexingtons CV-2 and CV-16*; *In Remembrance*; and *Reaper Leader: The Life of Jimmy Flatley*. With Robert J. Cressman he coauthored *A Glorious Page in Our History: The Battle of Midway, 4–6 June, 1942*, and with John Lundstrom he coauthored *Fateful Rendezvous: The Life of Butch O'Hare*, published by the Naval Institute in 1997. Dr. Ewing was a college professor for twenty years, and in 1988 he became senior curator at Patriots Point Naval and Maritime Museum, Charleston Harbor, South Carolina.

The Naval Institute Press is the book-publishing arm of the U.S. Naval Institute, a private, nonprofit, membership society for sea service professionals and others who share an interest in naval and maritime affairs. Established in 1873 at the U.S. Naval Academy in Annapolis, Maryland, where its offices remain today, the Naval Institute has members worldwide.

Members of the Naval Institute support the education programs of the society and receive the influential monthly magazine *Proceedings* and discounts on fine nautical prints and on ship and aircraft photos. They also have access to the transcripts of the Institute's Oral History Program and get discounted admission to any of the Institute-sponsored seminars offered around the country.

The Naval Institute also publishes *Naval History* magazine. This colorful bimonthly is filled with entertaining and thought-provoking articles, first-person reminiscences, and dramatic art and photography. Members receive a discount on *Naval History* subscriptions.

The Naval Institute's book-publishing program, begun in 1898 with basic guides to naval practices, has broadened its scope to include books of more general interest. Now the Naval Institute Press publishes about one hundred titles each year, ranging from how-to books on boating and navigation to battle histories, biographies, ship and aircraft guides, and novels. Institute members receive significant discounts on the Press's more than eight hundred books in print.

Full-time students are eligible for special half-price membership rates. Life memberships are also available.

For a free catalog describing Naval Institute Press books currently available, and for further information about subscribing to *Naval History* magazine or about joining the U.S. Naval Institute, please write to:

Membership Department
U.S. Naval Institute
291 Wood Road
Annapolis, MD 21402-5034
Telephone: (800) 233-8764
Fax: (410) 269-7940
Web address: www.navalinstitute.org